The Robotics Review 1

Editors

Oussama Khatib
Computer Science
 Department
Stanford University
Stanford, CA 94305

John J. Craig
SILMA Incorporated
1601 Saratoga-Sunnyvale Rd.
Cupertino, CA 95014

Tomás Lozano-Pérez
Artificial Intelligence
 Laboratory
Massachusetts Insitute
 of Technology
Cambridge, MA 02139

Contributing Editors

Haruhiko Asada
Massachusetts Institute
 of Technology

H. Harlyn Baker
SRI International

Jean-Daniel Boissonnat
INRIA (France)

Robert Bolles
SRI International

Rodney Brooks
Massachusetts Institute
 of Technology

John Canny
University of California
 Berkeley

Raja Chatila
LAAS-CNRS (France)

Mark Cutkosky
Stanford University

Paolo Dario
Università di Pisa

Hugh Durrant-Whyte
Oxford University

Roy Featherstone
Philips Laboratories

Malik Ghallab
LAAS - CNRS (France)

Christopher Goad
SILMA, Inc.

W. Eric L. Grimson
Massachusetts Institute
 of Technology

Gerd Hirzinger
DFVLR (West Germany)

Neville Hogan
Massachusetts Institute
 of Technology

John Hollerbach
Massachusetts Institute
 of Technology

Hirochika Inoue
University of Tokyo

Nenad Kircanski
M. Pupin Institute
(Yugoslavia)

Daniel Koditschek
Yale University

Jean-Claude Latombe
Stanford University

David Lowe
University of British
 Columbia

David Marimont
Xerox Palo Alto Research
 Center

Matthew Mason
Carnegie-Mellon University

J. Michael McCarthy
University of California
 Irvine

Joseph Mundy
General Electric R&D Center

Yoshihiko Nakamura
University of California
 Santa Barbara

David Orin
Ohio State University

Simeon Patarinski
Bulgarian Academy of
 Sciences

Jocelyne Pertin-Troccaz
Institut National
 Polytechnique de Grenoble

Marc Renaud
LAAS-CNRS (France)

Stanley Rosenschein
Teleos Research

J. Kenneth Salisbury
Massachusetts Institute
 of Technology

Victor Scheinman
Automatix Inc.

Jean-Jacques Slotine
Massachusetts Institute
 of Technology

Russell Taylor
IBM Thomas J. Watson
 Research Center

Roger Tsai
IBM Thomas J. Watson
 Research Center

Kenneth Waldron
Ohio State University

Michael Walker
University of Michigan

Tsuneo Yoshikawa
Kyoto University

Kamal Youcef-Toumi
Massachusetts Institute
 of Technology

The Robotics Review 1

edited by
Oussama Khatib, John J. Craig, and Tomás Lozano-Pérez

The MIT Press
Cambridge, Massachusetts
London, England

© 1989   Massachusetts Institute of Technology

All rights reserved. No part of this book may be reproduced in any form by any electronic or mechanical means (including photocopying, recording, or information storage and retrieval) without permission in writing from the publisher.

This book was printed and bound in the United States of America

Library of Congress Cataloging-in-Publication Data

The Robotics review 1.

   1. Robotics. I. Khatib, Oussama. II. Craig, John J., 1955-   . III. Lozano-Pérez, Tomás.
TJ211.R5682 1989   629.8'92   89-8044
ISBN 0-262-11135-7

Contents

Preface ix

Guest Editorial: Problems in Robotics 1
Michael Brady

I Programming, Planning, and Learning

Introduction 27

Articles

On the "Piano Movers'" Series by Schwartz, Sharir, and Ariel-Sheffi 33
John Canny

Triangle Tables in Execution Control and Robot Programming 41
Malik Ghallab

Motion Planning with Uncertainty: On the Preimage Backchaining Approach 53
Jean-Claude Latombe

Grasping: A State of the Art 71
Jocelyne Pertin-Troccaz

Reviews

Sensor-Based Control of Robotic Manipulators Using a General Learning Algorithm, by W. T. Miller III
Reviewed by Rodney Brooks 99

A Robust Layered Control System for a Mobile Robot, by R. A. Brooks
Reviewed by Raja Chatila 103

The Synthesis of Digital Machines with Provable Epistemic Properties, by S. J. Rosenschein and L. P. Kaelbling
Review by Christopher Goad 109

Planning for Conjunctive Goals, by D. Chapman
Reviewed by Stanley J. Rosenschein 113

MH-1, a Computer-Operated Mechanical Hand, by H. A. Ernst
Reviewed by Russell H. Taylor 121

II Sensing and Perception

Introduction 129

Articles

Tactile Sensing for Robots: Present and Future 133
Paolo Dario

Synopsis of Recent Progress on Camera Calibration for 3D Machine Vision 147
Roger Tsai

Reviews

4D-Dynamic Scene Analysis with Integral Spatio-Temporal Models,
by E. D. Dickmanns
Reviewed by H. Harlyn Baker 161

An *n*log*n* Algorithm for Determining the Congruity of Polyhedra,
by K. Sugihara
Reviewed by Jean-Daniel Boissonnat 167

Generating an Interpretation Tree from a CAD Model for 3D-Object Recognition
in Bin-Picking Tasks, by K. Ikeuchi
Reviewed by Robert C. Bolles 169

Interpretation of Contact Geometries from Force Measurements, by J. K. Salisbury
Reviewed by Mark R. Cutkosky 175

On Optimally Combining Pieces of Information with Application to Estimating
3-D Complex-Object Position from Range-Data, by R. M. Bolle and D. B. Cooper
Reviewed by Hugh F. Durrant-Whyte 181

Epipolar-Plane Image Analysis: An Approach to Determining Structure from
Motion, by R. C. Bolles, H. H. Baker, and D. H. Marimont
Reviewed by W. Eric L. Grimson 185

Machine Interpretation of Line Drawings, by K. Sugihara
Reviewed by Deepak Kapur and Joseph L. Mundy 179

TINA: The Sheffield AIVRU Vision System, by J. Porrill et al.
Reviewed by David G. Lowe 195

Three-Dimensional Model Matching from an Unconstrained Viewpoint,
by D. W. Thompson and J. L. Mundy
Reviewed by David H. Marimont 199

III Kinematics, Dynamics, and Design

Introduction 203

Articles

A Survey of Kinematic Calibration 207
John M. Hollerbach

A Review of Research on Walking Vehicles 243
Vijay R. Kumar and Kenneth J. Waldron

A Bibliography on Robot Kinematics, Workspace Analysis, and Path Planning 267
J. M. McCarthy and R. M. C. Bodduluri

Design and Control of Direct-Drive Robots—A Survey 283
Kamal Youcef-Toumi

Reviews

Computationally Efficient Kinematics for Manipulators with Spherical Wrists Based on the Homogeneous Transformation Representation, by R. P. Paul and H. Zhang
Reviewed by Roy Featherstone 303

Development of Holonic Manipulator, by M. Hirose, Y. Ikei, and T. Ishii
Reviewed by Hirochika Inoue 307

Numerical Simulation of Time-Dependent Contact and Friction Problems in Rigid Body Mechanics, by P. Lötstedt
Reviewed by Matthew T. Mason 311

Analysis of Multifingered Hands, by J. Kerr and B. Roth
Reviewed by Yoshihiko Nakamura 315

Efficient Parallel Algorithm for Robot Inverse Dynamics Computation, by C. S. G. Lee and P. R. Chang
Reviewed by David E. Orin 319

A Study of Multiple Manipulator Inverse Kinematic Solutions with Application to Trajectory Planning and Workspace Determination, by P. Borrel and A. Liegois
Reviewed by Marc Renaud 323

Mechanical Design of Robots, by E. I. Rivin
Reviewed by Victor Scheinman 327

The Calculation of Robot Dynamic Using Articulated-Body Inertias, by R. Featherstone
Reviewed by Michael W. Walker 331

IV Motion and Force Control

Introduction 335

Articles

Stability Problems in Contact Tasks 339
Neville Hogan and Ed Colgate

Robot Planning and Control via Potential Functions 349
Daniel E. Koditschek

Adaptive Trajectory Control of Manipulators 369
Jean-Jacques E. Slotine and John J. Craig

Reviews

Compliance and Force Control for Computer-Controlled Manipulators, by M. T. Mason
Reviewed by Haruhiko Asada 379

Stability and Robustness of PID Feedback Control for Robot Manipulators of Sensory Capability, by S. Arimoto and F. Miyazaki
Reviewed by John J. Craig 385

Satellite-Mounted Robot Manipulators--New Kinematics and Reaction Moment Compensation, by R. W. Longman, R. E. Lindberg, and M. F. Zedd
Reviewed by Gerd Hirzinger 389

Optimal Trajectory Planning for Industrial Robots, by R. Johanni and F. Pfeiffer
Reviewed by Nenad M. Kircanski 391

Extended Kinematic Path Control of Robot Arms, by E. Krustev and L. Lilov
Reviewed by Simeon Patarinski 395

Experiments in Force Control of Robotic Manipulators, by J. A. Maples and J. Becker
Reviewed by Kenneth Salisbury 401

Adaptive Control of Mechanical Manipulators, by J. J. Craig, P. Hsu, and S. S. Sastry
Reviewed by Tsuneo Yoshikawa 405

Preface

The undertaking of this annual has been motivated by the rapid growth of the still-young field of robotics. With the ever-larger volume of publications in journals and conference proceedings, it is difficult for those involved in robotics, particularly those who are just entering the field, to stay abreast of developments. This task is made even harder by the multidisciplinary nature of the field.

The Robotics Review is an annual serial publication that brings together reviews of selected papers, synopses of selected technical areas, and annotated bibliographies. The contents are organized in four parts, devoted to the following areas: Programming, Planning and Learning; Sensing and Perception; Kinematics, Dynamics, and Design; and Motion and Force Control.

The *Review* is the result of the efforts of an interdisciplinary group of Contributing Editors who themselves are involved in robotics research in various countries around the world. It is an enormous task to insightfully provide coverage of all the areas of robotics research, and a large, versatile, and evolving group of individuals committed to such an endeavor is vital in achieving this goal.

Research work to be reviewed is selected by the contributing editors on the basis of its significance and its contribution to the field. The reviews (which are not simply reprints of abstracts) attempt to emphasize new results and provide clear statements of how the work relates to previous efforts. The *Review* also contains surveys summarizing particular subareas of research. These synopses, along with the annotated bibliographies, tie together the results of many researchers and help place them in historical perspective.

It is our intent to provide a broad, in-depth, yet concise view of the field. We believe that this first *Robotics Review* constitutes a step toward this goal, and we hope it proves useful to the research and academic communities.

Oussama Khatib
John Craig
Tomás Lozano-Pérez

The Robotics Review 1

Problems in Robotics

Michael Brady

Robotics Research Group
Department of Engineering Science,
University of Oxford,
Oxford OX1 3PJ
England

Introduction

This article offers a personal view of some of the problems that need to be solved if we are to develop a *Science of Robotics*.

Robotics is a broad, inter-disciplinary subject that involves various facets of sensing and effecting, as well as many different levels of thinking that integrates sensing with action. "Thinking" takes place at many different levels. At its lowest levels, thinking is largely reflexive. Controlled responses to sensed changes in the environment are necessary, for example, for protective autonomy of a robot (Steer 1988) and they typically need to be fast and reflex-like. Naturally, reflexive thinking involves only rudimentary descriptions of the environment. At higher levels, intelligent thought enables the robot to cope with a complex, uncertain, changing world. To be intelligent, Robotics must rely upon Artificial Intelligence. An earlier essay (Brady 1985) explored this relationship. Here our emphasis is less on the state of the art in Robotics and more on outstanding problems.

The problems have been grouped (in no particular order) under the following ten headings: sensing (other than vision), vision, mobility, design, control, generic operations, reasoning, geometric reasoning, and system integration.

Sensing

Problem 1: The continuing need for better sensors

Tactile sensors that overcome basic sensitivity and ruggedness problems, and that compute shear, stress, and strain have been slow to emerge from the research laboratory. Until better sensors are available, theories for interpreting touch data will continue to languish.

As robots begin to move around factories (Brady et. al. 1988) or outdoors, they confront an awkward hiatus in direct range sensing. Almost all structured light range sensors have been developed for the range 0 – 2 metres. Because the magnitude of the speed of light implies high (Gigahertz) rates for time-of-flight sampling, robust, active electromagnetic range sensors are currently only available for ranges in excess of 50 – 100 metres. One partial exception is the ERIM sensor used in the CMU Navlab project; but it has several other problems (see Hebert and Kanade (1989)). Factory mobile robots need range sensors that operate in the range 2 – 10 metres, and they are currently not available. Clearly, there is a limit to the amount of laser power than can be beamed around a primarily metallic environment that contains people. Bell Laboratories have experimental prototypes of an infrared sensor that measures range up to 15 metres.

Continued development of all kinds of sensors, including TV cameras, is a problem for Robotics.

Problem 2: Sensor models

Currently, the choice of which sensor(s) to use in an application is based on experience or non-technical considerations such as availability or cost. More rational choice demands sensors models that use a better description of a sensor's abilities and limitations. Similarly, statistically combining uncertain sensor measurements requires sensor models to be developed that make explicit the type of information provided by the sensor, the error distribution in that measurement, and the sensor's characteristic failure modes. Sensor models may be quantitative, such as the corrupted Gaussian probabilistic models championed by Durrant-Whyte (1988), or they may be qualitative, such as the logical sensor concept pioneered by Henderson and Shilcrat (1984). In general, models of both sorts are required, together with functions that relate them, but how this should best be done remains an open problem.

Problem 3: Sensor control

reliably all of the time. This is the single most important reason why edge detection, for example, continues to be avoided in industrial vision. Speed is of lesser importance; it can always be increased if it is considered worthwhile and to do so is financially feasible.

Problem 6: Seeds of perception

Not all visual information is created equal (Brady 1987). Intensity changes provide more information than points in the middle of regions of smoothly changing or constant brightness. Since they provide more information, they offer tighter constraint. Corners offer even more information than edges, and so they in turn offer tighter constraint. Of course, it is harder to determine reliably image locations where the intensity changes significantly than locations where it barely changes at all. With care, however, it can usually be done. In turn, it is harder to determine image corners than edges; but again it can mostly be done. As Nagel (1983), Gong (1988), and Harris (1988) have shown, it is worth the effort to do so. We refer to corners, and similar locations of tight constraint as *seeds of perception*, and they can be found in processes as diverse as stereo, motion, shape from contour, and lightness.

Seeds of perception suggest a process in which interpretations are grown from locations of tightest constraint to those of decreasing constraint. Such a regime is implicit in a number of algorithms; but it currently falls some way short of a theory. The problem is to develop seeds of perception so that it involves (i) a precise model of the information or constraint associated with a feature; (ii) robust techniques for computing seeds; and (iii) a mechanism for combining the constraints associated with spatially distributed seeds.

Problem 7: Model-based vision

Current model-based vision systems are narrow minded. Some argue that since objects may be overlapped, the visibility of local features cannot be guaranteed. This leads to the pessimistic stance that assumes that those features will never be visible, and programs based on such a view fail to take advantage of distinct local features if they are visible. In consequence of their pessimism, such programs are ineffective when there is too much extraneous clutter.

Uncertainty is only one of many problems associated with sensors. An equally important problem stems from the fact that many sensors (including force, touch, and even certain visual processes) only deliver information that is local relative to the overall size of the object being sensed. This raises the unsolved sensor control problem: how to compute the most useful place to look next given what one has sensed previously and given prior expectations in the form of models.

Problem 4: Architectures for real-time sensory processing

Sensors are distributed in space, and they yield information at different rates, of different sorts, that needs to be interpreted in the light of past sensing and expectations. For all but the simplest sensors and tasks, serial hardware is too slow to support real-time sensor-based control. Parallel architectures are needed; but few compelling suggestions have been proposed. For example, early blackboard architectures depended critically on the time to process information being long compared to the time to write or access that information from a shared block of memory. This is unlikely to be true for a sensory processes distributed over a local area network. It has been suggested that MIMD architectures based on processors such as Transputers are appropriate for implementing schemes for sensor integration; but the suggestion has yet to be realised in practice.

Vision

Problem 5: Applied optics

There is a considerable body of heuristic knowledge about what might be called applied optics. This includes the placement of light sources and cameras, and the kinds of effects that are clearly visible (even to a computer) in images of various wavelengths and states of polarisation. An application that seems hard to make completely reliable or efficient may often be solved easily if polarised or ultraviolet light is used, or two images are superimposed to overcome shadowing effects.

A corpus of knowledge does not constitute a theory, and the largely *ad hoc* nature of such knowledge means that it is rarely written down in books. All too often the tricks have to learned afresh by engineers. Vision would be more widely adopted in Robotics if processes worked

Others argue for image structures such as parallelism, junctions, or salient features, and their programs collapse unceremoniously if those features are not visible. Graceful degradation in performance on the one hand, coupled with opportunism and the ability to recognise model instances in a sea of clutter on the other, may be possible if there is a hierarchy of representations of image structure, at one level making parallelism, symmetry and the like explicit, at a lower level simply consisting of a piecewise linear approximation to image edges. It is an open problem whether opportunistic programs that degrade gracefully can be built, and whether or not such a hierarchy of image structure representations holds a part of the key.

Problem 8: Dynamic vision

Few computer vision programs have a continuing perceptual existence. Instead, they gaze at one, or at most a small number, of images until their processing is done. This often takes time, though the author often consoles us that the processing could, in principle, be done in "real time". Even research in visual motion has worked on short image sequences, often consisting of as few as two images. Recently, several programs that aim at a continuing existence have been constructed, including (Khatib, 1986; Andersson 1988; Dickmanns 1988; Ayache and Faugeras 1988; Charnley and Blissett 1988). Andersson, for example, argues persuasively that building a vision program that had to work continuously and in real time qualitatively changed the problems that he confronted.

Many of the systems described in the current literature are based on ideas from control theory, particularly the extended Kalman filter. More generally, the relationship between control theory and vision needs to be explored more thoroughly, including (high-bandwidth) hardware architectures for achieving vision-based control. The term "dynamic vision" has recently been coined for this overlap between control based on vision and continuing vision based on ideas from control theory. There is no easy off-the-shelf solution to dynamic vision: ideas about hierarchical control are rudimentary; and since the matrices involved in EKFs are not banded it is difficult to implement EKF schemes in real time. In any case, the optimality results of the Kalman filter do not apply to the EKF. The development of dynamic vision systems, together with appropriate architectures is timely in view of the increasing availability of parallel hardware.

Problem 9: Shape

As Winston (1985) has noted, the primary role of a representation is to make information explicit and/or to expose the constraints that are exploited in recognition and reasoning. A system capable of performing a variety of tasks needs to make a variety of sorts of information explicit, hence needs to be equipped with a variety of representations. Nearly all vision programs have been constructed for a single goal, and they have used/championed a single representation that was chosen to be more or less well suited to the information required to achieve that goal.

One well-studied goal has been to determine the position and orientation of an object. The position and orientation of a rigid object can be specified by giving the position of one point, for example the center of mass, and a set of three characteristic directions, for example the principal axes of the object. Since the center of mass and principal axes are defined by integrals over the object, representations that accumulate information can be used to achieve the goal. Examples include the Hough transform and the Extended Gaussian Image.

More detailed representations are required if the goal is to recognise, inspect, or grasp an object, or to navigate amongst stationary objects. Yet more is required if the object is flexible or articulated with movable parts, for example a robot arm. Polyhedral approximations have been used for recognising objects, even when they are overlapped, and for some automatic assembly operations. The later section on Reasoning explains why polyhedral representations are of limited use in automatic planning of assemblies that involve parts such as shafts, gears, keyways, and cams.

Other representations try to make explicit commonly occuring geometric regularities, such as translational, reflectional, or rotational symmetry; surface discontinuities; and local geometric or topological surface features. Yet others attempt to represent the hierarchical decomposition of a shape into parts, for example to link function directly to form. A useful set of criteria by which to judge shape representations has been identified by Marr (1982), and augmented by other authors.

Still, there is no elixir for shape, no single representation that is useful and effective in all situations. It is probably illusory to chase such an all-embracing representation given the diversity of tasks that might confront an intelligent robot. However, no vision-guided system can mobilise a variety of representations, intelligently choosing between them to achieve different goals. And though we have progressed far in vision, no system can compute rich and useful descriptions of the complex, curved, textured, multicoloured objects that populate our everyday world.

Mobility

Problem 10: Systems architecture

Mobile robots focus attention on many of the problems discussed elsewhere in this article: sensor integration; dynamic sensing and vision; sensor control; and navigation. These components aside, mobile robots also force attention on systems integration (see section on System Integration). The appropriate hardware and software architecture for a multisensor, purposive, mobile vehicle remains an open problem, though there has been some preliminary work (Brooks 1989, Giralt et. al. 1984, Harmon 1987).

Problem 11: High bandwidth control

Advanced autonomous guided vehicles, such as that developed and marketed by GEC-Caterpillar, are based on low bandwidth industrial controllers. As noted earlier, the GEM-80 controller applies a Kalman filter to a sequence of trajectory set points using sensed data from the vehicle's odometers and laser ranging system.

As AGVs move into more complex, less structured environments, typified by outdoors applications such as stockyarding, the sensors will, of necessity, need to provide richer information about the environment. Vision and range sensors produce information at much faster rates than sonar and infrared sensors used in current AGVs. In consequence, high bandwidth controllers will need to be developed, based on more sophisticated nonlinear filters than the EKF. The design of such filters and controllers will be a major challenge for advanced Robotics, and for Computer Vision.

Problem 12: Sensor-based steering and foot placement

Legged robots barely know where to place their feet. Notwithstanding the extraordinary achievements by Raibert's group in making a biped robot perform gymnastics and a quadruped robot trot, no legged robot could reliably go for a jog through a park. Wheeled robots don't fare much better. They may be able to (slowly) circumnavigate obstacles; but they cannot judge the quality of the terrain and so which way to steer locally.

There are several dimensions to local steering control. First, the sensing problems of locating small rocks and holes and estimating the mechanical surface properties of the ground. Second, given a perceptual representation of the terrain immediately in front of the vehicle, deciding in real-time which way to steer or where to place the feet to provide stability, traction, and support.

Problem 13: Energy and legged locomotion

Why does a horse change from a trot to a canter, from a canter to a gallop? Part of the answer seems to be to minimise the energy expended for a given speed, terrain, and payload. Such energy considerations have yet to be studied in research on legged robots.

Design

Problem 14: Designing a head, knuckle, or knee

Roth's (1989) constraint equations, Jacobsen et. al.'s (1989) acuator model and parametric design, Asada's (1989) generalised inertia ellipsoid, and Salisbury's (1989) grasp stability all point the way toward a theory of design for robot arms, end effectors, and legs. Necessarily the theories are, to varying extents, abstract. The power of Roth's approach, for example, precisely derives from postulating idealisations such as instantaneous quantities at a point.

Though each of the authors referred to in this section have designed and built innovative robot structures, applying the currently fragmentary theory of design to produce real devices poses a considerable challenge. The title of the problem enumerates three specific devices to which the theory of design might be applied:

- the control of attention is a problem for visually guided control of robots as there is too much to look at in an image. Directing attention involves controlling head and eye movements. A number of "head" mechanisms have been built for this purpose but they are clumsy when compared with the beautiful structures implemented by the authors referred to above. A study of head designs for two (and three) camera vision systems would be of immediate interest to vision researchers. Coincidentally, the physiology and control of the human eye-head system has been closely studied, though this should not necessarily constrain the design.

- The fingers of a human hand approximate RRRR kinematic chains. This is not quite so, as the distal knuckle joint can not move independently of the proximal joint. Wood, for example, has investigated the intricate tendon structure of human fingers. Knuckles are largely driven remotely by an extraordinary tangle of tendons; though finger muscles do provide additional actuation. While Jacobsen, Wood, and their colleagues have done great work on robot hand design, the knuckle remains a paradigm of the challenge of building joints and (see below) actuators that are light, small, accurate, and fast.

- The knee is one of the most complex joints in the human body. It achieves remarkable performance in active suspension during running, jumping, and carrying loads. Prosthetic knees have been developed with some success, though their designs are but pale imitations of real, human knees. The mechanical design of a human knee equips it for the dynamic challenges that it must achieve; yet even the geometry of how it does this is scantily understood. The knee joint is a paradigm of relating mechanical design to complex control and dynamics for active suspension.

Problem 15: Actuators and Mechatronics

Actuation is all too-often the Achilles' heel of Robotics. Available actuators are either too big, too difficult to control (most hydraulics), too slow, not powerful enough, or have to overcome enormous friction from backlash, cogging and other non-linear effects of gearing (see

(Hollerbach 1989)). Choosing an actuator tends to be driven heuristically by a single facet of the design; then the further consequences of that choice have somehow to be overcome. For example, if strength or power-to-weight ratio is a key design requirement, hydraulic or pneumatic actuation is often chosen; but then the limited (relative to electric motors) possibilities for fine control have to be faced.

The deficiencies of current actuators have retarded the development of high performance robot arms. Hollerbach (1989) notes "the advent of direct drive technology has been a big boost to experimental robotics, since meaningful control studies incorporating dynamic models are being carried out". But direct drive motors are also plagued with problems, for example the significant levels of torque ripple (despite manufacturer's published claims to the contrary). Torque ripple severely degrades signal-to-noise ratio in the signal emanating from the motor, hence limits control. Other actuator technologies have been explored, including arrays of tiny electrostatic motors implemented on silicon chips and shape memory alloys.

A recurrent theme in the papers referred to in this section is the need to design a mechanical structure in tandem with the control. This is in essence one of the key ideas of *Mechatronics*. Conventionally, a motor is connected to a computer that is at some distance removed from the motor. The electrical characteristics of the motor are fixed: it is the computer's thankless task to try to make the motor appear other than it really is, in effect trying to make a silk purse out of a sow's ear. A fundamental idea of Mechatronics is to put the computer *inside* the motor, and to make the electronics of the motor define its performance characteristics. The idea is that simply by changing EPROMs, and the software they contain, one might change the velocity or torque profiles of the motor. In view of critical role played by actuators, realising this dream is a key problem for Robotics.

Problem 16: Parallelism, flexibility, and Robotworld

The potentially enormous space of designs for Robotics has barely been charted. The vast majority of arms are serial kinematic chains. Key contributions will be made by studying relatively uncharted areas of the design space:

- Asada (1989) discusses some of the advantages of parallel structures such as pantograph mechanisms, as does Raibert (1989) in reviewing Hirose's quadruped. Asada's parallel direct drive arm should encourage further efforts.

- Most arms are supposed to be rigid, though all have some (unmodeled, uncontrolled) flexibility in both joints and links. Flexible arms can be lighter and faster than rigid arms, and they can support much greater payload-to-weight ratios. The downside is that flexible arms are considerably more difficult to control. Daniel (1987, 1988) has designed flexible one, two, and three degree-of-fredom arms. Torque ripple of the direct drive motors powering the three degree-of-freedom Rotabot currently poses a problem for the effective control of the device. The design and control of flexible arms remains an open problem.

- Scheiman (1987) has proposed and implemented a totally integrated system of sensors, actuators, and fixtures called *Robotworld*. As the name implies, the entire design was oriented toward the needs of the robot and away from the needs or biases of the human designer putting it together. In essence, it works out the Gedanken exercise: "if I were a robot, what would I need to know, to what accuracy?".

Control

Problem 17: Experimental evaluation of modern control techniques

Too many papers that are backed up only by software simulation continue to be written on the application of modern control techniques to Robotics. To be sure, simulation has an important place in the development of control algorithms, for example in analyses of stability, and in checking one's proper use of numerical methods. However, the real world is hard to simulate. For example, Gaussian noise models are only an approximation (occasionally poor) to reality.

In view of the problems for control design facing the robot system designer, it is important that extensive experimental evaluation be made of the relevance and applicability to Robotics of the techniques of modern control theory. Hollerbach (1989) notes that this is not at all easy; he expresses the opinion that direct drive motors have only recently made feasible experimentation with control based on dynamic models. Nevertheless, experimental evaluation is imperative if the claims for modern control techniques are to be taken seriously and become routine in Robotics practice. Here are two specific examples:

- *on-line parameter estimation*: High performance arms with discontinuous changes in inertial loading implies accurate, on-line parameter estimation. Arimoto (1989) emphasises the importance of "learning" for accurate trajectory following. Hollerbach (1989) reports some encouraging initial results for the MIT direct-drive arm. On-line parameter estimation is essential to self-tuning adaptive control. Daniel (1987) reports some initial findings with the GPC algorithm. Clearly, considerably more work needs to be done.

- *computed-torque control*: is a promising technique for achieving force control of a single joint with a colocated force sensor. It has yet to be convincingly demonstrated in practice, however.

- *Multiple input - multiple output control*: joints 2 and 3 of the Unimation PUMA robot, and joints two and three of a SCARA robot have parallel axes. Interaction torques experience at joint 2 (for example), caused by the motion of joint 3, can be the dominant considerations in its motion. The independent joint control techniques that predominate in Robotics simply reject the interaction torque as a disturbance. An alternative approach is to view joints 2 and 3 as a two-input/two-output system, in which the interaction torques are (approximately) linear. Multivariable control techniques have, however, rarely been applied to robots.

Problem 18: Multiprocessor architecture for controlling a mobile vehicle or a robot hand

In the later section on system integration, we note that the software architecture of robot systems inevitably amounts to a real-time operating system. A number of multiprocessor architectures for control have been designed and implemented, for example based on recursive formulations of the dynamics of an open kinematic chain, a custom built bus, or on Transputer arrays. Brooks (1989) argues for a different approach to systems architecture for robot control that he calls a *subsumption architecture*. He presents some initial results with an implementation of the approach.

It is too early to know which of these multiprocessor architectures will be most appropriate for which aspects of robot control. Multiprocessor control of a sensor-guided vehicle, or of a multi-fingered hand, would provide good experimental data for judging the merits of different architectures.

Generic Operations

Problem 19: Increasing the repertoire

Operations with a reasonable claim to be generic that have barely been studied include: (i) *impacts* such as tapping, striking, dropping, and chopping; and (ii) *twisting* such as screwing, spinning, and drilling. No doubt others can be identified. More complex geometries such as L-shaped parts and T-sections pose additional problems.

The RCC is designed to accommodate two different events that may occur during chamfer crossing: a sideways translation and pure rotation. That is, the RCC blends the *two* manipulation operations of pushing and twisting, in the context of chamfered insertion. The axis of the twist is orthogonal to the direction of push. Bayonet fitting fixtures (such as some lightbulbs) also involve the simultaneous application of a push and a twist; but here the axis of the twist is parallel to the direction of push. In short, there is a hierarchy of operations in manipulation and assembly. However, it is unclear how different representations of the sort proposed by Mason (1989) can be combined (subject to temporal constraints) to automatically build representations for these "higher order" processes, in the way that pushing leads to squeeze-grasping.

Problem 20: Adding sensing

Passive compliance has the advantage of speed, as computation and sensing are significantly reduced and occasionally eliminated. The price paid is adaptability: software control algorithms are easier to change, perhaps on-line, than are mechanical structures. A descendant of the RCC was instrumented for force control to make the device applicable to a wider range of tasks. Recently, Mason and his colleagues have begun to investigate the trade-offs between (i) a sequence of sensor-less operations such as allowing a part to collide with the sides of a tray, and (ii) using a sensor to constrain the configuration of the part. Such analyses have great potential throughout Robotics. The heart of the problem is to relate the constraint on the state of an object as a result of a sensing step with the result of a sensor-less operation. This is relatively straightforward in a two-dimensional, rectangular environment. More generally it involves several other problems discussed in this essay, for example sensro models, logical sensors, and shape representations.

Problem 21: Qualitative process models

The applied mechanics of most practical instances of any manipulation operation are likely to be too complex to carry out analyses of the sort described by Mason (1989) and Whitney (1989). Significantly, Mason's work emphasises representations that form the basis for automatic planning of pushing, squeezing, and orienting parts. This suggests that qualitative accounts be developed for the manipulation operations, perhaps based on the idea of qualitative reasoning (naive physics).

Reasoning

Problem 22: Real-world planning

Early work on AI planning concentrated on off-line planning of simple actions in an idealised, static, toy world. Although this represents a considerable abstraction of planning, improvising, and reasoning about the real world, a number of important issues were uncovered and studied, including the frame problem, the need for hierarchical planning, and the need to represent dependencies and constraints. After something of a hiatus in planning research there is a resurgence of interest, this time on issues that are clearly important in any practical application:

- *temporal planning* of time-critical events (in particular exploring the relationship between planning and classical control). Temporal planners must satisfice in the sense of being able to offer their best suggestions for a plan when a resource such as time is exhausted;
- *planning in the face of uncertainty* including sensor-based planning; and
- *reactive planning and situated action*, in which a planner responds quickly to a changing world in which changes invalidate or render irrelevant a plan conceived in a previous world state. Agre and Chapman (1987) describe a program that plays a video game in which the hostile environment is continually changing. They suggest an indexical-functional representation for variables as opposed to that used in conventional robotics programs.

Despite this resurgence of effort, each of these topics requires considerable research.

Problem 23: Qualitative process theory and naive physics

Hopcroft and Krafft (1984) point out that conventional engineering is founded upon a variety of representations, each one of which supports some operations but makes others infeasible. Differential equations and finite element approximations have been used with great success to model heat and stress distributions in solids and in processes such as injection moulding. Such models are inappropriate for macroscopic processes such as assembly. Classical mechanics, though non-deterministic because of friction, is the basis for most current work in assembly and robot dynamics and control. Geometric modeling systems support computer-aided design and the simulation of the kinematics of robot motion. But none of these representations is appropriate for reasoning about whether or not a rusty screwdriver with a slightly chipped blade might be adequate for removing a partly worn screw.

The lesson is that modeling the physics of a situation is not necessarily the way to solve a problem of Robotics. On occasion, perhaps more often than we might initially suppose, a *qualitative*, symbolic, "naive" representation might be more appropriate. The past few years have seen a rapid growth in work on qualitative reasoning, together with its attendant technologies of truth-maintenance systems and sign algebras. (Bobrow, 1985) is a good introduction.

Qualitative reasoning occasionally needs to be refined to be quantitative but the existing qualitative reasoning techniques support this poorly or not at all ((Williams 1988) addresses this issue in his work on sign algebras). A second area in need of research is covered in the next topic, shape and dynamics: existing qualitative reasoning systems use only the simplest representations of objects and so their application to practical problems such as arise, for example, in production engineering is extremely limited.

Problem 24: Shape (and dynamics)

Some time ago we introduced a project called the *Mechanic's Mate* (Brady et. al. 1984) that aimed to reason about simple problems of assembly and disassembly, for example those that arise in woodwork. It turns out that there are many different ways to extract a nail or screw; which method is used depends upon the fine details of the shapes of the objects involved. We proposed to apply the smoothed local symmetry representation of shape since, based on it, we had implemented a system that generated and learned semantic network representations of shape starting from images. The project has been in abeyance since the author moved to Oxford; but the issues that it raised have not diminished in importance or timeliness. In related work, Winston et. al. (1984) explored the relationship between function and form in Robotics reasoning, based on the generalized cylinder representation of shape.

By and large, however, Robotics reasoning systems ignore shape; at most they rely on polyhedral approximations. Yet shape and symbolic representation of motion and dynamics are central to our ability to reason about the world. "If you put that too near the edge of the bench it will topple over"; "if you put that on top of the structure it will likely collapse, unless you first shore it up with a scaffold" (Fahlman 1974); and "if you put that milk on top of the eggs they will break because they are too delicate".

Related problems arise in building (perhaps automatically) representations of large scale space for mobile robots. Representations of shape (for example, the symmetric axis transform, the Delaunay triangulation, and generalized cylinders) originally developed in computer vision have been used in geometric reasoning and path planning.

Work on shape representations for practical reasoning and for robot perception continues. Recently, for example, Fleck (1988) has noted a number of problems with representations of shape and time that correspond to subsets of $\Re^n$. She proposes instead a generalisation of regular cell complexes to include boundaries. Considerably more work on the development of shape representations for practical reasoning is required.

Devising representations of shape that suffice for reasoning about how to assemble or troubleshoot one of a family of gear boxes would raise most of the important issues.

Problem 25: Design-to-Product

In the vast majority of manufacturing applications, design takes little or no account of the processes of production. Production engineers complain about the often unnecessary challenges that result and, more often than not, change the specified design *force majeure* to make their own job simpler. Recent AI projects have aimed to close the loop between design and production.

One such project was conceived by Popplestone (1986), and involved modeling shafts, transmissions, and motors parametrically, and then propagating constraints between these components to derive a design. The project subsequently became a demonstrator project for the UK Alvey program (Fehrenbach and Smithers, 1987). Other work (Fox and Kempf 1985) has addressed important issues such as tolerancing.

Although such pioneering projects point the way toward closing the loop between design and production, they barely scratch the surface of what is required in order to transform one of the blackest arts of manufacturing into a science.

Geometric Reasoning

Problem 26: Knowledge of assembly

Despite the attention that has been paid to systems that reason about assembly, we still don't know much about how to plan automatically real assembly operations such as the construction of electric motors, gearboxes, televisions, and washing machines. Viewed as abstract geometric problems that involve such convoluted shapes as gears and cams, they are probably infeasible so far as general solutions to "findpath" (Lozano-Perez and Wesly 1979) are concerned.

Fortunately, we know a considerable amount about real assemblies: electronics boards have connecting sides that slot into backplanes; gears mesh with other gears and do so easily if they are spun together; and motors have housings with designated places that enable them to be fit to brackets. Some of our knowledge of assembly involves the shapes of objects and the constraints that shape representations suggest in practice for the motions of objects. Rotationally symmetric objects tend to be moved along or twisted about their axes: we know in practice that combining these motions is an effective technique for inserting shafts and adding sleeves. A keyway on a shaft must be lined up with the keyway on the hole. It is surely unnecessary to *formally* derive these motions from abstract geometric representations. In fact, the numerical complexity and inaccuracies of boundary representations of the surfaces of shafts, sleeves, and keyways may well preclude such motions from being derived at all.

It seems necessary to mobilise both (i) the general knowledge of geometry and dynamics that are the topics of Mason (1989) and Lozano-Pérez and Taylor (1989), and (ii) commonsense knowledge about real assemblies like that enumerated in the previous paragraph. The development of techniques that can effectively mobilise both sorts of knowledge poses a well-known challenge to AI; but such techniques are crucial if we are to develop useful assembly systems.

Problem 27: Fine motion planning

The work of Lozano-Pérez and Taylor (1989) and in Mason (1989) makes a good start at automating and reasoning about the crucial problems involved in fine motions. To an increasing extent, Robotics is concerned with controlled collisions, and this poses as much of a challenge to reasoning as it does to design and control. As we noted in our discussion of manipulation problems, there is much to do. Mason draws attention to the severe assumptions that need to be made for applied mechanics analysis to be feasible. Lozano-Pérez and Taylor draw attention to the need to add more realistic dynamics to the operations they discuss.

Problem 28: Clash detection and its progeny

Geometric problems other than findpath have been studied. For example, Cameron (1988) has proposed a fast, complete algorithm for *clash detection*, which asks if two possibly complex shapes overlap. Such a technique may be useful in path and assembly planning if crucial points in a rough, putative assembly plan can quickly be identified and checked for collisions. Demonstrating the feasibility of such an approach remains an open problem. More generally, a list of practically important, generic, geometric reasoning tasks would be extremely valuable.

System Integration

Problem 29: Communicating processes

From a software engineering perspective, there is an open problem concerning the type of asynchronous software architecture that is most appropriate to a given project. Some projects have explored object-oriented programming (Henderson and Shilcrat 1984), others have provided semaphore guards to a shared data structure (Thorpe, Hebert, Kanade, and Shafer 1988), while others have proposed using communicating sets of processes either based on the CSP-Occam-Transputer model (Dickenson 1988), on the Ada model (Volz, Mudge, Naylor, Brosgol, 1986), or on a parallel version of C (Cox and Gehani 1989). At the moment it is not known how to gauge the appropriateness of a software architecture for a robot system, or what the design principles should be. A problem consequent upon the development of such architectures is the need to develop debugging tools to speed their development.

Problem 30: Distributed hierarchical organisation

Albus (1988) has begun to investigate the distribution of control over large systems. Some processors, such as those cooperating at a workcell, need to communicate quickly, while others may communicate mostly at low bitrates, apart from alarms and exceptions. Mapping asynchronous software architectures onto networked hardware architectures is a difficult problem that faces new challenges in Robotics.

Concluding Remarks

Plainly there is much work to do if we are to realise a *Science of Robotics*. We have enumerated a number of what we consider to be important problems that need to be solved, or at least explored. To conclude in a light-hearted vein, we now enumerate a number of "non-problems", some (but by no means all) of them spoofs, that we hope to deflect people from working on. Any subject has its "pure scholarship" syndrome: problems that once were important or timely take on a life of their own and keep a band of dedicated scholars busy long after they become less important, even irrelevant. There are still far too few researchers working in Robotics for the effort to be diluted by non-problems. The following is a tentative list of non-problems:

Some non-problems

- *Yet another formulation of rigid body dynamics* applied, for its own sake, to an open kinematic chain such as a robot arm waving in the breeze. Of course a judicial choice of representation can critically determine the effectiveness of some process such as computing the dynamics of closed kinematic chains (eg a hand grasping an object, or the legs of a robot contacting the ground), or computing dynamics in parallel; but it is the use of the representation that is of primary interest. Hollerbach (1989) discusses this non-problem in more detail.

- *kinematic trajectory planning* is of limited interest. The papers in (Brady et. al. 1984) pose a number of problems, particularly concerning knot placement in spline trajectories; but the entire approach has been overtaken by the invention of trajectory planning algorithms that incorporate the robot dynamics (see, for example, (Tan and Potts 1988)).

- *gratuitous complexity results* Complexity results about spatial reasoning problems (for example *the piano mover's problem* have provided deep insight into the intrinsic difficulty of many of the problems that are confronted regularly in Robotics. However, some problems have assumed a life of their own with no obvious application to Robotics. We caricature this with the spoof paper title: "the tethered piano tuner's problem I: a single tuning fork".

- *resurrecting well-known limited techniques* The needs of a particular application may dictate the use of a well-known technique whose (typically severe) limitations are also well known. Reasons for applying such techniques include speed, cost, processing power, or memory limitations. Applying such techniques may be expedient; but the fact that the application is in Robotics does not make it a contribution to Robotics Science. Examples include: (i) using the Sobel operator to find edges quickly or applying the Hough transform to inspect widgets; (ii) using (chronological) backtracking for robotic reasoning or planning.

- *applying a conventional SIMD architecture to a simple task in vision and/or control* This is another instantiation of the previous non-problem.

References

Agre, P., and Chapman, David. 1987. "Pengi: an implementation of a theory of activity", Proc. AAAI, Seattle

Albus, James S. 1988. "The central nervous system as a low and high level control system", in Dario ed. 1988. *Sensors and Sensory Systems for Advanced Robots*, NATO ASI series, Springer-Verlag.

Andersson, Russell L. 1988. *A Robot Ping-Pong Player*, MIT Press.

Arimoto, Suguru. 1989. "Control", in Brady (ed.) *Robotics Science*, MIT Press

Asada, Haruhiko. 1989. "Arm design", in Brady (ed.) *Robotics Science*, MIT Press

Ayache, Nicholas, and Faugeras, Olivier D. 1988. "Maintaining representations of the environment of a mobile robot", in Bolles and Roth (eds.), *Fourth Int. Symp. Rob. Res.*, MIT Press, 337 – 350.

Bobrow, Daniel G. 1985. *Qualitative Reasoning about Physical Systems*, MIT Press

Brady, Michael. 1985. "Artificial Intelligence and Robotics", Artificial Intelligence, 26, 79–121.

Brady, Michael. 1987. "Seeds of Perception", Proc. 3rd Alvey Vis. Conf. Cambridge, UK.

Brady et. al. 1984. "The mechanic's mate", O'Shea (ed.) *ECAI84: Advances in Artificial Intelligence*, North-Holland, 681–696.

Brady et. al. 1988. "Vision and the Oxford AGV", Proc. Image Processing '88 Conf., London.

Brooks, R. A. 1989. "The whole iguana", in Brady (ed.) *Robotics Science*, MIT Press.

Cameron, S. A. 1988. "Efficient Intersection Tests for Objects Defined Constructively", Int. J. Rob. Res. 8.

Charnley D. and Blissett R. J. 1988, "Surface reconstruction from outdoor image sequences", Proc. Alvey Vision Conf. 153-158.

Cox, I. J., and Gehani, N. J. 1989. "Concurrent programming and robotics", Int. J. Rob. Res. (to appear)

Daniel, R. W., Irving, M. A., Fraser, A. R., and Lambert, M. 1987, "The control of compliant manipulator arms", in Bolles and Roth (eds.), *Fourth Int. Symp. Rob. Res.*, MIT Press, 119 – 125.

Daniel, R. W. 1988 "A compliant direct drive industrial robot arm", Proc. RoManSy, Sept, Udine, Italy.

Dickenson, Mark. 1988. "A parallel-processing controller architecture for force control of an industrial robot", Proc. Second Wkshop on parallel processing and control, Bangor, UK.

Dickmanns, E. D. 1988. "4D-dynamic scene analysis with integral spatio-temporal models" in Bolles and Roth (eds.), *Fourth Int. Symp. Rob. Res.*, MIT Press, 311 – 318.

Durrant-Whyte, H. F. 1988. *Integration, Coordination, and Control of Multi-sensor Robot Systems*, Kluwer Academic Publishers.

Fahlman, Scott. 1974. "A planning system for robot construction tasks", Artif. Intell. 5(1), 1 – 49.

Fehrenbach, Paul, and Smithers, Tim. 1987. "Design and sensor-based robotic assembly in *design to product* project", in Bolles and Roth (eds.), *Fourth Int. Symp. Rob. Res.*, MIT Press, 391 – 399.

Fleck, Margaret M. 1988 *Boundaries and Topological Algorithms*, PhD thesis, MIT Artificial Intelligence Laboratory.

Fox, B. R., and Kempf, K. G. 1985. "A representation for opportunistic scheduling", in Faugeras and Giralt (eds.), *Third Int. Symp. Rob. Res.*, MIT Press, 109 – 116.

Gelb, Arthur. 1974. *Applied Optimal Estimation*, MIT Press.

Giralt et. al. 1984. "An integrated navigation and motion control system for autonomous multisensory mobile robots", in Brady and Paul (eds.), *First Int. Symp. of Robotics Research*, MIT Press, 191 – 214.

Gong, S. 1988. "Improved local flow", Proc. 4th Alvey Vis. Conf., Manchester, UK, 129 –134.

Harris, C., and Stephens, M. 1988. "A combined corner and edge detector", Proc. 4th Alvey Vis. Conf., Manchester, UK, 147 – 152.

Harmon, S. Y. 1987. "The ground surveillance robot (GSR): an autonomous vehicle designed to transit unknown terrain", IEEE J. Rob. and Aut. RA-3 (3), 266 – 279.

Hebert, Martial, and Kanade, Takeo. 1989. "3-D vision for outdoor navigation by an autonomous vehicle", in Brady (ed.) *Robotics Science*, MIT Press.

Henderson, T. C., and Shilcrat, E. 1984. "Logical sensor systems", J. Robotics Systems, 1(2), 169 – 193.

Hollerbach, John M. 1989. "Kinematics and dynamics for control", in Brady (ed.) *Robotics Science*, MIT Press.

Hopcroft, John E., and Krafft, Dean B. 1984. "The challenge of Robotics for Computer Science", Cornell University Internal Memo.

Jacobsen, Stephen C., Smith, Craig C., Biggers, Klaus B., and Iversen, Edwin K. 1989. "Behavior based design of robot effectors", in Brady (ed.) *Robotics Science*, MIT Press

Khatib, O. 1986. "Real time obstacle avoidance for manipulators and mobile robots", Int. J. Rob. Res., 5(1), 90 – 98.

Lozano-Pérez, T. and Wesley, M. A. 1979. An algorithm for planning collision-free path among polyhedral obstacles. Comm. of *ACM*, Vol. 22, No. 10, pp. 560-570.

Lozano-Pérez, T and Taylor, R. H. 1989. "Geometric issues in planning robot tasks", in Brady (ed.) *Robotics Science*, MIT Press

Marr, D. 1982. *Vision* Freeman.

Mason, Matthew T. 1989. "Robotic manipulation: mechanics and planning", in Brady (ed.) *Robotics Science*, MIT Press

Nagel, H.-H. 1983. Displacement Vectors Derived from Second-Order Intensity Variations in Image Sequences. Computer Vision, Graphics, and Image Processing 21, pp. 85-117.

Popplestone, Robin J., 1986. "An integrated design system for engineering", in Faugeras and Giralt (eds.), *Third Int. Symp. Rob. Res.*, MIT Press, 397 – 404.

Roth, Bernard. 1989. "Design and kinematics for force and velocity control of manipulators and end-effectors", in Brady (ed.) *Robotics Science*, MIT Press

Salisbury, Kenneth J., Brock, David., and O'Donnell, Patrick. 1989. "Using an articulated hand to manipulate objects", in Brady (ed.) *Robotics Science*, MIT Press

Steer, B. 1988. "Protective autonomy", Second Int. Workshop on manipulators, sensors, and steps toward mobility, Univ. Salford, UK, Oct 24-26

Tan, H. H., and Potts, R. B. 1988. "Minimum time trajectory planner for the discrete dynamic robot model with dynamic constraints", IEEE J. Rob. and Aut. RA-4 (2), 174 – 185.

Thorpe, C., Hebert, M. H., Kanade, T., and Shafer, S. A. 1988. "Vision and navigation for the Carnegie-Mellon Navlab", IEEE Trans. PAMI, 10(3), 361 – 372.

Volz, R. A., Mudge, T. N., Naylor, A. W., Brosgol, B. 1986. "Ada in a manufacturing environment", Proc. Control Eng. Conf. and Expo., Rosemont Ill, 433 – 440.

Whitney, Daniel E. 1989. "A survey of manipulation and assembly: development of the field and open research issues", in Brady (ed.) *Robotics Science*, MIT Press

Williams, Brian C. 1988. "MINIMA: A Symbolic Approach to Qualitative Algebraic Reasoning", Proc. Seventh AAAI Conf., St. Louis, 264 – 269.

Winston, Patrick Henry, et. al. 1984. "Learning physical descriptions from functional definitions, examples, and precedents", in Brady and Paul (eds.), *First Int. Symp. of Robotics Research*, MIT Press, 117 – 135.

Winston, Patrick Henry. 1985. *Artificial Intelligence*, 2nd Ed. Addison Wesley.

I

Programming, Planning, and Learning

Robotics has been defined as "the intelligent connection between perception and action." This section concerns the "intelligent connection" part of this definition. Programming, planning and learning are three approaches to establishing such a connection which involve different distributions of responsibility between the human and the robot.

In *programming*, the robot is viewed as the mechanical equivalent of a computer: a device that reliably executes a set of instructions. The human programmer combines these basic instructions to carry out tasks. In this approach, all the intelligence lies in the programmer.

One broad area of research in robotics concerns the development of robot programming systems that attempt to simplify the process of writing robot programs for complex tasks. Progress in this area happens first by expanding the basic "instruction set" – for example, by adding Cartesian straight-line motions or force-controlled motions or multi-fingered coordination to the robot's repertoire. Progress also comes about by providing language mechanisms that make the program development process more manageable – for example, by adding mechanisms to update automatically the position of the grasped object in the computer memory when the robot is moved (affixment).

One set of limits to robot programming are the fundamental ones due to the definition of the instruction set and the performance of the underlying hardware and control systems. Another set of limits are due to the practical difficulties of actually transforming the specification of the task into detailed instructions for the robot. These practical problems loom very large in robotics programming because of the difficulty of constructing general-purpose robot programs that can serve as building blocks for more sophisticated programs. In traditional computer programs, one can limit the effects of one computation on another. Most of the advances in computer programming concern ways of hiding information so as to control the propagation of side-effects. Unfortunately, all robot programs operate in a shared world, and a motion in one program affects the world for all subsequent programs. Because of this interdependence, it is nearly impossible to construct a robot program that is reusable from one task to another. The difficulties of robot programming have motivated the second area of research included in this section.

In *planning*, the objective is to construct systems that can transform high-level goals, such as the specification of a mechanical assembly for a robot arm or a target location for an outdoor mobile robot, into the sequences of commands necessary to achieve the tasks. A planning system can be said to be more "intelligent" than a programming system insofar as the specification that a human must provide to have the robot carry out a task is much simpler than a robot program.

A number of areas of robotics planning have received substantial attention in recent years:

- Gross motion (path) planning — Find a sequence of motions for a robot to reach a specified location without colliding with any obstacles.

- Fine motion planning — Find a sequence of (compliant) motion commands to achieve an assembly in the presence of position and model uncertainty.

- Grasping — Find the motion commands to achieve a stable grasp of an object even in the presence of uncertainty in the location of the object.

- Action sequence planning — Given a symbolic characterization of a set of actions (for example, an action for stacking one block on another), find the sequence of actions required to achieve a goal in a (partially) specified initial state.

The first three of these areas center on problems that involve detailed modeling of the geometry and mechanics of interaction between the robots and the task. Progress in this area has resulted in a fairly well-developed formal framework in each of these subareas.

In *learning*, the objective is to construct systems that can develop much of the expertise that they need to carry out tasks in the world directly from experience. These systems would take an even greater share of the burden from the human programmer. Interest in learning systems predates artificial intelligence and robotics. Unfortunately, progress in learning as applied to robotics has not been rapid. The current resurgence of interest in learning was motivated by the rebirth of interest in "neural networks."

In part I we have three survey articles, by Canny, Ghallab, and Latombe, each of which discusses four to ten published papers in some targeted subareas of planning research. We also have an extensive annotated bibliography on grasping, by Pertin-Troccaz. In addition, we have five reviews of selected articles in programming, planning, and learning. Below, we give a brief synopsis of these pieces, arranged in an order suggested by the introduction above.

Taylor reviews what is probably the earliest textual robot programming system, developed by Ernst in his 1961 Ph.D. thesis at MIT. This early work grappled with many of the fundamental issues that still challenge robot programmers today, notably how to incorporate sensing into the specification of action.

Canny reviews a series of papers by Schwartz, Sharir, and Ariel-Sheffi on the "piano movers' problem," which encompasses several variants of the gross-motion-planning problem mentioned above. These papers, together with an earlier paper by Reif, represent the beginning of a formal study of the computational complexity of motion-planning problems.

McCarthy's annotated bibliography on robot kinematics, workspace analysis, and path planning, included in part III, is also relevant here. This bibliography includes additional references to work on algorithmic motion planning that follows the work of Reif and Schwartz and Sharir. It also includes additional references to other, less formal work on path planning that has led to a number of practical implementations.

Koditschek's survey of potential field methods for robot planning and control, included in part IV, is also relevant. Potential fields are often cited as a local alternative or supplement to global gross-motion-planning methods.

Latombe reviews a sequence of papers on fine-motion planning using the "preimage backchaining" approach. These papers represent one of the few methodologies that have been developed for planning guaranteed strategies based on explicit modeling of geometry, mechanics, and uncertainty in measurement and control.

Pertin-Troccaz presents an annotated bibliography of research in grasping. This bibliography spans from collision-avoidance considerations for parallel-jaw grippers to stability considerations for multi-fingered hands.

The area of sequence planning, which had been essentially static for nearly ten years, has recently experienced a resurgence, partly as a result of some of the work reviewed here.

Rosenschein reviews Chapman's paper on planning for conjunctive goals. This paper is interesting partly because it "completes" an important line of development in action-sequence planning that was started in Sacerdoti's NOAH program. The other, perhaps more important reason that Chapman's paper is interesting is that it provides a detailed analysis of the computational complexity of this type of planning. Chapman's complexity results point out that – even for very simple domains – some of the basic subproblems of planning are intractable.

The combinatorial explosion inherent in traditional sequence planners is but one of the limitations of these planners in robotics domains. The other major limitation is the assumption that one can write programs to reliably implement the "primitive" actions in a domain – for example, grasping, stacking, and moving. In our discussion of programming we pointed out that this is the case only in very trivial domains. The traditional separation between planning sequences of actions and execution of these actions breaks down in practical situations. These limitations have prompted several new lines of work in planning systems, some of which are hardly recognizable as planners at all.

Ghallab reviews the use of triangle tables in robot programming and planning. Triangle tables arose in the context of one of the earliest of action planners, the STRIPS system developed at SRI. The triangle table was a mechanism for recovering from execution time failures without having to replan the whole task. Ghallab reviews more current work that extends this notion into a full-fledged mechanism for programming tasks in the presence of uncertainty.

Goad reviews a paper in which Rosenschein and Kaelbling propose a radical restructuring of the way we build robot systems. They propose to do away with the division between planning and execution altogether. Instead they develop systems that interact with the world whose semantics can be specified formally. The designer of such systems can then prove that the system can deal with some required class of tasks. But the system itself does not attempt to manipulate symbolic descriptions of its goals or of the state of the world.

Chatila reviews a paper in which Brooks proposes an architecture for mobile robot control systems that shares many of the same goals as that proposed by Rosenschein and Kaelbling. Once again the focus is on a tight connection between perception and action. Brooks makes a case for organizing robot systems as a number task-oriented modules that simultaneously monitor the sensors and all of which have access to the actuators. His subsumption architectures provides constraints on how individual modules can be implemented and how the modules can be connected.

Brooks reviews a paper in which Miller develops a learning algorithm that is an extension of Albus's CMAC model. The task considered is that of tracking objects moving on a conveyor belt with a robot-held camera. The procedure described by Miller learns the parametric representation of the control function that relates measurements to robot commands.

On the "Piano Movers'" Series by Schwartz, Sharir and Ariel-Sheffi

John Canny
EECS Department
University of California
Berkeley, CA 94720

Abstract

In 1983 and 1984, a series of five articles on robot motion planning appeared under the title of "The Piano Movers' Problem". The title was suggestive of the difficulty of the problem being considered, namely the navigation of some irregularly shaped object among obstacles. The class of problems considered included the navigation of three dimensional objects, like pianos, as well as rotary and sliding joint robots. This article reviews this series of papers.

1. Introduction

The approach taken by the authors of the "Piano Movers'" series differed from earlier work, with the notable exception of Reif (1979), in two respects. First of all, they considered motion planning in complete generality, and allowed the robot to have an arbitrary number of degrees of freedom. Secondly, they considered only exact methods, which always found a path if one existed, and which had well-characterized time bounds. This work formed a bridge between robotics and computer algorithms, and is probably responsible for a wave of algorithmic robotics papers which have followed since. From now on we will use the abbreviation PM for "Piano Movers'", and PM-n for the nth article in the series.

From an algorithmic point of view, the authors were successful, and in PM-II (Schwartz and Sharir 1983b), they were able to show that it is possible to plan an obstacle-avoiding path for virtually any robot in polynomial time (polynomial in the complexity of the environment). Even for simple robots however, this general algorithm was "catastrophically inefficient" to quote the authors in PM-V (Schwartz and

Sharir 1984). Thus while PM-II served as an existence proof for potentially efficient algorithms, the other papers in the series were detailed explorations of particular planning problems, with better performance than the general method.

In these other papers, the authors achieved mixed success. They made use of some very intricate geometric constructions, but the bounds they obtained were generally disappointing. Most of the polynomial time algorithms they gave had impractical exponents and large running time constants. However, theirs was the first attempt at some of these problems, and the bounds were improved significantly later on. The best bounds were for coordinating the motion of two disks in the plane with polygonal obstacles, in PM-III (Schwartz and Sharir 1983c). There they obtained an $O(n^3)$ algorithm for this problem, which is close to the best known bound of $O(n^2)$.

The practical impact of the whole line of research that might be called "algorithmic motion planning" is still unclear. But it is clear that there have been significant improvements in asymptotic running times since the "piano movers'" series first appeared. These improvements have often lead to much simpler algorithms. Presently several of these algorithms are being implemented and their typical running times explored. It seems likely that at least some algorithms from this diverse collection will prove useful in applications.

The papers are readily accessible to readers with some familiarity with analysis or topology. Even without this, the papers are self-contained and the topological arguments used in some proofs (there are not many) can be skipped by the reader who is only interested in the algorithmic details. We next give brief reviews of the papers in the series individually.

2. Piano Movers' I

Piano movers' I (Schwartz and Sharir 1983a) treats the problem of navigating a line segment or "ladder" in the plane among polygonal obstacles. The line segment is free to rotate and translate, and therefore has three degrees of freedom. Its configuration is specified by the x-y position of one endpoint (call it p) and the angle of the rest of the segment. The basic idea of the algorithm, like most motion planning algorithms, is to find a connected path from an initial configuration to a final configuration that remains entirely within the set of free configurations. The free configurations are those where there is no intersection between the segment and obstacles.

Rather than trying to represent the three-dimensional set of free configurations directly, they first project it into the plane, using the idea of "critical curves". The critical curves partition the plane into regions within which the possible motions in the rotation direction are qualitatively the same, and consist of a set of intervals for each placement of p. These motions are characterized by the types of contacts between the segment and obstacles as it rotates. That is, if we place the endpoint p at any point q_1 in a particular region, then as we rotate the line segment about p, it makes a series of transitions from overlapping to being clear of all obstacles, and vice versa. The sequence of features (edges or vertices) that contact the line segment when it makes these transitions describes the qualitative behaviour in the rotation direction, and the sequence is used to label the region.

So for example, suppose (e_3, v_2, e_7, e_4) is a description of a region. This means that with p fixed at any point in the region, as the segment is rotated clockwise from the horizontal, a transition to overlap occurs when the endpoint of the segment touches edge e_3, a transition back to non-overlap occurs when some part of the segment touches vertex v_2, etc.

The critical curves that separate neighboring regions are defined as the set of placements of p such that for some orientation of the segment, two types of contact occur simultaneously. For example, regions (e_2, e_4) and (e_2, v_3) if adjacent, will be separated by a critical curve of placements of p where some orientation of the segment contacts both e_4 and v_3.

With the two-dimensional partition in hand, a representation of the three-dimensional configuration space can be readily constructed. The three-dimensional set can be partitioned into volumes, and each volume is a "cylinder" over one of the two-dimensional regions. The notion of a cylinder will be described in more detail soon, but for now we note that the sequence of features defining the region give sufficient information to construct the cylinder over it. So using the geometric information from the two-dimensional partition and appropriate rules for jumps between volumes, a complete description of the configuration space can be computed.

Since critical curves are generated by pairs of features, there are $O(n^2)$ of them. This implies that they partition the plane into $O(n^4)$ regions. Each region may have $O(n)$ intervals in rotation space covering it, so the complexity of the partition of the three-dimensional configuration space is $O(n^5)$. In fact this is the running time that the algorithm achieves.

Some of the analysis in PM-I applies to the problem of navigating a general polygon in the plane with rotation. However, the authors stop short of a complete analysis, and no algorithm description or time bounds are given for the harder problem.

3. Piano Movers' II

Piano Movers II (Schwartz and Sharir 1983b) uses a very general cell-decomposition method for motion planning. Once again, the goal is to find a continuous path through the set of free configurations. The method assumes that this set can be described as a "semi-algebraic" set. A semi-algebraic set is the multidimensional generalization of a constructive solid geometry model. It is built from primitive semi-algebraic sets using intersection, union and complement. Each primitive semi-algebraic set is the set of points satisfying $f(x) \geq 0$, where $x \in \Re^n$ is a vector of real values, and f is a polynomial. In PM-II, Schwartz and Sharir show that under very reasonable assumptions, the configuration space of any conventional robot can be described as a semi-algebraic set.

To find a path through the set of free configurations when it is defined by a semi-algebraic set, the authors decompose the free set into simple pieces or cells. Each cell is a set which is topologically equivalent to an open ball of some dimension, or equivalently, to $\Re^m$ for some m. The cell decomposition is based on Collins' cylindrical algebraic decomposition (Collins 1975). To determine if there is a collision-free path between two given configurations, their algorithm searches for a sequence of cells $c_0, c_1, \ldots, c_m$ such that c_0 contains the start configuration, c_m contains the final configuration, and every pair c_i, c_{i+1} are adjacent, that is, their union is connected. Such a sequence exists if and only if there is a collision-free path between the configurations.

The general cell decomposition technique is not unlike the critical curve method described in PM-I. For a semi-algebraic set in $\Re^k$, one constructs a partition of $\Re^{k-1}$ (a set of polynomials in $x_1, \ldots, x_{k-1}$ is computed, and the zero sets of these polynomials partition $\Re^{k-1}$ in much the same way as the critical curves partition the plane in PM-I) with the property that the semi-algebraic set in $\Re^k$ is "cylindrical" over the partition of $\Re^{k-1}$. This means that within each region S_i

of the partition of $\Re^{k-1}$, the surfaces that define the set in $\Re^k$ form a covering space of S_i. The cylinder over S_i consists of a sequence of real values $(v_1(x_1,\ldots,x_{k-1}),\ldots,v_m(x_1,\ldots,x_{k-1}))$ where each real value gives the height (x_k coordinate) of one of the m surfaces that cover S_i.

In PM-I, the cylinder is defined by the sequence of features that define a region. For each transition contact, it is possible to determine the angle of the line segment for that contact as a function of the x and y coordinates of the endpoint. Each such function gives one of the values $(v_1(x_1,\ldots,x_{k-1}),\ldots,v_m(x_1,\ldots,x_{k-1}))$. Unlike the PM-I construction however, cell decomposition does not stop at dimension 2 but proceeds inductively from dimension k to dimension $k-1$ until dimension 1 is reached.

In the cell decomposition algorithm, we suppose that the partition of $\Re^{k-1}$ is a cell decomposition. Assuming inductively that this is true, it is then easy to generate a cell decomposition of $\Re^k$. We get a natural cell-decomposition of $\Re^k$ by taking as cells all sets of the form $S_i \times v_j$ or $S_i \times (v_j, v_{j+1})$, where (v_j, v_{j+1}) is the open interval with endpoints v_j and v_{j+1}.

Collins' algorithm gives a constructive method for cell decomposition, although it is not quite sufficient for motion planning, because adjacency information between cells is absent. In PM-II, Schwartz and Sharir gave a method for computing adjacency between cells of dimension k and $k-1$. This is enough to find a collision-free path so long as that path does not require the object to be moved through a region where its number of degrees of freedom is reduced. This could happen for example if a metal plate were moved though a slot with zero clearance. However, this requires contact between object and obstacles, and the physics of motion in contact must be taken into account, and the planning problem becomes much more difficult. Since the piano movers' problem is a gross motion planning problem, where it is assumed that the object can be moved without contacting the obstacles, the limitation to k and $k-1$ dimensional cell adjacency is no limitation at all.

While the cell decomposition technique is very general, this generality comes with a high price tag: a running time which grows doubly-exponentially with the number of degrees of freedom of the robot. This puts the algorithm out of the realm of feasibility for even the simplest of problems. However, if the number of degrees of freedom is a constant, the cell decomposition approach gives a solution in polynomial time. Although the exponent of this polynomial may be truly enormous, it did serve as a proof of the existence of polynomial time algorithms, and gave hope that the exponent could be greatly reduced in special cases.

4. Piano Movers' III, IV and V

Several cases where such improvements are possible were described in Piano Movers' III, IV and V. PM-III (Schwartz and Sharir 1983c) deals with coordinated motion of disks in the plane. The method uses a critical curve partition of the plane similar to PM-I, except that the configuration space has dimension at least 4 instead of 3, and the projection is considerably more complicated. For two disks, the system configuration is described by the coordinates of the centers of both disks. The configuration space for one disk is partitioned by critical curves generated by multiple contacts of the other disk with obstacles or with the first disk. For this problem, they obtain an $O(n^3)$ bound, where n is the number of environment edges, which is much better than direct application of the PM-II result. However, this technique still seems to exhibit double exponential growth with the number of degrees of freedom, and for three disks, the running time grows to $O(n^{13})$.

Piano Movers' IV (Sharir and Ariel-Sheffi 1984) considers a set of line segments joined at a common endpoint (a "spider") moving amid polygonal obstacles in the plane. Once again, a critical curve method is used. However, for this particular problem it is possible to simultaneously project the configuration spaces for all the arms of the spider into the plane, which greatly improves the running time. The method described in this paper runs in time $O(k^2 n^{k+4})$, where k is the number of arms on the spider, and n is the number of obstacle edges in the environment.

Piano Movers' V (Schwartz and Sharir 1984) considers a line segment (or "ladder") moving in three dimensions with polyhedral obstacles. The configuration of the object can be specified by the position of one endpoint, and by two angles specifying the direction of the segment. Thus the configuration space is five-dimensional. Once again the critical curve method is used, but instead of projecting into the x-y plane, projection is done onto the two-dimensional space of rotation angles. The bound eventually derived is $O(n^{11})$. They also give a generalization to the problem of moving an arbitrary polyhedron in three dimension with rotations, which has a six-dimensional configuration space. Here the bound is $O(n^{15})$.

5. Conclusions

The Piano Movers' series exemplifies one approach to robot collision avoidance: the environment is precisely known, and exact algorithmic bounds are given for computing a path in this environment, or deciding that one does not exist. As such, the series formed an important bridge between robotics and the study of algorithms. The papers in the series introduced quite sophisticated techniques, like cell decomposition and critical curve analysis, to find the connectivity of the set of free configurations. The algorithms were theoretically "fast" i.e. polynomial time, but almost all had impractical exponents. However, since the series appeared there have been significant improvements in running time for many of the problems they considered, and these later papers were clearly influenced by piano movers'.

The practical impact of algorithmic motion planning is still unclear, but there have been significant improvements and simplifications to planning algorithms since the initial "Piano Movers' " series. The final verdict depends on the results of experiments with the algorithms on real robots, which are currently underway in a number of robot labs.

References

Collins, G. E. 1975. "Quantifier elimination for real closed fields by cylindrical algebraic decomposition", *Lecture Notes in Computer Science, 33*, Springer-Verlag, Berlin, pp. 134-183.

Reif, J. 1979. "Complexity of the Movers' Problem and Generalizations" *Proceedings of the 20th IEEE Symposium on Foundations of Computer Science*, pp. 421-427.

Schwartz, J. T. and Sharir, M. 1983a. "On the Piano Movers' Problem: I. The case of a Two-Dimensional Rigid Polygonal Body Moving Amidst Polygonal Barriers", *Communications on Pure and Applied Mathematics*, vol. 36, pp. 345-398.

Schwartz, J. T. and Sharir, M. 1983b. "On the Piano Movers' Problem: II. General Techniques for Computing Topological Properties of Real Algebraic Manifolds", *Advances in Applied Mathematics*, vol. 4, pp. 298-351.

Schwartz, J. T. and Sharir, M. 1983c. "On the Piano Movers' Problem: III. Coordinating the Motion of Several Independent Bodies: The Special Case of Circular Bodies Moving Amidst Polygonal Barriers", *International Journal of Robotics Research*, vol. 2, no. 3, pp. 46-75.

Sharir, M. and Ariel-Sheffi, E. 1984. "On the Piano Movers' Problem: IV. Various Decomposable Two-Dimensional Planning Problems", *Communications on Pure and Applied Mathematics*, vol. 37, pp. 479-493.

Schwartz, J. T. and Sharir, M. 1984. "On the Piano Movers' Problem: V. The case of a Rod Moving in Three-Dimensional Space Amidst Polyhedral Obstacles", *Communications on Pure and Applied Mathematics*, vol. 37, pp. 815-848.

Triangle Tables in Execution Control and Robot Programming

Malik Ghallab
LAAS - CNRS
31077, Toulouse, France

Introduction

Triangle Tables have been proposed as data structures for representing and reasoning on sequences of actions. Their historical development follows closely that of mobile robot projects at SRI. Initially introduced in the context of the Shakey project [2], they remained quiescent for a decade until the Flakey project triggered a renewed interest. An extended formalism was then proposed as a general robot programming language [13], this in turn initiated more work on the subject [14 - 17]. Along a brief review of the above cited references, a discussion of Triangle Table formalism, its advantages and weak points, together with its potential uses in Execution Control and Robot Programming is proposed here.

Execution Control in Robotics aims at establishing an intelligent closed-loop control on the actions performed by a robot from what is perceived of their actual effects and of other events in the world. Reactive Planning shares exactly these same purposes but with a specific emphasis on the *unstructured environment* class of applications. There, a large set of possible goals can be pursued by (or given to) a robot. In each situation, it has to elaborate the correct plan, and to react in closed-loop, at the planning level, to events and effects. In the *structured environment* class of applications, a restricted set of possible goals and events is known in advance, for which a family of *flexible* plans is supposed to be available (either programmed or automatically generated). The controller task is to map permanently the current situation to the correct plan, or more precisely to an instance of a part of a plan, and monitor its taking place. Triangle Tables are involved in exactly that task.

Representation

Reasoning on actions requires a good representation of the concept of *action*. Such representation should among other things:
- give all possible conditions that permit the performance of the action and those that may forbid it (the qualification problem);
- provide or enable to easily find all possible effects of an action (the ramification problem);
- indicate efficiently all things that do not change and are not affected by an action (the frame problem).

STRIPS-based representation, on which Triangle Tables rely, addresses the last problem, but puts restrictions concerning the first two. The state of the world in such representation is modeled by a set of 1st order wffs. Some predicates are declared to be primitives. Deduced or non-primitive predicates have their support set (on which their validity depends) explicitly maintained.

An action **A** is a transition from a state **S** to a state noted **do(A, S)**, described by:
- its preconditions **p(A)**: a set of wffs such that **A** is a transition doable from state **S** iff the conjunction of formulas of **p(A)** has an instance that follows from **S**;
- its expected effects defined by 2 sets:
  . **add(A)**: clauses expected to be valid in the state **do(A, S)**; and
  . **del(A)**: clauses that may no longer be true in this new state.

If **add(A)** and **del(A)** give explicitly all primitive predicates that are affected by **A**, then the complete effects of **A** on non primitive predicates can be correctly deduced.

This representation enables one to express a general formula such as:

$$\forall x, y, z : \textbf{at}(x, y) \wedge \textbf{near}(y, z) \Rightarrow \textbf{closed-to}(x, z)$$

where **closed-to** is a non primitive predicate.
It cannot however deal with non atomic formulas that are not always true, such as:

$$\forall x : \textbf{tool}(x) \Rightarrow \textbf{at}(x, \textit{workbench})$$

Indeed, moving a particular tool out from the *workbench* cannot be handled by only updating its actual position (see [10] for a detailled discussion of this issue). Most systems relying on such a representation circumvent this problem by assuming that all wffs in the description of the world and actions are atomic formulas (but this is a slightly more restrictive assumption than necessary). In that case the state **do(A, S)** is computed by simple set operations, after unification and variable substitutions. Notice that variables in the description of an action have a special status (they are called schema parameters) and are specifically treated by the unification process.

Finally, such representation of an action **A** can be conveniently drawn in a 2 by 2 triangular array where each cell is a set of wffs:

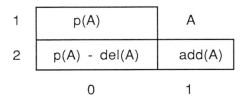

A Triangle Table is a generalization of the above representation to the description of a sequence of n actions. It is an (n+1) by (n+1) triangular array of cells. Rows are numbered from 1 to n+1, columns from 0 to n. Cells in row i (from column 0 to i-1) correspond to the preconditions of action A_i. Those in column i (from row i+1 to n) to the effects of A_i. More precisely the cell (i, j) contains the preconditions of A_i that result from A_j, but have not been destroyed by any other action from A_j to A_i (i.e. common elements of $p(A_i)$ and of $add(A_j)$ that do not belong to any $del(A_k)$, for $j < k < i$). In cell (i, 0) are the preconditions of A_i not provided by previous actions in the sequence. Cell (n+1, i) contains the effects of A_i that survive actions A_{i+1}, ..., A_n. Thus the last row of a triangular table describes the final outcome of the sequence of actions, whereas its column 0 contains formulas that must hold in a state **S**, starting from which the sequence can be entirely executed.

In summary, a Triangle Table is a formal representation of a sequence of actions and their interrelations. It shows clearly which effects of an action permit a subsequent action, which of its preconditions are not provided by previous actions in the sequence and are required to follow from the initial state of the world, and what will be the final effects of the sequence when it is carried out. The goal of a Triangle Table is precisely to achieve such effects in a state of the world.

A convenient concept toward that end is given by the *kernel* of a Triangle Table. The kernel of rank k, for k=1 to n+1, of a table **T** is the conjunction of wffs in the cells of the rectangular sub-array from row k to n+1, and from column 0 to k-1. For completion, the kernel of rank 0 is the trivial tautology. Thus, the kernel of rank n+1 (i.e. the last row) gives the final effects of the table; that of rank 1 (i.e. column 0) provides the preconditions necessary to the sequence of actions. One can easily see that the preconditions of the subsequence (A_i, ..., A_n) are given by the kernel of rank i : if it follows from some state S then this subsequence can be carried out.

The active kernel for a state **S** is the one with highest rank that follows from **S**. Starting from state **S**, to reach a state in which the goal of a table **T** is achieved, one has to compute the active kernel for **S**:

- if it is of rank n+1 then **S** already fulfils this goal;
- if it is of rank n then action A_n should be carried out ;
...
- if it is of rank i then the subsequence (A_i, ..., A_n) should be executed;
...
- if it is of rank 0 then the goal cannot be reached from **S** by table **T**.

Through the computation if its active kernel one can look at a Triangle Table as a goal-directed, ordered sequence of n+2 decision rules.

In addition to the extraction of its active kernel, several other formal operations can be defined on Triangle Tables. The following have been extensively analysed in [15]:

- the product of 2 tables: an associative operation that corresponds to the concatenation of 2 sequences of actions;
- the aggregation of a table: defines a single macro-action equivalent to the sequence, from the point of view of its preconditions and effects;
- the decomposition of a table: corresponds to the extraction of a particular subsequence;
- the hierarchization of a table: it is the substitution of a particular subsequence by the corresponding macro-action, it consists into a decomposition followed by an aggregation and a product.

An interesting generalization is considered in [17] where non sequential plans are examined: a structure corresponding to a Triangle Table for a DAG (directed acyclic graph) of actions can be defined, and similarly used. The algorithmic aspects of such generalization remain to be developed.

Possible utilizations

To actually use Triangle Tables in Robotics one has to bridge the gap from formal state transitions, on which the representation relies, to real world actions. Authors hopefully assume that such a large gap can be bridged by *lower level* control and perception systems. Once an abstract action (a state transition) has been selected and triggered, a corresponding complex program of commands is instantiated and run, most of which through sensory feedback closed-loops (e.g. motion actions, pick-and-place, part insertion , ...). A report on the outcome of such a program is brought back (assumed success, or failure and partial diagnosis). At this point the perception system is supposed to have translated more or less faithfully the actual state of the world into a set of wffs in a particular working memory. This state description, or updating, should include information about changes due to random events independent of what the robot does, and about those due to non intended or non modeled effects of its own actions. Admittedly those are strong assumptions, but they seem to be feasible for a well behave structured environment, on which a large body of knowledge is available.

A robot in such environment, trying to achieve a goal through a plan represented by a Triangle Table **T**, should not commit itself blindly to the entire sequence of actions. Once an action **A** has been triggered and terminated, its real effects and other events have to be monitored before considering the next action. This will be done by computing the active kernel of **T** relative to the updated state of the world. The procedure for controlling the execution of a plan **T** could be a loop over the following:

      compute the active kernel of **T** for the current state, let k be its rank;
      if   k=0   then exit with failure;
      if   k=n+1 then exit with success;
      else trigger action A_k, and wait until the outcome of A_k is reported

This is the basic control procedure proposed in [2]. It enables to take care of *negative events*, that force a robot to repeat one or several actions back in the sequence. It also handles *positive events* that bring closer to the goal, and permit to skip opportunistically some actions. It does not however guaranty termination; for that some kind of timeout process would be needed. It also fails for any state not in the scope of the table. Replanning, or chosing another plan, is then necessary. Replanning can be done toward the final goal, or, as proposed in [15], toward the kernel of the table closest to the actual state: one tries to reach back an available plan.

Worrying about the high cost of computing the active kernel, [14] distinguishes 2 classes of actions:
- low risk actions, basically those that are reversible in the real world, at the end of which only a monitoring of their own expected effects is performed, and, if successful, the next action in the sequence is triggered; and
- high risk or expensive actions for which the normal procedure of finding the active kernel is performed.

This reference proposes also some slight improvements on the algorithm of [2] for finding the active kernel (indexing a given predicate in order to match it to the working memory only once). The issue of compiling a Triangle Table has not been considered in the literature. A unification tree or RETE-like network could probably be designed to incrementally and efficiently compute the active kernel as the memory is updated.

The automatic synthesis of Triangle Tables can be done through planning for a specific instantiated goal, and a generalization of the found plan. This issue is pursued in [2] for the STRIPS planner. Plan generalization relies on changing all constants in the preconditions and effects of the plan actions into distinct variables, and redoing the proof of the sequence of preconditions to add constraints on variables that should be bound to the same object. A similar synthesis of tables is proposed in [15] on the basis of the ARGOS.II planner, that is more general than STRIPS, and uses a mixed chaining rule-based system where the preconditions and effects of actions are not directly given by rules but have to be found through their chaining.

Several authors have been very skeptical about the interest of a planning and execution control system for a robot at this abstract level: finding a sequence of state transitions that achieve some conjunction of predicates is the easiest part, they argue, compared with the formidable task of designing the program of commands corresponding to each action, and making it work in a reactive way, using sensors and actuators available today.

But Triangle Tables, as advocated in [13], can be seen to be of great help in precisely that programming task. Indeed, for such a programming one would require a system that permits:
- a goal directed behaviour,
- a hierarchical organization for implementing complex controllers with nested loops,
- short "sense-act" cycles for reacting to sensory data,
- modularity and granularity for ease of implementation and maintenance,
- parallel implementation by dedicated hardware for real-time performance.

Production Systems provide the last 3 features but are not convenient for the first two. In contrast Triangle Tables, seen as a programming tool, "maintain all advantages of Production Systems while allowing goal-directed behaviour and hierarchical organization", is argued in [13]. For that, the following scheme is proposed:
- when the robot is bound to run a program corresponding to a table **T**, it finds the current active kernel of **T**, and triggers the corresponding action,
- that action in turn has been decomposed into a sequence, programmed as a table **T'**, of simpler actions; current active kernel of **T'** is computed, an action is triggered leading to the call of another table,
- the search down a path in the hierarchy of tables continues until a primitive action is reached in some leaf-table. That action is executed.

Because the leaf-table surveys only some local conditions (that may more easily be provided by sensors), it won't be desirable to keep the control at this low level: the next active kernel and action in the leaf-table may no longer be relevant for the global state of the world. Control should be transferred back to the root caller.

Thus, there is a top level table that always keeps control. Primitive actions executed remain relevant for the current situation and goal, hence permitting a fast reaction to changes and short feedback loops.

A parallel implementation would improve response time. For that let us put in cell (n+1, 0) of a table **T** implementing the program of action **A** a proposition that checks that **A** is active. A table **T'** that triggers **A** will have to state that **A** is active by putting the right proposition in memory, and erasing it when its active kernel is not any more that of **A**. With an appropriate parallel hardware, all tables are interpreted concurrently and asynchronously. Each interpreter keeps on computing its active kernel and triggering the corresponding action if any. Primitive actions coming out of these computations are concurrently and asynchronously executed, after being checked by some conflict avoidance and resolution mechanism.

This programming scheme, as suggested in [13], could be at the basis of a general computational formalism: it enables to implement iteration (through repeatedly preforming a sequence of steps until a condition is met), recursion (when a call to a table appears in its own sequence), and conditional branching. To my knowledge, no experimentation of such programming scheme in robotics has been reported. It certainly deserves to be explored, although it raises several difficult open problems, such as:
- how to deal with timing and communication problems, e.g. from actuators and sensors to working memories of table interpreters, and communication between the interpreters,
- how to guaranty an acyclic hierarchy of tables, and deal with the halting problem even when it is acyclic,
- how to avoid and solve conflict between primitive actions,
- how to check progress toward the achievement of the task at hand.

Concluding remarks

In the design of an *intelligent connection from perception to action*, Execution Control is a central issue, for which there are today more unsolved problems than mastered techniques. Triangle Tables do provide some facilities for correctly representing and exploiting a sequence of abstract actions. They are indeed goal-oriented, reactive (to the projection in memory of the state of the world), and may support nicely a hierarchical distributed architecture. Those facilities should be helpful for a large class of simple robotics applications in well behaved structured environments. Although most of those facilities have not been yet fully explored and analyzed, one may say that Triangle Tables formalism will not be sufficient for supporting the design of a more general intelligent controller. That indeed may require, among other things, the ability to:

- represent and reason on time, durations, and temporal relations between events, actions and their effects (e.g. how long should one wait for the outcome of an action),
- deal with the inaccuracy and uncertainty inherent to sensors, actuators and a priori knowledge on the world,
- correctly find in a given situation all effects of an action and its preconditions (in addition to those explicitly stated),
- manage flexible non sequential plans with branching conditions and non deterministic choices solved opportunistically at execution time.

To conclude this brief survey, let us mention some additional works on Triangle Tables, and on few topics that arose along this discussion. The initial application of Triangle Tables, explored in [2], was in learning plans. That has been further considered in [15]. This author also developed an Explanation procedure for rule-based systems based on Triangle Tables [16]. The ramification and qualification problems in action representation are analysed in [6, 7], and dealt with through an interesting approach. Automated derivation of general recursive plans is explored in [11]. For Reactive Planning, see for example [8, 9]. The issue of Planning and Execution Control in *structured* vs *unstructured* environments is discussed in [5]. A comprehensive survey of Planning methods is presented in [4]. STRIPS and Mobile robot projects at SRI, Shakey and Flakey, are described in [1, 12, 3].

References

1. Fikes R.E. and Nilsson N.J. STRIPS: a New Approach to the Application of Theorem Proving to Problem Solving. Artificial Intelligence, 2, 189-208, 1971

2. Fikes R.E., Hart P. and Nilsson N.J. Learning and Executing Generalized Robot Plans. Artificial Intelligence, 3, 4, 251-288, 1972

3. Georgeff M.P., Lansky A.L., Schoppers M.J. Reasoning and Planning in Dynamic Domains: An Experiment with a Mobile Robot. In: Research on Intelligent Mobile Robots, 114-136, Technical Report, AI Center, SRI International, Menlo Park, California, May 1986

4. Georgeff M.P. Planning, Annual Review of Computer Science, 2, 359-400, 1987

5. Ghallab M., Alami R., Chatila R. Dealing with Time in Planning and Execution Monitoring. In: Robotics Research 4, Bolles and Roth (Eds.), 431-444, MIT Press, 1988

6. Ginsberg M.L and Smith D.E. Reasoning about Action I: A Possible World Approach. Artificial Intelligence, 35,-165-195, 1988

7. Ginsberg M.L and Smith D.E. Reasoning about Action II: The Qualification problem. Artificial Intelligence, 35,311-342, 1988

8. Hayes Roth B., Hayes Roth F., Rosenschein S. and Camarata S. Modeling Planning as an Incremental, Opportunistic Process. Proc. 6th IJCAI, 375-383 , 1979

9. Kaelbling L.P. An architecture for Intelligent Reactive Systems. In: Research on Intelligent Mobile Robots, 114-136, Technical Report, AI Center, SRI International, Menlo Park, California, May 1986

10. Lifschitz V. On the Semantics of STRIPS. In: Reasoning about Action and Plans, Georgeff and Lansky (Eds.), Morgan Kofmann, 1-9, 1986

11. Manna Z. and Waldinger R. How to Clear a Block: a Theory of Plans. J. of Automated Reasoning, 3, 343-377, 1987

12. Nilsson N.J. Shakey, the Robot. Technical Note 323, AI Center, SRI International, Menlo Park, California, April 1984

13. Nilsson N.J. Triangle Tables: A Proposal for a Robot Programming Language. Technical Note 347, AI Center, SRI International, Menlo Park, California, 1985

14. Picardat J.F. An AI Robot Programming Architecture: MUCAR. Proc. 16th Int. Conf. on Industrial Robots, 765-774, Springer-Verlag, Oct. 1986

15. Picardat J.F. Controle d'exécution, Compréhension et Apprentissage de Plans d'Actions: Développement de la Méthode de la Table Triangulaire. Thèse de Doctorat de l'Univ. P. Sabatier, Toulouse, Juin 1987

16. Picardat J.F. Reasoning about Reasoning: An Example of Two Level Explanation with the Triangular Table Method. Proc 2nd Int. Conf. on AI, 281-291, Marseille, Dec. 1986

17. Tenenberg J. Planning with abstraction. Proc. AAAI-86, 76-80, Philadelphie, Aug. 1986

Motion Planning with Uncertainty:
On the Preimage Backchaining Approach

Jean-Claude Latombe
Robotics Laboratory, Computer Science Department
Stanford University

Abstract

The goal of motion planning with uncertainty is to generate motion strategies for achieving specified spatial relationships among parts despite uncertainties in robot control, initial world model and sensing. Preimage backchaining is a general approach for planning such strategies. It was first proposed in 1983 by Lozano-Pérez, Mason and Taylor, and subsequently developed by several authors, mostly from the AI Laboratory at MIT. This paper surveys the series of papers and dissertations on preimage backchaining produced by the "MIT School" since 1983.

1. Introduction

Motion planning in the presence of uncertainty is one of the important problems one has to solve in order to create truly autonomous robots, i.e. robots which accept high-level task descriptions specifying what the user wants done rather than how to do it. The goal is to automatically generate motion strategies combining motion and sensing commands which can achieve an input goal, typically a spatial relationship between two parts, despite various kinds of uncertainty, namely uncertainty in robot control, uncertainty in the initial state of the world, and uncertainty in sensing.

Preimage backchaining is one of the few approaches which have been proposed to plan robot motion strategies in the presence of uncertainty. This approach has been first presented at the International Symposium on Robotics Research (Bretton Woods) in 1983 by Lozano-Pérez, Mason and Taylor (LMT paper). Since then, there have been several important contributions to this approach, mostly by researchers in the AI Laboratory at MIT, including Erdmann (1984; 1985; 1986), Buckley (1986) and Donald (1986; 1987; 1988). More recently, preimage backchaining has attracted the interest of researchers at several other places, e.g. (Natarajan 1988; Latombe 1988).

We believe that preimage backchaining is the cleanest and most rigorous approach to motion planning with uncertainty proposed so far, and that, despite some theoretical difficulties, it will have a major impact in the future. In this paper, we survey the series of papers and dissertations published by the "MIT School" from 1983 till 1987. We include in this series one paper by Mason (1984), which can be considered as a complement to the LMT paper of 1983.

In order to facilitate the survey, we first give a short presentation of the preimage backchaining approach (Section 2) and we compare it to other work (Section 3). Then, Section 4 reviews the original work on preimages (LMT paper and Mason's paper). Section 5 presents Erdmann's work reported in his S.M. thesis and in a couple of papers. Section 6 surveys Buckley's Ph.D. thesis. Section 7 presents extensions brought by Donald in his Ph.D. thesis and related papers.

A formal presentation of preimage backchaining requires a rather elaborate notational system. The notations used below are much simpler, but differ slightly from those used in the surveyed papers.

2. Preimage Backchaining

Let us consider a motion command of the form $\mathbf{M} = (\mathbf{CS}, \mathbf{TC})$, where **CS** is a *control statement* specifying a nominal trajectory and **TC** is a *termination condition* specifying when the execution of the motion should stop. For instance, **CS** may be a 'generalized damper' motion command specifying that the robot should move in a certain direction, while being compliant to forces perceived in perpendicular directions – see for instance (Mason 1981; Raibert and Craig 1981; Whitney 1985) for detailed presentations of compliant motions. **TC** is a logical expression whose arguments are both sensory inputs during the execution of the motion and the elapsed time since the beginning of the motion. When **TC** evaluates to *true*, the controller stops the motion.

Let the robot, $\mathcal{A}$, be a rigid object (e.g. the part held in a gripper or a mobile vehicle). $\mathcal{C}$ denotes the *configuration space* of $\mathcal{A}$, i.e. the set of all possible positions and orientations of $\mathcal{A}$ – see (Lozano-Pérez 1983) for a description of the notion of configuration space. Assume that the initial configuration of $\mathcal{A}$ is known at planning time to be within a given subset $\mathcal{I} \subset \mathcal{C}$. We want $\mathcal{A}$ to move to a goal configuration located in a given subset $\mathcal{G} \subset \mathcal{C}$. Typically, $\mathcal{G}$ corresponds to the achievement of a mating relation between two parts or the positioning of a mobile robot with respect to objects in its environment.

Let $\mathbf{M} = (\mathbf{CS}, \mathbf{TC})$ be a candidate motion command considered by the motion planner in order to achieve $\mathcal{G}$. Assume here that the shape and the location of the objects in $\mathcal{A}$'s environment are known exactly, but that both robot control and sensing are imperfect. Uncertainty in robot control means that $\mathbf{CS}$ specifies a nominal trajectory, but that any execution of $\mathbf{CS}$ will produce an *actual trajectory* that may be slightly different. Uncertainty in sensing means that the data read on the sensors are nominal values of physical parameters, but that the actual values may be somewhat different. All the papers surveyed here assume that errors in trajectories and sensing data are bounded, i.e. any actual data is contained in an 'interval', called *uncertainty*, centered at the nominal value. In probabilistic terms, this corresponds to modeling errors by uniform distributions over bounded intervals.

The *preimage* of $\mathcal{G}$ for $\mathbf{M}$ is defined as the subset of $\mathcal{C}$, denoted $\mathcal{P}(\mathcal{G}, \mathbf{M})$, such that: if the configuration of $\mathcal{A}$ is in $\mathcal{P}(\mathcal{G}, \mathbf{M})$ at the instant when the execution of $\mathbf{M}$ starts, then it is guaranteed both that the resulting motion will reach $\mathcal{G}$ and that $\mathcal{A}$'s configuration will be in $\mathcal{G}$ when the motion terminates.

This notion combines two more basic concepts, known as *reachability* and *recognizability*. Reachability concerns only $\mathbf{CS}$ and relates to the fact that any trajectory obtained by executing $\mathbf{CS}$ from $\mathcal{P}(\mathcal{G}, \mathbf{M})$ should be guaranteed to reach $\mathcal{G}$. Due to uncertainty in control, $\mathbf{CS}$ only describes a nominal trajectory and the planner must be certain that all the possible actual trajectories consistent with both $\mathbf{CS}$ and control uncertainty will ultimately traverse $\mathcal{G}$. Traversing $\mathcal{G}$, however, is not enough. The planner must also be certain that the termination condition $\mathbf{TC}$ will stop the motion of $\mathcal{A}$ in $\mathcal{G}$ (goal recognizability). This a much more subtle notion. One can regard $\mathbf{TC}$ as an observer of the actual trajectory being executed. Since sensing is imperfect, $\mathbf{TC}$ perceives the actual trajectory as an *observed trajectory*, which is likely to be neither the nominal one, nor the actual one. The problem for the planner is to do the following (in one way or another, implicitly or explicitly): (1) infer the set of all possible actual trajectories

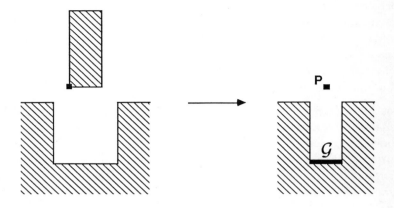

Figure 1: **The Peg-Into-Hole in Configuration Space**

from the nominal trajectory and the uncertainty in control; (2) infer the set of all possible observed trajectories from the possible actual trajectories and the uncertainty in sensing; (3) for every possible observed trajectory τ, verify that, when **TC** becomes *true*, all the actual trajectories compatible with τ have reached the goal. Then, assuming that the controller continuously monitors **TC** and that it can instantaneously stop the motion of $\mathcal{A}$, it is guaranteed that the execution of **M** will terminate in $\mathcal{G}$.

Example: Let us consider the task of inserting a peg (the robot $\mathcal{A}$) into a hole in a two-dimensional world. We assume that the peg can only translate, so that the task can easily be represented in configuration space as the problem of moving a point **P** into an edge $\mathcal{G}$ (see Figure 1).

The most common control law used in the framework is the generalized damper, whose nominal behavior is characterized by the following relationship:

$$\mathbf{f} = \mathbf{B}(\mathbf{v'} - \mathbf{v})$$

where **f** is the vector of forces acting on the moving object (represented as a point in configuration space), **v** is the commanded nominal velocity in free space, **v'** is the actual velocity, and **B** is a diagonal matrix known as the damping matrix. In free space $\mathbf{f} = 0$ and $\mathbf{v'} = \mathbf{v}$. When the point is in contact with an obstacle, **v'** is the projection of **v** on the tangent plane at the contact point.

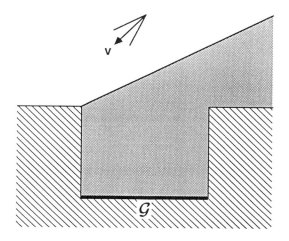

Figure 2: **Example of Preimage**

Figure 2 shows an example of preimage (grey area) of $\mathcal{G}$ for a generalized damper control statement with the commanded velocity **v**. Execution of the motion command is guaranteed to generate a trajectory that is contained in the uncertainty cone centered on the commanded velocity **v** as shown in the figure, and to slide over any encountered edge which is not orthogonal to a direction contained in the uncertainty velocity cone (we assume frictionless edges). Thus, **M** is guaranteed to reach $\mathcal{G}$ if the initial location of **P** is within the grey region.

The termination condition **TC** is

$$[\mathbf{c}(\delta t) \in \mathcal{G} \oplus \Sigma(0)] \text{ and } [angle(\mathbf{f}(\delta t), \nu(\mathcal{G})) \leq \theta]$$

where
- δt denotes the elapsed time since the beginning of the motion,
- $\mathbf{c}(\delta t)$ denotes the sensed configuration of $\mathcal{A}$ at instant δt,
- $\Sigma(\mathbf{c})$ is a sphere of constant radius ρ centered at $\mathbf{c}$ modeling the set of possible actual configurations of $\mathcal{A}$, when the sensed configuration is $\mathbf{c}$ ($\Sigma(0)$ is the same sphere centered at the origin),
- $angle(n_1, n_2)$ evaluates to the magnitude of the angle between two directions n_1 and n_2,
- $\mathbf{f}(\delta t)$ denotes the sensed reaction force on $\mathcal{A}$ at instant δt,
- $\nu(\mathcal{G})$ denotes the outgoing normal vector to the edge $\mathcal{G}$,

- θ denotes the maximal angular error on the sensed force direction,
- $\oplus$ is the Minkowski's operator for set addition[1].

The second conjunct in **TC**, $angle(\mathbf{f}(\delta t), \nu(\mathcal{G})) \leq \theta$, guarantees that the motion will terminate in contact with one of the three horizontal edges. The first conjunct, $\mathbf{c}(\delta t) \in \mathcal{G} \oplus \Sigma(0)$, allows **TC** to distinguish the bottom edge $\mathcal{G}$ from the two sides of the hole. ∎

Now, suppose that an algorithm is available for computing preimages. Given $\mathcal{G}$, $\mathcal{I}$ (the subset of possible initial configurations of $\mathcal{A}$), and the geometry of the environment, *preimage backchaining* consists of constructing a sequence of preimages $\mathcal{P}_1, \mathcal{P}_2, ..., \mathcal{P}_q$, such that
- $\mathcal{P}_i$, $\forall i \in [1, q]$, is a preimage of $\mathcal{P}_{i-1}$ for a selected motion command $\mathbf{M}_i$ (with $\mathcal{P}_0 = \mathcal{G}$);
- $\mathcal{I} \subseteq \mathcal{P}_q$.

The inverse sequence of the motion commands which have been selected to produce the preimages, $[\mathbf{M}_q, \mathbf{M}_{q-1}, ..., \mathbf{M}_1]$, is the generated motion strategy. This strategy is guaranteed to achieve the goal successfully, whenever the errors in control and in sensing remain within the ranges determined by the uncertainty intervals.

At the eventual expense of completeness, the problem of generating the sequence of preimages can be transformed into the combinatorial problem of searching a graph by selecting motion commands from a predefined discretized set. The root of this graph is the goal region $\mathcal{G}$, and each other node is a preimage region; each arc is a motion command, connecting a region to a preimage for this command. Construction of this graph requires discretizing the set of possible control statements. With generalized damper control, it requires discretizing the set of velocity orientations.

The preimage backchaining approach raises three kinds of issues:
- modeling issues (representation of geometry, physics, motions, uncertainty),
- computational issues (computation of the preimage of a goal for a given motion command),
- combinatorial search issues (construction and search of the preimage graph).

---

[1] Thus, $\mathcal{G} \oplus \Sigma_{\rho_c}(0)$ is the edge $\mathcal{G}$ grown by ρ_c.

3. Relation to Other Approaches

Research on robot motion planning became active in the mid-seventies, when the goal of automatically programming robots from a geometrical description of the task was first considered attainable. Since the early eighties, a great deal of effort has been devoted to this domain. Part of this effort was motivated, on the one side by the difficulties encountered in using explicit robot programming systems, and on the other side by the goal of introducing autonomous robots in hazardous or remote environments (e.g. nuclear sites, space, undersea, mines).

During the last ten years, most of the effort has been oriented toward solving the *path finding* problem, i.e. the problem of planning motions among obstacles without uncertainty. This effort has produced several major results, both theoretical and practical (e.g.: configuration space, complexity analysis, operational path planners). The problem of planning motions in the presence of uncertainty is conceptually more difficult than the path finding problem. It has attracted less attention so far, and less results have been produced. Three basic approaches to this problem have been developed to some extent.

The first has been proposed simultaneously by Lozano-Pérez (1976) and Taylor (1977), and is known as the *skeleton refining* approach. It consists of: first, retrieving a plan skeleton appropriate to the task at hand; and second, iteratively modifying the skeleton by inserting refinements (typically sensor-based readings). Refinements are chosen after checking the correctness of the skeleton, either by propagating uncertainty through the steps of the plan skeleton (Taylor 1977), or by simulating several possible executions (Lozano-Pérez 1976). Subsequent contributions to the approach were made by Brooks (1982), who developed a symbolic computation technique for propagating uncertainty forward and backward through plan skeletons, and by Pertin-Troccaz and Puget (1987), who proposed techniques for verifying the correctness of a plan and amending incorrect plans. Backward propagation of uncertainty in this approach can be regarderd as a particular case of preimage backchaining with known motion commands.

The second approach to motion planning with uncertainty has been proposed by Dufay and Latombe (1984), and is known as the *inductive learning* approach. It consists of assembling input partial strategies into a global one. First, during a training phase, the system uses the partial strategies to make on-line decisions and execute several instances of the task at hand. Second, during an induction phase, the system combines the execution traces generated during the training phase, and generalizes them into a global strategy. In fact, the training phase and the induction phase are interweaved. The generation of a strategy for the task ends when new executions do not modify the current strategy. A system based on these principles has been implemented, and tested successfully on several part mating tasks. Some aspects of this approach have been extended by Andreae (1986).

Both the skeleton refining and inductive learning approaches deal with uncertainty in a second phase of planning. The plan skeleton and the local strategies used during the first phase could be produced using path-finding methods assuming no uncertainty. The second phase takes uncertainty into account, either by analyzing the correctness of the current plan, or by directly experimenting with the local strategies and combining them into execution traces shaped by actual errors. In contrast, the rationale of the third approach to motion planning with uncertainty, i.e. preimage backchaining, is that uncertainty may affect the overall structure of a plan, in such a way that a motion strategy may not be generated by modifying or composing plans generated assuming no uncertainty.

In addition to being based on a different rationale, preimage backchaining is a much more rigorous approach to motion planning with uncertainty than the other two approaches. Consequently, it is natural to expect that preimage backchaining raises new theoretical issues, which were not considered in the other approaches. It does not mean that these issues are not present in the other approaches, but they are hidden by their ad-hocness.

In parallel with the research mentioned above, there has recently been an increasing interest in planning motions of objects in contact with other objects, e.g. (Hopcroft 1986; Valade 1984; Laugier and Théveneau 1986; Koutsou 1986). Planning motions in contact space differs from the traditional research on path planning in that it considers paths in contact space (which is a closed manifold whose dimension is lower than the dimension of the configuration space manifold) rather than paths in free space (which is an open manifold of the same dimension as configuration space). However, like the traditional research on path planning in free space, most of the research on motion planning in contact space has assumed so far accurate and complete knowledge of the world and perfect control of the robot motions. Nevertheless, an approach to motion planning with uncertainty has been developed recently by Desai (1987), which uses the concept of *contact formation* to partition the motion planning problem into a sequence of subproblems, each of lower dimensionality.

To terminate this section, it is worth noticing that preimages are also related to techniques and methodologies in other domains than Robotics. A goal may be regarded as the specification of a postcondition, and its maximal preimage (if it exists) for a certain motion command **M** as the 'weakest precondition' of this postcondition for the statement **M**. The notion of weakest precondition is a classical one in Computer Science for both program verification and program synthesis. Preimage backchaining is also related to the *goal regression technique* developed for the 'robot planning' problems traditionally considered in Artificial Intelligence (Waldinger 1975; Nilsson 1981).

4. LMT and Mason's Papers

The preimage backchaining approach was first presented by Lozano-Pérez, Mason, and Taylor at the First International Symposium on Robotics Research in 1983. The paper was reprinted in 1984 in the International Journal of Robotics Research. This paper sets up the important concepts related to the approach. It first gives an intuitive presentation of these concepts and then proposes a formal statement of the approach.

Part of the conceptualization developed in the LMT paper deals with modeling issues. The authors show how a task can be represented in configuration space. They model motions in this space as generalized damper motions along input velocities, and uncertainty in robot control by means by an uncertainty velocity cone as shown in the example of Section 2. Uncertainty in position sensing is modeled by a ball of fixed radius centered at the sensed configuration. Uncertainty in force sensing is specified by a maximal angular error on the orientation of the force vector. The LMT paper also introduces friction (in two-dimensional configuration spaces) and describes it my means of a cone whose axis is the normal at the contact point (this is basically Coulomb's law with equal static and dynamic friction coefficients). Due to friction, generalized damper motion may slide or stick along edges (a simple sticking/sliding test is defined in the paper). These various models have been re-used in most subsequent work on preimage backchaining.

The paper includes an analysis of the role of the termination predicate and a formal definition of preimages based on this analysis. Following these considerations, the formal statement of the approach is given as a general theoretical procedure (LMT procedure). However, this procedure is too abstract to be implemented directly. In particular, it includes non-deterministic statements applying to continuous sets. In fact, as we will see, due to theoretical difficulties, only limited subsets of the general framewok have been implemented so far.

Mason's paper (1984) brings several complements to the LMT paper. It mostly addresses the correctness and the completeness of the preimage backchaining framework formalized in the LMT paper. In particular, Mason analyzes several variants of the proposed procedure. Through this analysis, he shows that the procedure must be able to consider several intermediate goals (sub-goals) at a time in order to generate conditional strategies. This corresponds to the case where the planner (1) can plan a motion command whose execution is guaranteed to reach and terminate in one goal among n specified goals without being able to know which one at planning time, and (2) can construct a recognition condition that will permit the controller to recognize which of the goals has been achieved at execution time. For each individual goal, the conditional plan may contain a specific sequence of motion commands to be executed after.

Mason proves two important results:

- *Correctness of the LMT Procedure:* If the LMT procedure produces a motion strategy, this strategy is correct, i.e. its execution is guaranteed to achieve the goal (provided that all errors are within the uncertainty bounds).

- *Bounded-Completeness of the LMT Procedure:* If there is a motion strategy that can achieve a goal with a bounded number of generalized damper motions, then the LMT procedure is guaranteed to produce a motion strategy.

5. Erdmann's Work

Erdmann has investigated further the notion of preimage. His work is reported in his S.M. thesis (Erdmann, 1984) and in two papers (Erdmann 1985; Erdmann 1986). It concerns mainly computational and modeling issues underlying preimage backchaining.

One important contribution of Erdmann is his investigation of the computation of preimages. This investigation includes several facets:

- Erdmann analyzes the role of both the arguments in the termination condition and the knowledge incorporated in the condition predicate, and their impact on the 'power' of the condition, i.e. its ability to recognize goal achievement. He illustrates this analysis by many examples and he introduces several types of conditions – with(out) state, with(out) sense of time, with(out) history, ... – based on the data and knowledge available to them. The greater the power of the termination condition the larger the preimages will be, leading to easier graph searching.

- Unfortunately, maximal (i.e., largest) preimages may not exist in any situation and, when they exist, the problem of computing them remains open. Erdmann suggests to consider simpler (less powerful) termination predicate using essentially instantaneous sensing for recognizing the achievement of a goal. This leads him to introduce the concept of *backprojection*. The backprojection of a goal for a motion command is a region in $\mathcal{C}$ such that if the motion starts from within this region it is guaranteed to ultimately traverse the goal. Erdmann gives an algorithm for computing the backprojection of an edge in a two-dimensional configuration space

- In order to compute preimages as backprojections, Erdmann introduces the notion of *first entry set*. It is not clear however how this notion could be implemented in a program.

The other important contribution of Erdmann concerns the modeling issue. In his thesis, he investigates at length the representation of forces in configuration space and the modeling of friction when rotations of the moving object are allowed. Some readers may find a more recent paper by Rajan, Burridge and Schwartz (1987) a helpful preliminary reading before digging into the details of Erdmann's thesis.

6. Buckley's Work

The work of Buckley is reported in his Ph.D. thesis published in 1986 (Buckley 1986). It consists of the following contributions:

- In the previous papers, all motions were generalized damper motions described in configuration space. Buckley describes a representation of another kind of motion, *generalized spring* motions, in configuration space. Such a motion is aimed at achieving a certain configuration of the moving object, while complying to lateral external forces. Buckley also shows how uncertainty in generalized spring control can be modeled as a cylinder whose axis is the nominal path.

- Buckley's thesis describes the embedding of the preimage concept into two new frameworks. In one framework, *verification*, the motion strategy is given as an output to a program that verifies its correctness. In the other framework, *teaching*, a human user interactively builds a possibly conditional motion strategy by selecting motion directions; in this second framework, the computation of preimages is done by a program, which also keeps track of a AND/OR search graph, while the user is responsible for guiding the search by selecting the (intermediate) goals and the velocity directions.

- In the most original part of his thesis, Buckley describes an implemented planner based on the concept of *forward projection* (a better name might be 'postimage'). The forward projection of a given subset $\mathcal{I} \subset \mathcal{C}$ for a given motion command $\mathbf{M}$ is the subset of $\mathcal{C}$ in which the execution of $\mathbf{M}$ is guaranteed to terminate if it started from within $\mathcal{I}$. The planner considers three possible terminations of a motion: (1) termination by sticking on a surface, (2) termination by sensing the configuration in a predefined range, (3) termination by sensing the configuration and the external force in predefined ranges. The planner requires to discretize configuration space into atoms and builds a transition graph between the atoms. The implemented planner operates in a three-dimensional polyhedral configuration space. In such a space, the atoms considered by Buckley are free space and each of the obstacle faces, edges and vertices. The rationale underlying this particular decomposition is that the environment can produce the same reaction force at every configuration in the same atom.

7. Donald's Work

Donald's work is reported in his Ph.D. thesis in 1987 (Donald 1987a) and in related papers (Donald 1986; Donald 1988).

The two major contributions of Donald to the preimage backchaining approach are the following:

- Previous research on preimage backchaining assumed no uncertainty in the shape of the objects in the workspace (including the robot) and in their position in space (excluding the position of the robot). Donald introduces uncertainty in the model of the world and represents it by extending the notion of configuration space to the notion of *generalized configuration space*. The basic idea is the following: if a world parameter (e.g. the width of the hole in the peg-into-hole task) is not known exactly, a new dimension (i.e., a new axis) is added to the configuration space; this dimension corresponds to the uncertain parameter. Thus, the generelized configuration space is the original configuration space augmented by as many dimensions as there are uncertain world parameters. Donald allows objects to be pushed by the robot, so that when the robot is in contact with an object, it may push the object, slide along its surface, or a combination of both. Donald extends the representation of forces, friction, and generalized damper motions to generalized configuration space.

- As proven by Mason, the procedure proposed in the LMT paper (assuming it is implemented) generates a motion strategy that is guaranteed to succeed, and fails to generate a strategy if there is no guaranteed strategy (with a bounded number of steps). Donald argues that this might be a too strong requirement put on the motion planner. For example, in the peg-into-hole example, uncertainty in the width of the hole may be too large and, since the peg may sometimes be wider, no strategy is guaranteed to always succeed. In other examples, generating a guaranteed strategy might be too costly, due to the fact that it requires anticipation of very unlikely situations. In response to this, Donald introduces and develops the concept of *Error Detection and Recovery (EDR)* strategies, which are allowed to fail. However, such strategies either succeed or fail recognizably. This means that there is no possibility for the strategy to fail without the controller realizing it.

In his thesis, Donald describes an implemented planner generating EDR strategies in a limited domain.

8. Conclusion

We have reviewed a series of work on the preimage backchaining approach to motion planning with uncertainty. Since the approach was first presented in 1983, there has been much progress at the theoretical level. However, implementations of the approach remain quite limited despite some recent realizations (Buckley 1986; Donald 1987a) and propositions (Latombe 1988). In fact, as mentioned in Section 3, the preimage approach is more rigorous than other approaches, and makes explicit some difficult theoretical issued that are hidden in these other approaches because they are more ad-hoc. Conversely, dealing with these issues is a prerequisite to implementing the preimage backchaining approach, but not to implementing the other approaches. This underlines the fact that in general it is easier to build ad-hoc implementations of ad-hoc approaches than ad-hoc implementations of rigorous approaches. It is likely however that there will be more and more implementations in the near future. The increasing interest in both the problem and the approach is also likely to generate new conceptual developments.

Over the last nine years, the computational complexity of the path planning problem without uncertainty has been heavily investigated (both lower and higher bounds). The computational complexity of motion planning in the presence of uncertainty and of the preimage backchaining approach has received much less attention. The publications of Natarajan (1988), Donald (1987b), Canny and Reif (1987), and Canny (1987) should be noticed, however. Canny and Reif (1987) have proven that the three-dimensional compliant motion planning problem is non-deterministic exponential time hard. Donald (1987b) has shown that planning a guaranteed planar multi-step strategy with sticking termination conditions can be done in time polynomial in the number of vertices in the polygonal environment, and roughly simply exponential in the number of steps in the strategy. The lower bound given by Canny and Reif suggests, as with the more traditional path planning problems, to avoid the most general problem and to focus on more specific, but still interesting problems.

References

Andreae, P.M. 1986. Justified Generalization: Acquiring Procedures from Examples. Ph.D. Dissertation, Technical Report 834, Artificial Intelligence Laboratory, MIT.

Brooks, R.A. 1982. Symbolic Error Analysis and Robot Planning. *International Journal of Robotics Research*, vol. 1, no. 4, pp. 29-68.

Buckley, S.J. 1986. Planning and Teaching Compliant Motion Strategies. Ph.D. Dissertation, Department of Electrical Engineering and Computer Science, MIT.

Canny, J.F. 1987. The Complexity of Robot Motion Planning. Ph.D. Dissertation, Department of Electrical Engineering and Computer Science, MIT.

Canny, J.F. and Reif, J. 1987. New Lower Bound Techniques for Robot Motion Planning Problems. *Proceedings of the IEEE Symposium on the Foundations of Computer Science.*

Desai, R.S. 1987. On Fine Motion in Mechanical Assembly in Presence of Uncertainty. Ph.D. Dissertation, Department of Mechanical Engineering, University of Michigan.

Donald, B.R. 1986. A Theory of Error Detection and Recovery for Robot Motion Planning with Uncertainty. *Preprints of the International Workshop on Geometric Reasoning,* Oxford, UK.

Donald, B.R. 1987a. Error Detection and Recovery for Robot Motion Planning with Uncertainty. Ph.D. Dissertation, Department of Electrical Engineering and Computer Science, MIT.

Donald, B.R. 1987b. The Complexity of Planar Compliant Motion Planning Under Uncertainty. Technical Report 87-889, Department of Computer Science, Cornell University.

Donald, B.R. 1988. A Geometric Approach to Error Detection and Recovery for Robot Motion Planning with Uncertainty. To appear in *Artificial Intelligence.*

Dufay, B. and Latombe, J.C. 1984. An Approach to Automatic Robot Programming Based on Inductive Learning. *International Journal of Robotics Research,* vol. 3, no. 4, pp. 3-20.

Erdmann, M. 1984. On Motion Planning With Uncertainty. Technical Report 810, Artificial Intelligence Laboratory, MIT.

Erdmann, M. 1985. Using Backprojections for Fine Motion Planning with Uncertainty. *Proceedings of the IEEE International Conference on Robotics and Automation,* St. Louis, MO, pp. 549-554.

Erdmann, M. 1986. Using Backprojections for Fine Motion Planning With Uncertainty. *International Journal of Robotics Research,* vol. 5, no. 1.

Hopcroft, J. and Wilfong, G. 1986. Motion of Objects in Contact. *International Journal of Robotics Research,* vol. 4, no. 4, pp. 32-46

Koutsou, A. 1986. Planning Motion in Contact to Achieve Parts Mating. Ph.D. Dissertation, Department of Computer Science, University of Edinburgh.

Latombe, J.C. 1988. Motion Planning With Uncertainty: The Preimage Backchaining Approach. Technical Report No. STAN-CS-88-1196, Department of Computer Science, Stanford University.

Laugier, C. and Théveneau, P. 1986. Planning Sensor-Based Motions for Part-Mating Using Geometric Reasoning Techniques. *Proceedings of the European Conference on Artificial Intelligence*, Brighton, UK.

Lozano-Pérez, T. 1976. The Design of a Mechanical Assembly System. Technical Report No. AI TR 397, Artificial Intelligence Laboratory, MIT.

Lozano-Pérez, T. 1983. Spatial Planning: A Configuration Space Approach. *IEEE Transactions on Computers*, vol. C-32, no. 2, pp. 108-120.

Lozano-Pérez, T., Mason, M.T. and Taylor, R.H. 1983. Automatic Synthesis of Fine-Motion Strategies for Robots. *Proceedings of the First International Symposium on Robotics Research*, Bretton-Woods, NH, pp. 65-96. Reprinted in *International Journal of Robotics Research*, vol. 3, no. 1, pp. 3-24.

Mason, M.T. 1981. Compliance and Force Control for Computer Controlled Manipulators. *IEEE Transactions on Systems, Man, and Cybernetics*, vol. SMC-11, no. 6, pp 418-432.

Mason, M.T. 1984. Automatic Planning of Fine Motions: Correctness and Completeness. *Proceedings of the IEEE International Conference on Robotics and Automation*, Atlanta, GA, pp. 492-503.

Natarajan, B.K. 1988. The Complexity of Fine Motion Planning. *International Journal of Robotics Research*, vol. 7, no. 2, pp 36-42.

Nilsson, N.J. 1980. *Principles of Artificial Intelligence*. Morgan Kaufmann, Los Altos, CA.

Pertin-Troccaz, J. and Puget P. 1987. Dealing with Uncertainty in Robot Planning Using Program Proving Techniques. *Proceedings of the Fourth International Symposium on Robotics Research*, Santa-Cruz, CA, pp. 455-466.

Raibert, M.H. and Craig, J.J. 1981. Hybrid Position/Force Control of Manipulators. *Journal of Dynamic Systems, Measurement and Control*, no. 102, pp. 126-133.

Rajan, V.T., Burridge, R. and Schwartz, J.T. 1987. Dynamics of a Rigid Body in Frictional Contact With Rigid Walls. *Proceedings* of the IEEE International Conference on Robotics and Automation, Raleigh, NC, 671-677.

Taylor, R.H. 1976. Synthesis of Manipulator Control Programs from Task-Level Specifications. Ph.D. Dissertation, Technical Report AIM 228, Department of Computer Science, Stanford University.

Valade, J. 1984. Automatic Generation of Trajectories for Assembly Tasks. *Proceedings of the Sixth European Conference on Artificial Intelligence*, Pisa, Italy.

Waldinger, R. 1975. Achieving Several Goals Simultaneously. In Elcock, E. and Michie, D. (eds.), *Machine Intelligence 8: Machine Representations of Knowledge*, Ellis Horwood, Chichester, UK, pp. 84-136.

Whitney, D.E 1985. Historical Perspectives and State of the Art in Robot Force Control. *Proceedings of the IEEE International Conference on Robotics and Automation*, St. Louis, MO, pp. 262-268.

Grasping: A State of the Art

Jocelyne Pertin-Troccaz
LIFIA/IMAG Laboratory
Institut National Polytechnique de Grenoble
46, Avenue Félix Viallet
38031 Grenoble cedex - France

Over the last decade, grasping has evolved from a somewhat marginal topic to an important field of robotics research. This increasing interest in grasping is partly due to the evolution of industrial automation towards flexible automation. The transition from large batch size to medium and small sizes has led to the replacement of special purpose devices with more general purpose end-effectors enabling the manipulation of a broader class of objects. At the same time, more attention has been given to fine manipulation and assembly, especially precise assembly, e.g. electronics. This has pointed out the need for tools able to increase the robot's manipulative capacity with fine position and force control. Therefore, as end-effectors become more flexible, control becomes more complex, and a better understanding of grasping turns out to be a challenging issue.

Since grasping refers to research both on tool design and on action planning, we distinguish two main fields that we name *design and control of hands*[1], and *grasp planning*.

Design and control of hands

The common use of general but simple grippers requires a good understanding of the grasping process. Mostly geometrical approaches have been developed for such grippers. On the other hand, dextrous

---
[1] In the following we will use indiscriminately end-effector, gripper and hand even though end-effector is the very general term and hand is usually used to refer either to the human hand or to any end-effector with several articulated fingers; this last type of tool is also called a dextrous hand.

hands have forced researchers and engineers to focus on mechanical and kinematic problems; these hands were designed either to approximate a subset of human hand capabilities [2], or to achieve particular properties (for example, mobility). These quite complex articulated devices along with the common use of force control has raised new problems. In particular, all of the kinematic problems encountered for manipulators have had to be reconsidered; since articulated hands behave as several cooperating robots, they pose much more complex control problems than classical grippers.

Grasp planning

The action of grasping an object can be defined as the placement of the gripper relative to the object and is characterized by a set of contacts called a *grasp*. Grasp planning consists in choosing these contacts: location, type (point with friction, soft finger, ...), and applied forces and torques. This choice depends on the gripper, the object, the workspace, and the task to be performed while holding the object. Grasp planning must be done such that the contacts:

- are reachable, which implies both that the space occupancy in the object neighbourhood allows the contacts (there is no interference between the gripper and the nearby objects, including the object to be grasped) and that there is a collision free trajectory to reach the contacts;

- securely grasp and hold the object[3];

- allow the task to be performed (for example, screwing): this includes ensuring that both the appropriate forces and torques can be applied and that the contact locations do not prevent the task from being carried out (from an accessibility point of view);

- are "destructable", that is there exists at least one trajectory that permits the object to be dropped in a known configuration.

Since dealing with these four conditions at the same time is intractable, grasp planning is generally split into overlapping sub-problems.

---

[2] Even though the control capabilities are very different, hand design and control took advantage of earlier research in the field of prosthetics.

[3] We make a clear distinction between grasping and holding: grasping refers to the interval of time within which the contacts occur and the object lies on another object while holding supposes that the main action is to perform a particular task

Two main approaches to these different sub-problems exist: a *geometric* approach and a *mechanical* one. The former consists of considering the shape of the object and the spatial occupancy of the environment in order to deal with accessibility problems. The latter considers forces and torques in order to deal with both the mechanics of grasping and the mechanics of the task. The sub-problems are the following:

- *grasp generation and/or analysis*: this phase is generally restricted to the study of the object and the gripper; it consists in generating a grasp according to a particular criterion (shape matching between the object and the gripper, stability, and so forth); in some cases the generated grasp is a first approximation of the final solution (for example when two faces are chosen for grasping a part; in this case no precise location of the gripper is generated at this time);

- *reaching* considers the grasping environment in order to find a precise location of the contacts and to generate a collision-free trajectory for a particular grasp; sometimes the target location accessibility is also considered to find a grasping position;

- it may occur that no grasp that is compatible with the task can be found either because the object geometry, the task, or both are very constrained. In this case, *regrasping* has to be considered to go from some attainable intermediate grasp to a desired one. Very little work exists in this field. Geometric approaches analyze the compatibility of several grasps while mechanical approaches are based on dextrous manipulation in the hand.

Dextrous hands raise new problems related to grasp planning. In particular, because of the large number of degrees of freedom, it is impossible to generate contact locations from a description of the object, the hand, and the task. Therefore, *preshaping* is used as a pre-planner. It consists in determining a posture of the hand, i.e., a global arrangement of the fingers, as a function of the object shape and a description of the task. The fingers' fine positioning is done afterwards using some local sensing capabilities at grasping time. A lot of work related to human grasping is concerned with preshaping.

Preshaping implies the determination of a set of basic postures; many taxonomies exist in the literature. This review lists as exhaustively as possible the bibliography of grasping for robots. Because of the development of dextrous hands, some papers related to human grasping are also cited because they can offer many insights for both the design and the control of robot hands.

The following pages present a table which recapitulates this annotated bibliography; this table requires some preliminary comments. It is divided in fields and sub-fields. Zero, one or two "diamonds" are used:

- no diamond when the paper is not concerned with the field,
- one diamond to indicate the relevant sub-field,
- and two diamonds to indicate the most relevant sub-field (when more than one is involved in this paper).

The fields are defined as follows:

Field includes:

> **geometry of grasping** mainly spatial and morphological reasoning;
>
> **mechanics of grasping** including studies on stability, equilibrium, resilience to slipping and twisting, rolling and slipping at the fingertips, force-closure, form-closure, mobility analysis;
>
> **kinematics of hands** ;
>
> **human grasping** .

Overall topic includes *hand design and/or control*, *grasp planning*, and *human grasping*;

Tool can be a *human* hand, a tool, *dextrous* or not, with *two* or *three* fingers or *another* type of tool (vaccuum for example); among other types of tools we include what we call a *generalized* gripper when other objects are used with, or as an effector (for example, the palm of the gripper, a surface on which lies the object or the links of the robot itself);

Objects denotes what type of object model is used, if any; *2D* can be convex (*cvx*), *2,5D* represents prisms and *3D* can be polyhedra (*poly*), polyhedra and cylinders (*poly+*), parallelepipeds (*para*) or parallelepipeds with cylinders (*para+*);

Grasp planning can be divided in four different topics:

Preshaping the hand according to the object morphology and/or the task to be performed holding the object; a potential field (*pot. field*) may be used;

Reaching : the problem is to determine how to reach the grasp; some spatial reasoning techniques can be used in order to avoid collisions: configuration space (*cspace*), interference checking (*int. check.*), potential field method (*pot. field*) or some kind of *ray tracing*; some mechanical reasoning capabilities can also be used (*mech*) to determine how the object behaves during reaching the grasp when some contacts with friction are available (for example with the surface on which the object lies);

Grasp generation or analysis deals with generating or analyzing existing grasps according to geometric (*geo.*) and/or mechanical (*mech.*) criteria;

Regrasping is used when the initial grasp and the task to be performed with the object are incompatible and can rely on a geometric (*geo.*) or mechanical (*mech.*) analysis;

Sensing can be *vision* or *tactile*; *no* is used when specifically sensorless manipulation is performed;

Friction is basically defined with the Coulomb model except if it is specified otherwise (*no Clb*);

Contact type can be a point (*pt*), a polygon (*poly*), a soft gripper (*sf*); it can be *multiple* when one finger contacts on more than one point or *general* when it involves other surfaces than those of the fingertips; * is used when the shape of the contact is not specified.

References	Field				Overall topic			Tool				
	Geometry of grasping	Mechanics of grasping	Kinematics of hands	Human grasping	Hand design and/or control	Grasp planning	Taxonomy of grasps	Dextrous	Human	Bi-fingered	Tri-fingered	Other
[AHM85]		equilibrium				◊				◊		
[AIL83]				◊	control	◊	◊	◊	◊ ◊			
[Arb81]				◊	control	◊			◊			
[BDR81]	◊					◊				◊		vacuum
[Bea85]				◊	control	◊			◊			
[BFG85]		stability				◊					elastic	
[Boi82]	◊	stability				◊				◊		
[Bro85]		stability				◊				infinite		
[Bro88]		stability				◊				infinite		gene.
[BVD*86]		no slip./twist				◊				◊		
[CAHK87]		◊	◊					◊				
[Cut85]		◊ ◊	◊	◊	◊		◊					◊
[CW86]			◊		◊		◊ ◊		◊ ◊			
[Eds85]								◊				
[Fea84]		roll. stab.				◊		◊		◊		
[Fea86]		slip. & roll.				◊		◊				
[Fea87]		slip. & roll.			control	◊		◊				
[GBL84]		form-closure	◊		◊						articulated	
[GBL85]		stability	◊		control						articulated	
[HA77a]			◊		◊						elastic	
[HA77b]	◊	stability				◊					elastic	
[HC85]		equilibrium					◊			◊		
[Hol81]			◊		control				◊			
[HT83]			◊		control			◊	◊ ◊			
[HU77]			◊		◊							soft grip.
[HNW86]		◊	◊					◊				
[IBA85]			◊		control	◊	◊	◊	◊ ◊			
[Ibe87a]			◊		◊		◊		◊			
[Ibe87b]			◊			◊		◊	◊ ◊			
[Ibe87c]			◊				◊	◊	◊ ◊			
[IL84]			◊				◊	◊	◊ ◊			
[INH*86]	◊	stability				◊				◊		
[Ja84]			◊		◊			◊				
[Jea81]				◊		◊			◊			
[JIK*86]			◊		◊			◊				

Grasping 77

2D	2,5D	3D	Preshaping	Reaching	Grasp Generation/Analysis	Regrasping	Sensing	Friction	Contact type	References
◊					mech.					[AHM85]
			◊ ◊	◊			◊			[AIL83]
			◊	◊			◊			[Arb81]
		◊			geo.		◊	◊	point	[BDR81]
				◊			◊			[Bea85]
cvx ◊					mech.			no	point	[BFG85]
	◊ ◊	◊			geo & mech		vision	no		[Boi82]
		◊		mech.			no	◊	gene.	[Bro85]
		◊		mech.			no	◊	gene.	[Bro88]
		poly.			mech.			◊	poly.	[BVD*86]
							tactile	no Clb	pt,sf	[CAHK87]
								◊	*	[Cut85]
										[CW86]
										[Eds85]
◊					mech.			◊	point	[Fea84]
					mech.			◊	*	[Fea86]
					mech.		tactile	◊	*	[Fea87]
								◊	point	[GBL84]
										[GBL85]
										[HA77a]
		◊			geo & mech		vision	no	point	[HA77b]
		poly.			mech.			◊	point	[HC85]
										[Hol81]
										[HT83]
										[HU77]
										[HNW86]
			◊							[IBA85]
			◊	◊						[Ibe87a]
			◊							[Ibe87b]
			◊							[Ibe87c]
							◊			[IL84]
		◊		int. check.	mech.		vision			[INH*86]
										[Ja84]
			◊ ◊	◊ ◊	◊		◊			[Jea81]
										[JIK*86]

References	Field				Overall topic			Tool				
	Geometry of grasping	Mechanics of grasping	Kinematics of hands	Human grasping	Hand design and/or control	Grasp planning	Taxonomy of grasps	Dextrous	Human	Bi-fingered	Tri-fingered	Other
[JL86]		stability	◊			◊		◊				
[Kob84]	◊		◊								dextrous	
[KR86]	◊		◊				◊					
[Lau81]	◊					◊				◊		
[Lau87]	◊					◊				◊		
[LJM*87]	◊					◊				◊		
[Loz76]	◊					◊				◊		
[Loz81]	◊					◊				◊		
[LP83]	◊					◊				◊		
[Lyo85]	◊					◊	◊	◊				
[Lyo86a]	◊					◊		◊				
[Lyo86b]	◊					◊		◊				
[Mas82]	◊	friction				◊				◊		gene.
[Maz87]	◊					◊				◊		
[MB86]	◊	friction				◊				◊		
[MS85]		mobility			◊	◊		◊		◊		
[Nap56]			◊				◊		◊			
[Nap62]			◊					evolution				
[Ngu85a]		force closure				◊						elastic
[Ngu85b]		stability				◊						elastic
[Ngu87a]		force closure				◊						
[Ngu87b]		stability				◊						
[Oka79]			◊		◊						dextrous	
[Per86]	◊					◊				◊		
[Per87]	◊					◊				◊		
[PS86]	◊					◊				infini. thin		
[RO87]	◊					◊					◊	
[Sal85]					cont & prog.		◊					
[Sal87]		mobility										gene.
[SC82]		mobility	◊		◊		◊					
[TAP87]		sliding				◊				◊		gene.
[TBK87]						◊						
[TLM87]	◊					◊				◊		
[Win77]	◊					◊				◊		
[WVW84]	◊					◊				◊		

2D	2,5D	3D	Preshaping	Reaching	Grasp Generation/Analysis	Regrasping	Sensing	Friction	Contact type	References
					mech.			◊	pt, sf	[JL86]
										[Kob84]
										[KR86]
		poly+			geo.					[Lau81]
		poly+		cspace	geo.					[Lau87]
		◊		cspace, pot.	geo.	geo.	vision			[LJM*87]
		para+		cspace	geo.					[Loz76]
		poly		cspace	geo.					[Loz81]
		poly		cspace	geo.					[LP83]
			◊	pot. field						[Lyo85]
			pot. field							[Lyo86a]
			◊	◊						[Lyo86b]
◊				mech.			no	◊	gene.	[Mas82]
	◊			cspace, pot	geo.	geo.	vision			[Maz87]
							no	◊	gene.	[MB86]
							◊/no	◊	gene.	[MS85]
										[Nap56]
										[Nap62]
◊				mech.			no		point	[Ngu85a]
◊				mech.			no		point	[Ngu85b]
		poly		mech.			◊/no		pt,sf	[Ngu87a]
		poly		mech.			◊/no		pt,sf	[Ngu87b]
						◊				[Oka79]
		poly+		cspace	geo.					[Per86]
		◊		cspace			vision			[Per87]
◊				ray tracing	geo.				multi	[PS86]
				cspace			vision			[RO87]
							tactile			[Sal85]
										[Sal87]
										[SC82]
cvx					mech.					[TAP87]
						◊				[TBK87]
		poly			geo.					[TLM87]
		para		cspace						[Win77]
		poly		int. check.	geo, mech			◊	poly	[WVW84]

References

[AHM85] J.M. Abel, W. Holzmann, and J.M. McCarthy. On grasping planar objects with two articulated fingers. *IEEE Jour. of Robotics and Automation*, RA-1(4):211–214, December 1985. communication.

> 2D grasps - two articulated fingers - points of contact with friction - From a set of equilibrium equations parametrized by the magnitude of the applied forces, a curve of possible grasps is obtained from which an optimum can be chosen

[AIL83] M.A. Arbib, T. Iberall, and D. Lyons. *Coordinated control programs for movements of the hand.* COINS Technical Report 83-25, Laboratory for Perceptual Robotics, Dept of Computer and Information Science, University of Massachusetts, Amherst, MA 01003, August 1983. also published in Experimental Brain Research, Suppl 10, pp 111-129, Springer Verlag Berlin. Heidelberg 1985.

> Human grasping - Describes perceptual and control schemas for a dextrous hand

[Arb81] M.A. Arbib. *Handbook of Physiology – The Nervous System II*, chapter Perceptual structures and distributed motor control, pages 1449–1480. American Physiological Society, Bethesda, 1981.

> Human grasping: description of the human hand command from the physiological point of view

[BDR81] J.R. Birk, J.D. Dessimoz, and R.B.Kelley. General methods to enable robots with vision to acquire, orient and transport workpieces. In *Eighth NSF Grantees' Conference on Production Research and Technology*, Stanford, January 1981.

> Includes experiments with a contour adapting vacuum gripper connected with a vision system - A classic solution to the "bin-picking problem"

[Bea85]   D. Beaubaton. Motor plans and programs in a goal directed behavior: reaching to grasp an object. In *Congrès International d'Ethologie*, Toulouse, 1985.

> Human grasping - Uses the well-known task- object- effector- actuator hierarchy for human grasping and tries to determine the locus of the different control levels in the human brain

[BFG85]   B.S. Baker, S. Fortune, and E. Grosse. Stable prehension with a multi-fingered hand. In *IEEE Int. Conf. on Robotics and Automation*, St Louis, March 1985.

> 2D - Three spring-loaded fingers producing frictionless punctual contacts - Proof of the existence of stable grasps on any convex polygon and presentation of the algorithms - Discussion of [HA77a]

[Boi82]   J.D. Boissonnat. Stable matching between a hand structure and an object silhouette. *IEEE Trans. on Pattern Analysis and Machine Intelligence*, PAMI-4(6):603–612, November 1982.

> 3D - two or three planar fingers - Computes the stable possible grasps of a gripper along a silhouette obtained by a range-data sensor. Based on the search of typical geometrical features (arrangement of segments in 2D)

[Bro85]   R.C. Brost. *Planning robot grasping motion in the presence of uncertainty*. CMU-RI-TR 85-12, The Robotics Lab., CMU, July 1985.

> Prismatic objects - Infinite parallel jaws - Automatic planning of gripper motions allowing a grasp (offset, push or squeeze) despite bounded variation in the initial location of the object - Conditions for stability and convergence of the object orientation

[Bro88]   R.C. Brost. Automatic grasp planning in the presence of uncertainty. *Int. Jour. of Robotics Research*, 7(1), February 88.

Prismatic objects - Infinite parallel jaws - Automatic planning of gripper motions allowing a grasp (offset, push or squeeze) despite bounded variation in the initial location of the object - Conditions for stability and convergence of the object orientation

[BVD*86] J. Barber, R. Volz, R. Desai, R. Runinfeld, B. Schipper, and B.Wolter. Automatic two-fingered grip selection. In *IEEE Int. Conf. on Robotics and Automation*, San Francisco, April 1986.

Two parallel jaws gripping two parallel faces on a polyhedron - Polygonal contact surfaces with linear pressure variation - Analysis of stability and resilience to rotational slipping: extension of [WVW84]

[CAHK87] M.R. Cutkosky, P. Akella, R. Howe, and I. Kao. Grasping as a contact sport. In *4th Int. Symp. on Robotics Research*, Santa Cruz, August 1987.

Dextrous hands - Reveals the inadequacy of point contact and soft-fingertips contacts models with Coulomb friction model - Describes the computation of the stiffness of a grasp in presence of structural compliance in the fingers - Presents an improved friction model for elastic materials and relates it to tactile sensing

[Cut85]   M.R. Cutkosky. *Grasping and fine manipulation for automated manufacturing*. PhD thesis, Carnegie Mellon University, Pittsburgh, January 1985.

Design-directed analysis of industrial robots tasks - Design of both an instrumented, controllable wrist and an active hand for fine manipulation - Comparison between different types of grasps in terms of their stability, stiffness and resilience to slipping

[CW86] M.R. Cutkosky and P.K. Wright. Modeling manufacturing grips and correlations with the design of a robot hand. In *IEEE Int. Conf. on Robotics and Automation*, San Francisco, April 1986.

Presents a hierarchical taxonomy of grasps involving the type of objects, their appearance, the type of task (power/precision, sensitivity to forces and torques...), the number of fingers - Aimed at CAD of specialized industrial grippers, based on the type of both grasps and tasks more than on the shape of the objects

[Eds85] D.V. Edson. Giving robot hands a human touch. *High Technology*, September 1985.

Very brief overview of the on-going research on dextrous hands - The market of dextrous hands

[Fea84] R.S. Fearing. *Simplified grasping and manipulation with dextrous robot hands.* AI-Memo 809, Artificial Intelligence Lab., M.I.T. Cambridge, November 1984. also published in the IEEE Jour. of Robotics and Automation, RA-2, 4, December 1986.

2D - two fingers - Strategies using active compliance and slipping at the fingers for stably grasping objects with a dextrous hand when no model is available - Assumes single friction point of contact

[Fea86] R.S. Fearing. Implementing a force strategy for objects reorientation. In *IEEE Int. Conf. on Robotics and Automation*, San Francisco, April 1986.

Open-loop force strategy (twirling) for the full rotation of an object in a plane - Point contact with friction model; quasi-static grasping - A system of forces represented in a spherical reference frame is computed so that the object slipping at the fingertips is bounded - Describes the implementation on the Stanford/JPL dextrous hand - Analysis of the open-loop stategy limits

[Fea87] R.S. Fearing. Some experiments with tactile sensing during grasping. In *IEEE Int. Conf. on Robotics and Automation*, Raleigh, March/April 1987.

Extension of [Fea86] - Twirling experiments using a 8x8 sensor array covering a cylindrical finger - Describes preliminary tactile information interpretation (contact localization, object orientation on a finger...)

[GBL84] J.C. Guinot, P. Bidaud, and J.P. Lallemand. Mechanical analysis of a gripper manipulator. In *2nd Int. Symp. on Robotics Research*, Kyoto, August 1984.

Describes the design and control of a three fingered hand (each finger has three joints) allowing an isostatic gripping of objects - Point contact with friction

[GBL85] J.C. Guinot, P. Bidaud, and J.P. Lallemand. Modelization and simulation of force-position control for a manipulator-gripper. In *3rd Int. Symp. on Robotics Research*, Gouvieux (France), October 1985.

Extension of [GBL84] - Tri-fingered hand - Formulates an analytical model for active force control of this hand

[HA77a] H. Hanafusa and H. Asada. A robot hand with elastic fingers and its application to assembly process. In *IFAC Symposium on Information and Control Problems in Manufacturing Technology*, Tokyo, 1977. also published in Robot

Motion: Planning and Control, edited by Brady and al., MIT Press, 1982.

> Presents a tri-fingered hand with adjustable rigidity: its control scheme, its static and dynamic characteristics, the parameter adjustment with respect to a desired assembly task

[HA77b]   H. Hanafusa and H. Asada. Stable prehension by a robot hand with elastic fingers. In *7th Int. Symp. on Industrial Robots*, Tokyo, October 1977. also published in Robot Motion: Planning and Control, edited by Brady and al., MIT Press, 1982.

> 2D - Three equidistant elastic fingers producing frictionless contact points - Presents a heuristic for obtaining a stable grasp from a sensory profile - Based on the search for a local minimum of the potential function with the center of the hand near the centroid of the profile

[HC85]   W. Holzmann and J.M. Mc Carthy. Computing the friction forces associated with a three fingered grasp. In *IEEE Int. Conf. on Robotics and Automation*, St Louis, March 1985. also published in IEEE Jour. of Robotics and Automation RA-1, 4, December 1985.

> Tri-fingered hand - Point contacts with friction - Presents a procedure which allows the computation of the friction forces required to satisfy equilibrium for a known set of contact forces, and to check for a potential slipping

[Hol81]   J.M. Hollerbach. An oscillation theory of handwriting. *Biological Cybernetics*, 39(00):139–156, 1981.

> Analysis of the handwriting task - Extracts what is called functional DOFs (writing direction and letter height) and suggests a controller containing two coupled oscillators

[HT83]   S. Himeno and H. Tsumura. The locomotive and control mechanism of the human finger and its application to robotics. In *International Conference on Advanced Robotics*, Tokyo, September 1983.

> Human finger description - Control analysis: force, position and tonus (due to the simultaneous contraction of agonist and antagonist tendons) - Emphasizes the importance of tonus control for grasping objects of various weight under constant force and position

[HU77]   S. Hirose and Y. Umetani. The development of soft gripper for the versatile robot hand. In *7th Int. Symp. on Industrial Robots*, Tokyo, October 1977.

> Soft gripper design - Presents a singular type of gripper enabling the gripping and holding of various objects with uniform pressure - Describes the kinematics of soft gripping

[HNW86]   J.M. Hollerbach, S. Narasimhan and J.E. Wodd. Finger force computation without the grip Jacobian. In *IEEE Int. Conf. on Robotics and Automation*, San Francisco, April 1986.

> Dextrous hand (three or four fingers) - Point contact with friction - Presents a fast and efficient way of computing joint torques for a desired external force on a grasped object using special kinematic relationships of the articulated fingers in order to avoid the use of Jacobians (here, the Utah/MIT dextrous hand) - Efficiency is multiplied by four

[IBA85]   T. Iberall, G. Bingham, and M.A. Arbib. *Opposition space as a structuring concept for the analysis of skilled hand movements*. COINS Technical Report 85-19, Laboratory for Perceptual Robotics, Dept of Computer and Information Science, University of Massachusetts, Amherst, MA 01003, July 1985.

> Human prehension - Describes functional units, virtual fingers, which are groupings or real fingers - Presents the concept of opposition space as the area where three basic oppositions can take place, alone or in parallel, so that opposing forces can be exerted between virtual finger surfaces to effect a stable grasp - Relates to the preshape of the hand

[Ibe87a] A.R. Iberall. *A neural model of human prehension*. PhD thesis, University of Massachusetts, Amherst, January 1987.

> Neuro-physiology - In-depth analysis of human prehension offering many insights for both the design and the control of dextrous robot hands - Among which: a functional prehensile classification based on the concept of virtual fingers (grouping of real fingers) moving in the so-called opposition space

[Ibe87b] T. Iberall. Grasp planning for human prehension. In *Int. Conf. of Artificial Intelligence*, Milano, August 1987.

> Human prehension analysis - A two-level hierarchical planner is presented - The object and task representation is transformed into a grasp-oriented description of the task which is then mapped into an opposition space (it captures the available forces and DOFs of the hand and is related to the virtual fingers concept)

[Ibe87c] T. Iberall. The nature of human prehension: three dextrous hands in one. In *IEEE Int. Conf. on Robotics and Automation*, Raleigh, March 1987.

> Goal-directed classification of the prehensile postures (in terms of the available forces and DOFs) - Three basic, non exclusive, methods based on the virtual fingers concept: pad-opposition, palm-opposition and side-opposition - Comparison with well-known classifications of the literature

[IL84]   T. Iberall and D. Lyons. *Towards perceptual robotics*. COINS Technical Report 84-17, Laboratory for Perceptual Robotics, Dept of Computer and Information Science, University of Massachusetts, Amherst, MA 01003, August 1984.

> The human hand as a model for a perceptual robot hand - Suggests six basic postures of the hand based on the concept of virtual fingers (a hierarchical substructure in hand control)

[INH*86] K. Ikeuchi, H.K. Nishihara, B.K.P. Horn, P. Sobalvarro, and S. Nagata. Determining grasp configurations using photometric stereo and the prism binocular stereo system. *Int. Jour. of Robotics Research*, 5(1):46–65, Spring 1986.

> 3D - Two parallel jaws - Sensor-based system that grasps parts from a pile: locates the part, measures the attitude of the part, its elevation, determines legal stable grasps (grasp points eliminating 5 DOFs) and generates a collision-free grasp configuration (eliminates the sixth DOF: the approach direction)

[Ja84]   S.C. Jacobsen and al. The utah/mit dextrous hand: work in progress. *Int. Jour. of Robotics Research*, 3(4), Winter 1984.

> Anthropomorphic tendon-operated hand - Describes subcontrol systems and various hardware elements of the hand

[Jea81] M. Jeannerod. *Attention and Performance*, chapter Intersegmental coordination during reaching at natural visual objects, pages 153–168. Erlbaum, Hillsdale, 1981.

Human grasping - Proposes the open-loop, so-called ballistic, movement of the arm (U-shaped trajectory) combined with a preshaping of the hand with respect to the object properties (weight, size, shape, material,...)

[JIK*86] S.C. Jacobsen, E.K. Iversen, D.F. Knutti, R.T. Johnson, and K.B. Biggers. Design of the utah/mit dextrous hand. In *IEEE Int. Conf. on Robotics and Automation*, San Francisco, April 1986.

Discusses the motivations and presents the characteristics and performances of the hand version II

[JL86] J.W.Jameson and L.J. Leiffer. Quasi-static analysis: a method for predicting grasp stability. In *IEEE Int. Conf. on Robotics and Automation*, San Francisco, April 1986.

Dextrous hands - Point contact with friction and soft-finger contact - Prediction of grasp stability

[Kob84] H. Kobayashi. On the articulated hands. In *2nd Int. Symp. on Robotics Research*, Kyoto, August 1984.

Dextrous hand (three fingers and twelve joints) - Discusses kinematic issues for fine motion control - Computes the handling and grasping forces at fingertips

[KR86] J. Kerr and B. Roth. Special grasping configurations with dexterous hands. In *IEEE Int. Conf. on Robotics and Automation*, San Francisco, April 1986.

Analysis of hand kinematics for non nominal operating configurations (under/overconstrained object, redundancy, singular configurations) - Determination of the possible manipulations for these special circumstances

[Lau81]   C. Laugier. A program for automatic grasping of objects with a robot arm. In *11th Int. Symp. on Industrial Robots*, Tokyo, October 1981.

> 3D - Two parallel jaws - Presents an algorithm for grasps generation using an iterative filtering process based on morphological properties of the object to be grasp (local accessibility, mass distribution, mutual visibility, ...)

[Lau87]   C. Laugier. *Raisonnement géométrique et méthodes de décision en robotique. Application à la programmation automatique des robots*. PhD thesis, Institut National Polytechnique de Grenoble, LIFIA 46, Avenue Félix Viallet 38031 Grenoble Cedex, December 1987. in french.

> Description of morphological and spatial reasoning capabilities used in the SHARP object-level system

[LJM*87]   T. Lozano-Pérez, J.L. Jones, E. Mazer, P.A. O'Donnel, W.E.L. Grimson, P. Tournassoud, and A. Lanusse. Handey: a task-level robot system. In *4th Int. Symp. on Robotics Research*, Santa Cruz, August 1987.

> Overall description of the first operational object-level system for pick-and-place operations: HANDEY - Grasp planning using sensory data, a configuration space approach and a pseudo-potential field method - Regrasping

[Loz76]   T. Lozano-Pérez. *The design of a mechanical assembly system*. AI-TR 397, Artificial Intelligence Lab., M.I.T. Cambridge, December 1976.

> Specification of the LAMA object-level system - 3D grasping by two parallelepipedic jaws - Associates one or several sets of legal grasp positions (GSET) to each geometric feature (cuboid, cylinder) - GSET are bi-dimensional spaces - Prunes the GSETs by growing and projecting the obstacles lying in the space swept by the jaws onto this bi-dimensional space

[Loz81] T. Lozano-Pérez. Automatic planning of transfer movements. *IEEE Trans. on System, Man and Cybernetics*, SMC-11(10), 1981. also published in: Robot Motion: Planning and Control, edited by Brady and al., MIT Press, 1982.

> 3D - Two parallel fingers - Use of the configuration space approach in order to verify that (1) the internal faces of the jaws will overlap the grasp surfaces, (2) the manipulator will not collide - Considers the final environment in pick-and-place operations

[LP83] C. Laugier and J. Pertin. Automatic grasping: a case study in accessibility analysis. In *International Conference for advanced software in robotics*, Liege, May 1983. also published in: Advanced Software in Robotics, edited by A. Danthine and M. Geradin, North Holland, 1984.

> 3D - Two parallel jaws - Polyhedra - Global accessibility analysis for potential grasps performed by slicing the space swept by the gripper and projecting grown slices onto a suitable plane (the gripping plane) - An initial bi-dimensional zone depending on the features to be grasped is iteratively reduced

[Lyo85] D. Lyons. A simple set of grasps for a dextrous hand. In *IEEE Int. Conf. on Robotics and Automation*, San Francisco, April 1985.

> 3D - Dextrous hand - Preshape planning considering both the type of task to be performed (power /precision) and the size (small /long /large) and shape (flat /round) of the object to be grasped - Chooses among precision /lateral /encompass preshaping - Deforms the preshape for the purpose of obstacle avoidance using a potential field approach

[Lyo86a] D. Lyons. Tagged potential field: an approach to specification of complex manipulator configurations. In *IEEE Int. Conf. on Robotics and Automation*, St Louis, March 1986.

> Presents a general potential function (GPF) combining a position attractive field and a position repulsive field - Manipulator configurations can be described as systems of GPFs acting on the DOFs - Illustrated by some examples among which the description of three types of preshape configurations of the hand (precision/ lateral/ encompass)

[Lyo86b] D.M. Lyons. *RS: a formal model of distributed computation for sensory-based robot control*. PhD thesis, University of Massachusetts, Amherst, September 1986.

> In response to the increasing complexity of robot programming, presents a computational model of the robot domain - Uses the task of programming grasping and manipulation with a dextrous hands as an example domain

[Mas82] M.T. Mason. *Manipulator grasping and pushing operations*. PhD thesis, Massachusetts Institute of Technology, June 1982. also published as a technical report of the AI Lab., AI-TR-690.

> Prismatic objects - Sensorless reduction of uncertainty based on the passive compliance of the object with the gripper and the environment - Combination of frictional forces and geometrical constraints is analyzed to plan motions that reduce uncertainty

[Maz87] E.F. Mazer. *HANDEY: un modèle de planificateur pour la programmation automatique des robots*. PhD thesis, Institut National Polytechnique de Grenoble, LIFIA 46, Avenue Félix Viallet 38031 Grenoble Cedex, December 1987. in french.

> Detailed description of the first operational object-level system for pick-and-place operations: HANDEY - Grasp planning using sensory data, a configuration space approach and a pseudo-potential field method - Regrasping

[MB86] M.T. Mason and R.C. Brost. Automatic grasp planning: an operation space approach. In *6th Symp. on Theory and Practice of Robots and Manipulators*, Cracow (Poland), September 1986.

> Sensorless mechanical reduction of uncertainty by grasping - Extension of [Bro85] to fingers with finite width and motions with finite length for a push-grasp

[MS85] M.T. Mason and J.K. Salisbury. *Robot hands and the mechanics of manipulation*. Artificial Intelligence, MIT Press, 1985. Compilation of Mason's and Salisbury's Ph.D Theses.

> Central topic: friction - Analysis of the mechanics of constraint and freedom for two rigid bodies in contact - Used for the design of hands that grasp securely and for generalized grasping (when the motion of the object to manipulate is not completely constrained by the gripper - in particular when pushing)

[Nap56] J.R. Napier. The prehensile movements of the hand. *Jour. of bone and joint surgery*, 38B(4), November 1956.

> Human grasping - From anatomical and functional considerations, proposes only two patterns for prehensile movements: the power grasp and the precision grasp and one non-prehensile posture: the hook grasp

[Nap62] J.R. Napier. The evolution of the hand. *Scientific American*, 207(6):56–62, 1962.

Human grasping - Because of the discovery of a million-year-old man-ape's hand with man-made tools, hypothesizes the very old origin of the modern man's hand and supposes that man-ape was already capable of power grip

[Ngu85a] Van Duc Nguyen. *The synthesis of force-closure grasps in the plane.* AI-Memo 861, Artificial Intelligence Lab., M.I.T. Cambridge, September 1985. a condensed version has been published in the IEEE Conference of Robotics and Automation in 1986.

2D - From the shape of the object and the type of contacts (point or edge with/without friction contacts, soft-finger contact) a polynomial simple algorithm generates force-closure grasps including the number of contacts and their location (so that each finger can be positioned independently in its contact region)

[Ngu85b] Van Duc Nguyen. *The synthesis of stable grasps in the plane.* AI-Memo 862, Artificial Intelligence Lab., M.I.T. Cambridge, October 1985. a condensed version has been published in the IEEE Conference of Robotics and Automation in 1986.

2D - Fingers with linear stiffness and programmable compression producing frictionless contact points - Presents a linear algorithm, in the number of springs, that allows to make stable every force-closure grasp and to choose the compliance center and the stiffness matrix of the grasp

[Ngu87a] Van-Duc Nguyen. Constructing force-closure grasps in 3d. In *IEEE Int. Conf. on Robotics and Automation*, Raleigh, March/April 1987.

3D - Polygons - Presents algorithms to construct the whole set of force-closure grasps on a polygon (algorithm polynomial in the number of fingers) - Necessitates either two soft-finger contacts or four hard-finger contacts or seven frictionless point contact - Relation with equilibrium grasps

[Ngu87b] Van-Duc Nguyen. Constructing stable grasps in 3d. In *IEEE Int. Conf. on Robotics and Automation*, Raleigh, March/April 1987.

|Extension of [Ngu85b] to the 3D case (polygons)

[Oka79] T. Okada. Object handling system for manual industry. *IEEE Transactions on Systems, man and Cybernetics*, SMC-9(2), February 1979.

|Dextrous hand design - Description of the structure and kinematics of a three fingered hand (respectively two, three and three rotational joints)

[Per86] J. Pertin-Troccaz. *Modélisation du raisonnement géométrique pour la programmation des robots d'assemblage*. PhD thesis, Institut National Polytechnique de Grenoble, LIFIA, 46, Avenue Félix Viallet, 38031, Grenoble Cedex, March 1986. in french.

|3D - Two planar jaws - Detailed presentation of the geometric aspect of grasping - Presentation of a method: grasps generation based on local accessibility analysis and checking for global accessibility using 2D geometric tools

[Per87] J. Pertin-Troccaz. On-line automatic robot programming: a case study in grasping. In *IEEE Int. Conf. on Robotics and Automation*, Raleigh, March/April 1987.

|3D - Two parallel fingers - Applies a configuration space approach to data obtained by a laser-based sensor - Explains specific optimizations of the basic algorithms allowed by this experimental environment - Path planning viewpoint: complementary to [RO87])

[PS86]   M.A. Peshkin and A.C. Sanderson. Reachable grasps on a polygon: the convex rope algorithm. *IEEE Jour. of Robotics and Automation*, RA-2(1):53–58, March 1986. communication.

> Two parallel, infinitely thin fingers - Extrusion objects - Generation of grasps made of several contacts (for example, a finger against two vertices), computed using a linear algorithm in the number of vertices of the cross-section - Determines the range of angles from which a vertex is externally visible, therefore reachable, and combines compatible vertices to produce reachable grasps

[RO87]   G. Roth and D. O'Hara. A holsite method for parts acquisition using a laser rangefinder mounted on a robot wrist. In *IEEE Int. Conf. on Robotics and Automation*, Raleigh, March/April 1987.

> 3D - Two parallel fingers - Applies a configuration space approach to data obtained by a laser range-finder mounted on the robot wrist - Detailed explanation of the grasp generation from the viewpoint of sensing: complementary to [Per87]

[Sal85]   J.K. Salisbury. Integrated language, sensing and control for a robot hand. In *3rd Int. Symp. on Robotics Research*, Gouvieux (France), October 1985.

> Stanford/JPL dextrous hand - Describes a LISP based programming environment for coordinating the finger motions - Presents a sensor fingertip enabling the determination of location, magnitude and direction of a pure force exerted through a point contact - Sample program for the construction of a map from the sensed contacts

[Sal87]   J.K. Salisbury. Whole arm manipulation. In *4th Int. Symp. on Robotics Research*, Santa Cruz, August 1987.

> Generalized grasping and manipulation using the whole arm surfaces - A categorization of mechanisms in terms of their whole arm manipulation potential - Design of a prototype link whose control allows a broad range of interaction modes with the environment

[SC82] J.K. Salisbury and J.J. Craig. Articulated hands: force control and kinematics issues. *Int. Jour. of Robotics Research*, 1(1), Spring 82.

> Dextrous hand (Stanford/JPL) - Design principles - Structural design based on desired properties of the hand - Mobility analysis, singularity avoidance, optimization of hand kinematics based on parameter optimization techniques - Tendon-level control potentially allowing active position and force control

[TAP87] J.C. Trinckle, J.M. Abel, and R.P. Paul. Enveloping, frictionless, planar grasping. In *IEEE Int. Conf. on Robotics and Automation*, Raleigh, March/April 1987.

> 2D - Planner/simulator for a generalized grasping (using the surfaces of the hand) of convex polygons when the friction forces are not large enough to prevent sliding - Mechanical analysis for form-closure grasps generation

[TBK87] R. Tomovic, G.A. Bekey, and W.J. Karplus. A strategy for grasp synthesis with multi-fingered robot hands. In *IEEE Int. Conf. on Robotics and Automation*, Raleigh, March/April 1987.

> Anthropomorphic grasping: rule-based system under development aimed at preshaping a multi-fingered hand - Based on sensory data

[TLM87] P. Tournassoud, T. Lozano-Pérez, and E. Mazer. Regrasping. In *IEEE Int. Conf. on Robotics and Automation*, Raleigh, March/April 1987.

> 3D - Polyhedra - Two parallel jaws - Regrasping: construction of a sequence of ungrasping and grasping operations based on the search of a path in a grasp-placement space (a placement is a stable position of the object)

[Win77] M. Wingham. *Planning how to grasp objects in a cluttered environment*. Master's thesis, University of Edinburgh, 1977.

> 3D - Block world - Accessibility analysis: 2D computation allowed by the projection of the space swept by the gripper onto the plane of the parallelepipedic jaws

[WVW84] J.D. Wolter, R.A. Volz, and A.C. Woo. *Automatic generation of gripping position*. RSD-TR 2-84, Center for Robotics and Integrated Manufacturing, Ann Arbor (Michigan), February 1984.

> 3D - Polyhedrons - Selection of parallel grasping surfaces - Accessibility analysis based on a 2D geometrical processing - Ranking according to the grasp's resilience to translational and rotational slipping and to twisting (makes use of heuristics)

Sensor-Based Control of Robotic Manipulators Using a General Learning Algorithm[1]

W. Thomas Miller, III

Reviewed by

Rodney A. Brooks
MIT Artificial Intelligence Lab
545 Technology Square
Cambridge, MA 02139

This paper presents a working, implemented system which learns the kinematics of a manipulator and the optics of a camera well enough to use visual feedback to guide a manipulator to reliably and accurately track objects on a conveyor belt. The learning technique uses a very clever updating rule to modify weights in a state-space table that follows Albus' (1975) work on Cerebellar Models.

Although it is not explicitly mentioned, the techniques used in the paper have a lot in common with modern *neural networks* or *parallel distributed processing (PDP)* (Rumelhart and McClelland, 1986) approaches to learning. Miller does not use thresholds in his approach however.

Miller gives a method for learning an approximation to a control function f where

$$g = f(s)$$

maps $s \in S$, a multidimensional discrete input state vector, to g, a multidimensional output vector. In the case of tracking objects on a conveyor belt the space S has six dimensions; the positions of the three rotary joints of the manipulator (there are actually five, but the wrist joints are constrained to keep a gripper-held camera pointing downwards), the x and y image coordinates of the center of mass of the image of the moving part, and z, a measure of the size of the part in the image. The output vector g is three dimensional giving changes in the three joint angles. The goal is to drive x and y to 0 and z to some fixed z_0 thus keeping the part on the conveyor belt centered in the image, and to keep the camera a fixed height above the conveyor.

---

[1] IEEE Journal of Robotics and Automation, Vol RA-3, No. 2, April 1987, 157–165.

The Cerebellar Model provides an efficient way of representing f, and capitalizing on its smoothness so that information learned about it at one point can be interpolated at nearby points. Furthermore it provides a compact way of representing an approximation to f over a high dimensional space S. The idea in essence is to cover the space S with many overlapping subsets R_i and then associate a weight w_i with each of these subsets, and finally to approximate f linearly as

$$f(s) = \sum w_i \chi_{R_i}(s)$$

where χ_R is the characteristic function of the set R (i.e., valued 1 for points in the set R and 0 elsewhere). The Cerebellar Model simply provides a way of choosing sets R_i which ensure that continuity is approximated in all directions. Furthermore it maps distant parts of the space to common subsets in such a way that the "noise" from one part of the space should be low in the approximation of another part of the space. This assumes that overall we are only interested in values of f over some relatively sparse subset of S.

The key point in the paper is a very clever way of learning f from unsupervised trials. The strategy relies on being able to identify for each observed state of the system s, some desired change Δs in the observed variables. In the case of the conveyor belt, the desired change at each control step is to move the three image parameters x, y, and z halfway to their target values. It is not very likely that the system will be observed doing exactly the desired behavior very often during learning, nor is there a model of the correct behavior to learn from as in classical neural networks.

Miller's clever idea is to embed S as a manifold in a higher dimensional space S'. For the conveyor belt example the extra dimensions of the space are the observed changes in the parameters x, y, and z. The manifold arises from only considering points in S' where these new dimensions take on exactly the desired changes in the state space given the other six dimensions. Miller learns the function f' on the whole pf S' where f' is the obvious lifting of f into the higher dimensional space. A set of commands g_0 to change the position of the robot joint angles is computed by evaluating f' on the appropriate point $s \in S'$ of the embedded manifold, where s includes the desired changes in the image coordinates of the moving part. After each motion of the robot, $s_0 \in S'$ is formed by the state of the robot and the camera image before the motion, and the actual changes observed in image coordinates. Having observed the system with the command g_0, ideally it should be the case that

$$g_0 = f'(s_0)$$

if f' were known exactly. Note that s_0 may not be on the embedded manifold. Now a learning rule is used to update the weights w_i to improve the appoximation of f' at s_0. Each w_i involved at point s_0 is incremented by

$$\delta = \beta \times \frac{g_0 - f'(s_0)}{C}$$

where C is the total number of weights w_i involved at s_0, and β is a learning factor between zero and one (set at 0.5 in the experiments reported by Miller). Thus the approximation for f' is improved at points in S' at which the exact kinematic effects of changes in joint angles have been observed relative to the moving workpiece on the conveyor. Hopefully many of these points will be near to the embedded manifold of interest and so over time the interpolations of f' on the manifold improve.

The empirical results presented by Miller are quite impressive. Starting with a very naive control law and a 0.5 second update rate, the manipulator always loses sight of a moving part on the conveyor belt within 4 seconds. But even in the first trial with learning turned on, the manipulator successfully keeps the part in view for the full 10 seconds possible, lagging about 15 centimeters behind the part. After ten trials, the robot was able to track the part within a centimeter in the x and y directions and within 5 centimeters in the vertical direction (the visual information is somewhat poorer in that direction).

There are some important questions not adressed in the current paper. For instance the paper does not give any feel for what class of functions f and embeddings f' into higer dimensional spaces can be learned by this procedure. Nor does it tell us what the requirements are on the distributed representation of f for the learning procedure to work.

The technique certainly appears to be useful for a wide variety of applications. In particular it should be well suited to constructing self-calibration procedures to relate actuator to sensor coordinate systems.

References

Albus, J. S., 1975 (Sept). A New Approach to Manipulator Control: The Cerebellar Model Articulation Controller (CMAC). *Trans. ASME, J. Dynamic Systems, Measurement, and Control.* Vol. 97, 220–227.

Rumelhart, D. E., and McClelland, J. L. 1986. Parallel Distributed Processing. *MIT Press*, Cambridge.

A Robust Layered Control System For A Mobile Robot[1]

Rodney A. Brooks

Reviewed by

Raja G. Chatila
LAAS-CNRS
7 Avenue de Colonel Roche
31077 Toulouse, France

Designing a control structure for an autonomous robot system (*i.e.*, a robot possessing multiple functions for sensing, decision, and action) is a fundamental question, and not merely the secondary integration problem of several functionalities. The control structure determines the robot's behavior and performance when it acts in the *real world*, confronted by a range of situations. This area of research has only recently become very active. Previously research groups focused their efforts on specific aspects of a robot system (*e.g.*, vision and modelling, path planning) rather than investigating how various systems achieving these functions can be integrated into a single entity producing robust behavior. Interest in this problem is increasing as sub-systems are better understood and achieved.

Two basic requirements of a control structure are:

1. *The operation of the robot system as such* for achieving tasks and monitoring the execution of actions, and

2. *Reactivity* to asynchronous events. Changes in the environment due to various factors such as other agents, or an incomplete or inaccurate model used by the robot, must be detected and dealt with when necessary.

In this paper Brooks proposed a control architecture for autonomous mobile robots, (or to put it in his own words "creatures"), based some observations from which we excerpt:

---

[1] IEEE Journal of Robotics and Automation, Vol. RA-2, No.1, March 1986.

1. Complex behavior does not necessarily result from complex internal structures (whatever an external observer might think).

2. The system design should be simple. When a system is composed of several elements, the interfaces should be simpler than the components, or the decomposition is incorrect. If a subsystem is designed to solve an ill-defined problem, it will not be robust.

3. Sensors should self-calibrate.

4. Robots should be able to survive by themselves as long as possible without human intervention.

The basic idea proposed by Brooks is to decompose the robot system not on the basis of the system's internal operation (based on a functional decomposition for example), but rather of its **external behavior**. A level of competence is then defined as an informal specification of a class of behaviors. Eight levels are introduced, where level i embeds level $i - 1$.

To each *level of competence* corresponds a *layer of control*. These layers are built incrementally, and therefore, not all functionalities of the robot system are necessary at one time to enable its operation. The proposed architecture is thus based on an evolutionary approach. Higher levels can subsume lower levels' operation, while these lower levels still continue to function. Some important consequences of this approach are the following:

- Each layer can deal with a different goal, and thus the robot can cope with multiple goals.

- Multiple sensors can be used according to the needs of each layer without necessarily building a central representation.

- The global system is robust. Debugging is facilitated because each layer can run independently before adding new layers. This is an important point in system development which helps avoid the burden of debugging a large complex system.

- Clearly, the structure is easily extensible, and can correspond to a simple implementation.

The levels of competence proposed by Brooks are:

0. Collision avoidance.
1. Random motion.
2. Environment exploration (set goals and reach them).
3. Environment map building and path planning.
4. Noticing environment changes.
5. Reasoning on objects in the world and achieving tasks.
6. Generation and execution of plans that tranform the world into a desired state.
7. Reasoning on the behavior of objects and adapting plans.

Of course, one could question this decomposition, but it is likely that Brooks intended it only as an example. The underlying principle remains the same, whatever the decomposition.

A given level is realized by processes linking sensing and action through a set of simple modules. Modules are finite state machines, and are loosely coupled: they run asynchronously and exchange messages without any protocol. Messages can be lost: to every input line is associated a single buffer containing the most recent message.

The basis of the subsumption architecture is that a given module can suppress or inhibit the inputs (outputs) of another module for a given time period, thus actually replacing its usual inputs (outputs). A higher layer of control can thus be very easily integrated with a previous structure through this mechanism.

The following example (figure 1) describes levels 0 and 1.

Level Zero: Collision Avoidance

This level does not achieve general collision avoidance (*i.e.*, based on an analysis of the known obstacle configuration), but rather it uses a local avoidance technique (namely, the potential field method). This lowest level makes use of ultrasonic sensors and acts as a basic reflex loop. The modules composing this level are:

- FEELFORCE: builds a force field from ultrasound readings.

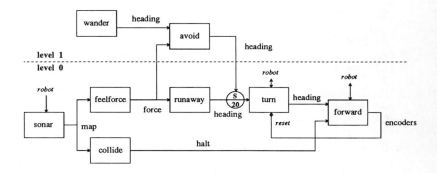

Figure 1: Levels zero and one. (from [1])

- RUNAWAY: computes an escape command (heading) from the force field.

- TURN: receives **turn** (in-place rotation) and **move** (forward motion) commands and executes them.

- FORWARD: execute a straight line motion command and then resets TURN to receive new commands.

- COLLIDE: Detects possible collisions and sends halt messages.

A force field is created and used to generate commands to TURN that execute rotations and FORWARD which executes straight line motions. TURN reads its input line when FORWARD finishes because FORWARD resets TURN to its initial state: read input line. When an obstacle is detected by COLLIDE, FORWARD receives a halt, and thus TURN is also reset so that it may receive a command from runaway.

Level One: Random Motion

Level One is simple. The modules composing this level are:

- WANDER: periodically produces a random heading.
- AVOID: integrates level 0 obstacle data with WANDER.

AVOID uses indications from level zero of obstacles and the random heading generated by WANDER to compute a resulting heading that is sent to TURN by suppressing the output of RUNAWAY.

So far the levels of competence actually built do not exceed level two (**explore**).

This control architecture is very appealing, elegant, simple and without doubt meets the requirements of reactivity and robustness. Obviously, long term research is needed to reach higher levels.

However, the fact that the approach results in a *fixed* wiring of the system has its advantages in terms of efficiency, but does not enable a flexibility, in terms of programmability, in the reactions.

This architecture was met with some criticism that mainly focused on its ability to grow in complexity when including the upper levels, while keeping the purity of its initial assumptions. However, this criticism has generally not been based on theoretical grounds, but on the fact that the increasing complexity of the system would result in great difficulties in adding new modules while guaranteeing a consistent behavior. But this is just what Brooks wants to demonstrate. A good answer would be to *prove* the properties of such a system, not only observe them via *experiments*.

In more recent papers [2,3,4], this architecture is used for arm and walking robot control, and a tracking experiment, with no major changes in the approach, except for the suppression schema which replaces timers by a continuous flow of messages from the suppressor node. These new applications make use of a larger number of layers (15 levels of a simpler structure for the arm controller). Whereas this tends to show the architecture's ability to grow (in terms of number of modules), inclusion of actually *higher level* capacities is still a step beyond.

Sooner or later in the evolution of this system, the ability to *plan actions* and try to execute them (and not only to react to external stimuli) will have to be incorporated to achieve increased autonomy. At that time the question will arise as to whether the known problems of planning (*e.g.* goal decomposition, sub-goal interactions, time, etc.) can be solved through this architecture.

References

[1] R. A. Brooks, "A Robust Layered Control System for a Mobile Robot", IEEE Journal of Robotics and Automation, Vol. RA-2 No. 1, March 1986.

[2] R. A. Brooks, "A Robot that Walks: Emergent Behaviors from a Carefully Evolved Network", September 1988.

[3] J. H. Connell, "A Behavior-Based Arm Controller", MIT AI Lab, AI Memo 1025, June 1988.

[4] I. D. Horswill and R. A. Brooks, "Situated Vision in a Dynamic World: Chasing Objects", 1988.

The Synthesis of Digital Machines with Provable Epistemic Properties[1]

Stanley J. Rosenshein and Leslie Pack Kaelbling

Reviewed by

Christopher Goad
SILMA, Inc.
1601 Saratoga-Sunnyvale Road
Cupertino, CA 95014

This paper starts with a formal analysis of what it means for a machine to know something about its environment, and develops on this basis an approach for programming machines which interact with the external world. The ideas are presented at a level of generality relevant to programming any such machine, but have been applied principally to robotic programming - in particular, to both the control and vision aspects of the mobile robot project at SRI with which the authors are associated. In what follows, I will start by sketching this programming approach, and then will return to the formal analysis of knowledge upon which it is based.

The methods used for robot programming - in research as well as production environments, - normally differ only slightly from those of the mainstream of programming practice. Sequential or modestly parallel algorithms are designed and then coded in close relatives of standard non-robotic languages. The approach presented in this paper differs from conventional programming in that the programs constructed admit massively parallel implementation, and in that there is a level of indirection in the process: the human programmer writes programs which write programs.

The approach finds concrete implementation through the language REX, designed by the authors. Roughly speaking, a REX program is a description of an abstract machine - given by a set of memory

---

1. J.Y. Halpern (editor), "Proceedings of the Conference on Theoretical Aspects of Reasoning About Knowledge", 1986.

locations, together with rules which govern low the state of the machine evolves through time under external inputs. REX machines are built up hierarchically. A typical REX subroutine will build a class of machines with properties parameterized by the inputs to the subroutine. Complex machines are built by combining simpler components constructed by subroutines.

As indicated earlier, the design of REX arose out of an analysis of a machine's knowledge of its environment. Briefly stated, the analysis runs as follows. Suppose that after some sequence of interactions with the external world, a machine is in state S. Then the machine is said to know a proposition P if P is true in every state of the world which is consistent with the machine having arrived in S. This is referred to by the authors as the "situated automata" approach to knowledge. Similar definitions have served as a basis for applying logics of knowledge to distributed systems (Halpern 1986). The situated automata definition of knowledge applies to parts of machines as well. The paper begins with a formal language for describing machines, and an axiomatization of the relevant primitive notions, including knowledge. The analysis is very general, in the sense that it applies to any machine whose internal state may be viewed as a set of memory locations, and whose behavior can be expressed in terms of constraints between the contents of memory locations. A machine description in this formal language may or may not be effective; it may or may not provide sufficient information to construct a machine with the specified properties. Now, REX may be regarded as a programming language which generates a particular class of effective machine descriptions in this formal language. So, REX is a hardware description language designed for simplicity of formal reasoning about the described hardware.

This work makes important contributions to robotics along two lines. First, REX taken by itself represents a radical approach to robotic programming which may prove practically effective. One might doubt the feasibility of the approach for development of complex algorithms in robotics and vision, since general experience indicates that designing highly parallel specialized hardware is far more difficult than designing conventional sequential algorithms. (REX is still a significant step of abstraction above a "real" hardware design language; layout for example finds no place in REX.) Let me quickly indicate just one reason why this initial fear may be unjustified. In REX, the basic composition method involves hooking simpler machines together, and the simplicity of the primitives involved appear to make this kind of composition flexible and easy to deal with. This method is appropriate for example, in combining submachines which "track" different parts of the external environment to form a machine which tracks a larger part of this environment at an appropriate level of abstraction. Hence, the kind of composition involved in REX may ease rather than complicate the construction of certain kinds of robotic algorithms. Beyond this there is the general consideration that massive parallelism is likely to be necessary for the solution of the most difficult problems in robotics and vision, and this justifies exploration of parallel techniques, even if the initial difficulties are substantial.

Second, the axiomatization of machine behavior and knowledge constitutes a real advance in formalizing reasoning about machines interacting with a complex environment. Its success derives from the underlying definition of knowledge used - a definition which does not restrict attention to machines which code statements about the external world in any particular way, and from its hierarchical nature.

In the current exposition, the formalization of knowledge and REX are linked by matters of motivation and the use of the formal system in verifying REX programs. Nonetheless, the two aspects of the work can be understood independently. An interesting direction - not pursed in this paper - would be the use of the formal system in synthesis, rather than just verify of REX programs. The work described here has its roots in the extensive study of formalization of reasoning about knowledge and action which as been carried out in philosophy and artificial intelligence over more than two decades, beginning with the work of Hintakka (Hintakka 1962) and Kripke (Kripke 1963). In particular, Kripke semantics constitutes the core of the approach. The paper launches quickly into a quite technical exposition of the formal system - an exposition which assumes basic familiarity with previous work in model logics of knowledge. Fortunately for the reader who is not familiar with this literature, the volume in which the paper appears starts with an excellent introduction to the field by Joseph Halpern. This introduction assumes little prior knowledge, and provides definitions of the basic notions which Rosenshein and Kaelbling assume. The examples of REX and formal reasoning about REX given in the paper are very simple, and give little flavor for how the approach would apply to complex situations. Wells (Wells 1987) describes vision algorithms used in the mobile robot project at SRI - algorithms which were partly implemented in REX. However, the topic of the latter paper is the algorithms themselves rather than the manner of their implementation. I look forward to the publication of description this approach as applied to robotic application of realistic complexity.

References

Halpern J.Y., 1986. Reasoning about knowledge: An Overview, Proceedings of the Conference on Theoretical Aspects of Reasoning About Knowledge (ed. J.Y. Halpern), Morgan Kaufmann.

Hintikka J., 1962. Knowledge and Belief, Cornell University Press, Ithaca.

Kripke S. 1963. Semantical Analysis of Modal Logic, Zeitschrift fur Mathematische Logik und Grundlagen der Mathematic 9, pp. 67-96.

Wells, William W, 1987. 1987 AAAI Conference.

Planning for Conjunctive Goals*

David Chapman

Reviewed by
Stanley J. Rosenschein
Teleos Research
Palo Alto, California

Chapman's paper is concerned with the correctness and efficiency of domain-independent planning systems. Although planning has been a central topic in artificial intelligence (AI) approaches to robot programming since the late 1960s, its logical and algorithmic foundations have been poorly understood. This is especially true for *nonlinear* planning systems, that is, systems that operate by constructing a graph representing a partially ordered set of actions. Chapman has taken an important step toward rectifying this situation by developing an elegant nonlinear planning system called TWEAK and carrying out a detailed analysis of its mathematical properties. In this paper he presents several interesting results: (1) TWEAK's planning procedure is provably correct and complete; (2) previous nonlinear planning systems can be viewed as variants of the general method embodied in TWEAK; and (3) the formal planning problem for even moderately rich domains is computationally intractable. Although the last conclusion will hardly startle AI planning researchers (many of whom openly embrace heuristic methods for this very reason), Chapman's analytic insights are illuminating and raise important questions about future research directions in planning.

Historically, the field of AI planning arose as a response to the complexity of robot programming. Rather than explicitly program the robot's reaction to every conceivable situation, the aim was to have the robot "program" itself at run time by manipulating symbolic descriptions of actions and their effects. Early suggestions along these lines were made by John McCarthy and form the basis for his work on the situation calculus [3] and for the

---

* *Artificial Intelligence*, **32**(1987), pp. 333-377.

STRIPS planner used as part of the control system for SHAKEY [2]. The model implicit in these and later studies can be described briefly as follows: The *planner* is assumed to have available a symbolic description of the current state of the world and a description of the desired goal state, supplied either by a human supervisor or by online perceptual processes operating on a stream of low-level sensory data. The output of the planner is a *plan*, a program-like data structure describing a sequence of actions which, if executed starting in the current state, would cause the goal to be achieved. Responsibility for carrying out the plan lies with the *execution module*, which produces a stream of low-level actions. The execution module is also responsible for monitoring dynamic perceptual information and detecting anomalous conditions that might require replanning.

In this model, planning is a form of *reasoning*, i.e., drawing conclusions from premises. As in other areas of automated reasoning in AI, much of the research has focused on striking a suitable balance between the logical richness of the representation language and the computational complexity of the derivation process. For certain specialized forms of planning, such as motion planning, algorithmic efficiency can often be guaranteed by exploiting domain-specific (e.g., geometric and physical) constraints. Domain-independent planning, on the other hand, has tended to use formalisms resembling those in the formal theory of programs and has often fallen into the same computational difficulties as automated program synthesis and verification. To insure tractability while preserving domain independence, researchers have been forced to simplify the problem in a variety of ways: (1) modeling actions as discrete state transformations; (2) considering only transformations that can be described in a simple "syntactic" fashion, e.g., by adding and deleting assertions from highly structured state descriptions; (3) restricting the logical form of goal statements; and (4) limiting plans to action sequences, thus excluding complex action-forming operators such as iteration, conditionals, or concurrency.

Even with these simplifications, it is hard to find semantically interesting yet computationally tractable domain-independent planning methods.

Early research in planning concentrated on methods that search for an action sequence either by working forward from the initial condition or backward from the goal. At each stage, a symbolic situation description is transformed to reflect the effects of candidate action steps until eventually a sequence of steps is found that transforms the initial condition into one satisfying the goal. These methods are sometimes called *linear* because the planning process mimics the temporal order of the resultant plan steps. The logic of linear planning is well understood [5,8], but its combinatorial properties are usually unsatisfactory, since the search process is exponential in the length of the action sequence generated.

Nonlinear planning methods were developed with the aim of improving search efficiency. These methods work by adding plan steps and constraining their temporal order and the bindings of variables until the plan can be shown to satisfy the goal. The advantage of this style of planning is that the search process only imposes constraints as necessary rather than arbitrarily choosing orderings and backtracking on failure. The hope was that this would make a larger class of planning problems tractable, and, indeed, some of the most successful experimental work on planning (e.g., Sacerdoti [6], Wilkins [9], Stefik [7]) has been based on this style of planning. However, this experimental work has proceeded largely without the benefit of formal analysis, either of the inherent logic of nonlinear planning or of its combinatorial properties. Chapman's work, along with other recent work [4], provides the beginnings of such an analysis.

Chapman begins by carefully defining the version of the planning problem he intended TWEAK to solve. In Chapman's formulation, propositions are represented by negated or unnegated tuples of constants and variables (e.g., ~(on a x)). A plan step, corresponding to an individual action, has a set of pre- and post-condition propositions, as illustrated in Figure 1. A set of plan steps can be restricted by constraints of two varieties: temporal constraints that govern the relative order of plan steps and codesignation constraints that restrict the denotations of constants and variables. A plan is simply a collection of plan steps and associated constraints.

Preconditions: (on x z)
(clear x)
(clear y)

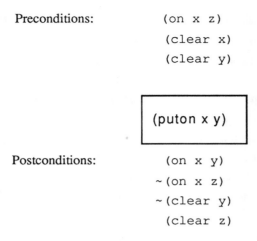

Postconditions: (on x y)
~(on x z)
~(clear y)
(clear z)

Figure 1. A plan step

Since a plan is only a partial description of an action sequence, there are ordinarily many possible "completions" of the plan, and important plan properties must be defined relative to this set of completions. The most important such property is what the author calls the Modal Truth Criterion (the modalities in question being *necessarily*, meaning "in all completions," and *possibly*, meaning "in some completion.") This criterion defines the precise circumstances under which a proposition is necessarily true in a situation. To paraphrase the author: A proposition is necessarily true in a situation if there is a situation necessarily previous (or equal) in which the proposition is necessarily asserted and for every possibly intervening step and every possibly codesignating proposition which the intervening step denies, there is a further necessarily intervening step which necessarily re-establishes it by asserting a codesignating proposition. (Unfortunately, the same modal precision that renders the statement barely interpretable is central to the author's key technical points.) Having defined what it means for a proposition to be true in a situation, the author is able to formally define the planning problem: Given an initial situation and a set of goal propositions, find a plan in whose final situation all the goal propositions are necessarily true.

Having formally defined the problem, the author proceeds to present an algorithmic solution based on an exhaustive analysis, using the modal truth criterion, of the various ways a proposition can come to be true in a situation. Each clause of the criterion gives rise to operations the planner might perform to make a goal proposition true, namely inserting a plan step, constraining it to be before or after another, constraining terms to codesignate or not, and so on. This analysis leads to a relatively simple non-deterministic procedure for exploring the space of plans. The author has implemented a system called TWEAK that uses this procedure together with a dependency-directed breadth-first search control strategy to systematically construct plans.

By separating the logic of nonlinear planning from the control, Chapman has made both easier to understand. He proves several theorems establishing important mathematical properties of his algorithm, notably the polynomial decidability of "truth in a situation," the undecidability of the overall planning problem, and the correctness and completeness of his planning algorithm. This last result guarantees that if TWEAK, given a problem, terminates claiming a solution, the plan it produces does in fact solve the problem; and if TWEAK returns signalling failure or does not halt, no solution exists. Having given these results, Chapman proceeds to survey past work on planning, classifying various heuristic approaches according to which clauses of the modal truth criterion they happened to exploit and which elements of his own more general algorithm them embody. He argues quite convincingly that most of the main ideas in nonlinear planning are effectively subsumed by TWEAK.

The polynomial-time complexity result for checking truth in a situation depends crucially on the restricted nature of the action-description language. Chapman proves that for action representations even slightly more expressive than those used by TWEAK, the problem of deciding truth in a situation is intractable. The kinds of extensions that bring on intractability include the representation of conditional actions, the dependence of effects on the input situation, and derived side-effects (i.e., indirect effects of an action that must be deduced by the planner.) In all these cases the problem of determining whether a proposition is necessarily true in a nonlinear plan is NP-hard.

These results are important for two reasons. On the positive side, they help structure and extend previous results in AI planning. They also serve as a paradigmatic example of the trade-offs between expressiveness and tractability. On the negative side, they clarify just how little expressiveness is required to push domain-independent reasoning systems over the combinatorial cliff. Derived side-effects and actions with conditional effects are necessary to formalize all but the most trivial domains. For example, even an action as simple as pushing a block will, in the real world, have effects that are contingent on the weight of the block, whether the block is nailed to the table, and so on. Chapman acknowledges these issues fully and even goes so far as to raise doubts about domain-independent planning as a fruitful area of investigation. He points out that in his own research he has been exploring alternatives to classical planning, chiefly the detection of "cliches," or problem-solving abstractions that occur across many domains and a more reactive approach to behavior generation known as "situated activity" [1].

References

[1] Agre, Philip E. and David Chapman. "Pengi: An Implementation of a Theory of Activity," *Proceedings of Sixth National Conference on Artificial Intelligence*, Seattle, Washington, July 1987, pp. 268-272.

[2] Fikes, Richard and Nils J. Nilsson. "STRIPS: A New Approach to the Application of Theorem Proving to Problem Solving," *Artificial Intelligence*, Vol. 2, Nos. 3,4, 1971.

[3] McCarthy, John and P. Hayes. "Some Philosophical Problems from the Standpoint of Artificial Intelligence," in *Machine Intelligence 4*, B. Meltzer and D. Michie (eds.) Edinburgh University Press, Edinburgh, Scotland, 1969.

[4] Pednault, Edwin P.D. "Extending Conventional Planning Techniques to Handle Actions with Context-Dependent Effects." *Proceedings of Seventh National Conference on Artificial Intelligence,* St. Paul, Minnesota, August 1988, pp. 55-59.

[5] Rosenschein, Stanley J. "Planning: A Logical Perspective," *Proceedings of the International Joint Conference on Artificial Intelligence*, Vancouver, B.C., August 1981, pp. 331-337.

[6] Sacerdoti, Earl D. *A Structure for Plans and Behavior*, Elsevier North Holland, New York, 1977.

[7] Stefik, Mark. "Planning with Constraints (MOLGEN: Part 1). *Artificial Intelligence,* Vol. 16, No. 2, May 1981.

[8] Waldinger, Richard. "Achieving Several Goals Simultaneously," in *Machine Intelligence 8,* E. Elcock and D. Michie (eds.), Ellis Horwood, Edinburgh, Scotland, 1977.

[9] Wilkins, David. *Practical Planning: Extending the Classical AI Planning Paradigm,* Morgan Kaufmann Publishers, San Mateo, CA, 1988.

MH-1, A Computer-Operated Mechanical Hand[1]
Heinrich A. Ernst

Reviewed by

Russell H. Talyor
Manufacturing Research Department
IBM T. J. Watson Research Center
Yorktown Heights, New York 10598

Introduction

This 1961 Ph.D. thesis describes a pioneering effort to develop a general purpose, programmable, sensor-controlled manipulator. The outlook expressed is very much that of early Artificial Intelligence research, and much of the content may seem rather quaint after almost thirty years, especially to those whose main interests are the engineering or analytical aspects of robot control. Nevertheless, Ernst's work can provide an invaluable perspective into many current problems in manipulator programming. A careful reader will recognize many familiar concepts in nascent form and may profitably consider whether their subsequent evolution was the only possible way things might have turned out.

Experimental Equipment

Ernst added motors and position-feedback potentiometers to a commercial mechanical master-slave manipulator designed for handling radioactive materials. This system was interfaced to the TX-0 computer through a custom electronic interface. This interface supported velocity commands to the seven motors through a multiplexed 6-bit DA converter and up to 32 analog inputs through a multiplexed AD converter with unspecified precision. Position loops were closed through the computer every 100 ms.

The heavy emphasis on sensing in this early system is especially noteworthy. Indeed, MH-1 was rather better equipped than were most

---
[1] Ph.D. Thesis Massachussetts Institute of Technology December, 1961.

subsequent research systems or, indeed, than the vast majority of present-day industrial robots. The gripper was fitted with eight binary touch sensors, photodiode proximity sensors in each fingertip, and 16 conductive elastic pressure sensors. Interestingly, he reports that manipulation forces were better obtained by monitoring motor inputs, and that the finger force sensors were most commonly used to detect contacts. There seems to have been little attempt to measure sensor or motor performance, but the system was evidently capable of moving about 10-20 inches/second.

Ernst goes to some lengths to explain his choice of computer. He had two candidates: an IBM 709 located in the MIT Computation Center and the TX-0, an early transistorized "minicomputer." He chose the TX-0 largely because of interfacing ease and because "whole weekends" were available for him to use it. The computation power was stunningly small by today's standards. The machine's basic instruction time was 12 microseconds, there was no hardware multiply or hardware interrupt, and memory consisted of 8K 18-bit words. Ernst's basic control system seems to have required about 2K words, and he reports never coming close to the 8K memory limitation. This is consistent with my own experience a few years ago in putting together a simple manipulator controller based on an early 16-bit microprocessor. Such systems are, in fact, quite adequate, so long as the underlying manipulator is sufficiently precise so that extensive computation is not required to correct for misalignments and so long as the total task complexity is low. These restrictions were fine for Ernst, who did not care much about performance and was interested in getting *any* sort of interesting behavior from the robot. Going beyond them has been a primary motivation driving the development of multiple subsequent generations of robot control system, each of which seems almost Parkinsonian in its hunger for more memory and computation power.

Explicit Programming: Mechanical Hand Interpreter

The Mechanical Hand Interpreter (MHI) was developed primarily as a a debugging and familiarization vehicle for the underlying manipulation system. The description of MHI is maddeningly vague, and one is left with the impression that Ernst does not consider the details to be very important. This inattention to clarity is unfortunate, since

MHI introduced the fundamental "guarded move" primitive that has been the basis of manipulator programming ever since. A typical MHI "statement" consists of one or more motion specification clauses with the general form

>   MOVE <direction & speed specification>
>   UNTIL <sensor condition 1>
>   UNTIL <sensor condition 2>
>   ...
>   UNTIL <sensor condition n>

followed by conditional clauses specifying a transfer of control if one of the sensor conditions is satisfied. The sensor conditions include both binary tests and analog thresholds. For example,

>   b2, MOVE z don 240
>   UNTIL s3 100 rel
>   UNTIL s8 200 lol cst
>   UNTIL s15 200 lol cst
>   UNTIL s27 200 lol cst
>   IFGOTO f2,b3
>   IFGOTO f3,b4
>   ...
>   IFCONT t,b2

moves the hand down at speed 240 until position sensor s3 is 100 less than its starting value or at least one of sensors s8, s15, or s27 reports a value less than 200. If the second until clause "trips," control is transferred to label b3, and so forth.

Ernst describes two simple MHI programs. The first one searches the work table for a box. It then searches the table for wooden blocks, which it picks up and drops into a box. The second program works interactively with the machine operator to stack blocks one on top other. These programs can be viewed as crude prototypes of two important classes of robotic application (sensor-based programmable automation and semi-autonomous "smart" teleoperation) that have motivated much subsequent research. Ironically, neither application, in itself, appears to be of particular interest to Ernst, who is primarily concerned with questions of whether a computer can be made "understand" and interact intelligently with the real world.

Despite the relative success of MHI, Ernst has several interesting criticisms of explicit programming that remain relevant even for more modern languages and systems. The first is that explicit programming is tedious. Although complex motion sequences, once written, can be called many times,

> ... these sequences must still be written originally. There should be a way to specify a whole action, for instance ... "grasp the object you are touching," so that the decomposition into elementary steps, including (this is most important!) the assignment of parameters such as speed or sense input can be performed automatically by the computer.

In Ernst's view, this problem arises inherently from the amount of information that must be specified; improvements in the language's conciseness are no help. If the computer can get more of the requisite information from the real world, then the job of the human programmer will be greatly simplified.

Ernst's second criticism is more philosophical. An explicitly programmed robot has no real "understanding" of the world. Although it may be more flexible than a tape-controlled manipulator, its behavior is still that of a "normal automoton," wholly dependent on a human programmer's judgement to allow it to respond appropriately to real-world uncertainities. If the human forgets to include the appropriate sensory checks, the machine will persist in highly inappropriate, even destructive, behavior.

Although Ernst perhaps doesn't give sufficient credit to the practical significance of large subroutine libraries or to the importance of language design in promoting easy customization and avoidance of careless errors, his criticisms are largely sound. They have been the justification for a large and growing body of work on "model based" or "task level" programming.

Implicit Programming: Teleological Hand Interpreter

Ernst's assessment of the limitations of explicit programming led him to propose a "Teleological Hand Interpreter" (THI), in which motions are specified in terms of desired sense element values, rather than of the motors to be used. Unfortunately, his description of THI is, if anything, even less exact than his description of MHI. Nevertheless, a careful reading and some informed guesswork yield several very interesting ideas.

THI has three kinds of statement. *Symbolic* statements specify motions required to achieve desired touch sensor values, *explicit* statements specify positioning motions, and *transfer* statements which specify where control is to go when sensor conditions are satisfied. Blocks of these statements are executed in parallel, and control is transfered when conditions are fulfilled. For example,

SYMBOL 0,0,btm,lbt,v10
IFG a0
EXPLIC s3 370 rel stp,v10
IFC b0

would cause motion of the gripper at speed 10 in an appropriate direction to bring the wrist bottom sensor into contact with something while also moving with speed 10 so that position feedback sense element s3 increases in value by 10 units. If the wrist bottom sensor touches something then control is transferred to statement a0. Otherwise, control passes to statement b0 after s3 changes by 10 units.

A second idea is an unusual "interrupt" feature. If a sensor that is not mentioned in the current active block statemets changes value, the system searches nearby blocks for a statement that would be expected to make this change. If such a statement is found, then control is passed to the corresponding block. Ernst reports that use of this facility significantly improved the robustness of a program for putting blocks into a box. If the gripper hits the box after it is moved unexpectedly, the program eventually recognizes it from the pattern of sensor "events" and traps to code associated with putting the box back at a standard location. There seem to be several assumptions (e.g., that behavior can be specified primarily as a graph of desired sensory conditions, that lexical closeness in a sequential program can be used to encode intentions, that any accidental change in a sensor value should preempt present behavior) that seem to be of limited generality. Nevertheless, some variation on this theme might well be worth further investigation to see where it leads.

The third idea was the introduction of a relatively small set of higher-level subroutines for such common actions as "go to," "scan," "take," and "place," together with a standard data structure (called a "model") for storing important object attributes such as location and size. The subroutines used parameters from these models and updated appropriate slots. Evidently, there were several different versions of each subroutine, and the human was expected to select which one was appropriate for a particular case.

Ernst reports that THI programs were shorter and ran perhaps twice as fast as the corresponding MHI programs. He clearly finds them more interesting, although (as he recognizes) they were no more reliable. The use of a runtime model, albeit a very simple one, together with subroutines whose semantics are defined in terms of model variables was clearly a very important step. The subroutines themselves are reported to have been quite difficult to write for many of the same reasons that MHI programming was hard, i.e., the human must anticipate and provide for many future runtime situations. Sensing must be used to distinguish different states and appropriate actions selected. Once the decision to build the subroutines has been taken, the tradeoff between MHI and THI paradigms for the internals becomes less critical.

Conclusions

This review has focused primarily on the system and programming aspects of Ernst's thesis, since these seem to me to have the greatest interest for present-day robotics researchers. Paradoxically, these aspects may have been less interesting to Ernst than the Artificial Intelligence implications of building a machine that could interact with the physical world. He is very concerned with *qualitative* descriptions of behavior, and is very impatient with (indeed, dismissive of) more *quantatative* descriptions. For example, the survey of the (very limited) prior art in 1961 concludes with this statement:

> D. Bobrow undertook a study of a certain way of controlling a mechanical hand by means of a digital computer. He described the motions of the hand in a coordinate system fixed to the hand. His use of an abundance of transformations and similar mathematical operations on the coordinates describing the motions of the hand violates some of the basic conditions which I feel should be imposed on the work described here. His study will therefore not be considered in any detail here.

Although Ernst does not specify which conditions Bobrow violates or why Bobrow's approach violates them, this judgement seems rather hasty, even without the benefit of hindsight. My own view is that rigorous mathematical formulations can play an invaluable role in the specification of complex behavior, and we should take advantage of them wherever possible to simplify the engineering problems associated with providing a suitable "vocabulary" of system capabilities. *At the same time,* it is important to recognize that human programmers are not very good at mathematical specifications except in the most constrained circumstances. Thus, while one should not ignore mathematical formulations of behavior, their best role would seem to be as a suitable intermediate language, or "object code" for higher level planning systems.

Overall, this thesis represents an important landmark in the development of Robotics. Although it is often difficult to read and occasionally tendentious, it nevertheless can be read profitably by anyone with a serious interest in understanding the origins of current research. Many key concepts in manipulator programming have their roots in this work, and it was a long time before a more powerful system was developed.

II

Sensing and Perception

A robot's ability to deal with a variable and uncertain world depends largely on its ability to sense it. Therefore robotics has motivated the development and application of a wide array of sensors: tactile, force, ultrasonic, infrared, and visual. But sensors are not enough in themselves. The raw data from sensors must be transformed into information about the world. Paraphrasing our earlier definition of robotics as the intelligent connection of perception to action, we can define perception as "the intelligent connection of sensory data to 3D models."

We can classify (roughly) research in the area of sensing and perception into the following categories:

1. Contact sensing, specifically, tactile and force sensing.

2. Non-contact sensing, specifically, visual sensing. This category can be further subdivided as follows:

    (a) Image understanding — the process of converting visual data from one or more images to 3D measurements.

    (b) Scene understanding — the process of converting disparate geometric measurements into a coherent scene description, possibly with the aid of geometric models of objects.

Actual research tends to span these categories, but they serve as a useful organization principle.

Much of the work in sensing and perception is devoted to image understanding, although traditionally this work has been perceived to be distinct from robotics. This is due partly to the perception that much of this work is not yet ready to meet the real-time requirements of robotics applications. Much of the perception work in robotics has focused instead on localizing specific objects from contact and non-contact data while attempting to minimize the general-purpose processing of the sensing data. Many of the papers reviewed in part II are in this vein. Improvements in algorithms and computers, however, are beginning to enable sophisticated image processing within the context of robotics tasks.

In part II we have two survey articles, by Dario and Tsai, each of which discusses four to ten published papers in some subareas of sensing and perception. In addition, we have nine reviews of selected articles spanning the categories discussed above. Below, we give a brief synopsis of all of these pieces, arranged in an order suggested by the introduction above.

One of the surveys and one of the reviews focus on contact sensing:

Dario's synopsis considers a wide variety of work in tactile sensing, explores the types of tasks where they may be useful, and examines the requirements on sensors for the different tasks.

Cutkosky reviews a paper by Salisbury that suggests how fingertip force sensors may be used to obtain contact information that has been traditionally regarded as the domain of matrix tactile sensors.

Two of the reviews, within the area of image understanding, focus on retrieving information in image sequences:

Baker reviews a paper by Dickmanns that addresses some of the visual problems encountered in the autonomous visual control of a vehicle. This work has been demonstrated in driving a van at 75 km/h on a highway. The focus of the paper is on Kalman filtering techniques for integrating predictions and observations.

Grimson reviews a paper in which Bolles, Baker, and Marimont show how to use the known motion of a camera and the tracking of points within rapidly sampled images to reconstruct the positions of objects in the environment.

Several other reviews span the boundary between image understanding and scene understanding:

Durrant-Whyte reviews a paper in which Bolle and Cooper employ statistical techniques to segment and classify segments of range data into primitive surface descriptions. These segments are then used (singly and together) to construct a maximum likelihood estimate of object location.

Tsai reviews a number of papers on camera calibration. Calibration is a problem that arises in many 3D vision techniques that require knowing the parameters that govern the conversion from external 3D coordinates to image coordinates.

Lowe reviews a number of papers by the Sheffield AIVRU vision project started by Mayhew and Frisby. The members of the Sheffield team have been engaged in developing a complete 3D vision system. In particular, they have focused on the problem of matching 3D models to the depth data obtained through binocular stereo.

Bolles reviews a paper by Ikeuchi that addresses the problem of locating an object from depth data obtained through the use of "photometric stereo." Ikeuchi's system automates the creation of a localization program, starting with a CAD model of the part and a model of how the sensor values are related to the object's orientation.

Marimont reviews a paper by Thompson and Mundy that locates a polyhedral model in a two-dimensional image by matching pairs of vertices in the model with vertices identified in the image.

Boissonnat reviews a paper in which Sugihara develops a very efficient algorithm for deciding whether two polyhedra (or parts of polyhedra) are congruent; this is a common subproblem in model-based matching systems. Sugihara's result is notable for being one of the few complexity results available on the problem of matching images to models.

Kapur and Mundy review a monograph by Sugihara that concerns the problem of interpreting line drawings as three-dimensional scenes. This is one of the earliest areas of research in computer vision. Sugihara introduces an algebraic approach to the problem and succeeds in deriving necessary and sufficient conditions for a two-dimensional image to represent a drawing of a three-dimensional scene. This is a long-standing problem in computer vision.

Tactile Sensing for Robots: Present and Future

Paolo Dario
Scuola Superiore S. Anna
and
Centro "E. Piaggio", University of Pisa
56100 Pisa, Italy

Much attention and substantial research has been devoted in the recent past to the study of tactile sensing for robots. Thus, it is now perhaps time to try to answer a fundamental question that has not been posed with the emphasis it deserves. The question is: what is the real application domain for robotic tactile sensing? The answer to this question bears an important relationship to the main purpose of this synopsis: to put accomplishments in this field into perspective and to identify key areas for future research.

It might appear surprising that real interest in tactile sensing technology and its applications to robotics dates to only a few years ago. In fact, the very notion of a robot, viewed as the physical interface between a computer and the outer world, seems to imply some degree of manipulation ability, and thus the capability of sensing and controlling the contact conditions between the robot end effector and a manipulated object. However, our experience is that industrial robots, although inherently unable to manage any but very simple types of contact with their surroundings, have been widely successful.

This seeming contradiction can be explained by observing that many problems related to the role of tactile sensing originated in the fact that until recently the paradigm of the "real" robot has been the industrial robot, the only one with a market which justified investments and extensive research. Advanced robotics, a field in which touch is unanimously considered as essential for the robot not only to manipulate objects but also for exploring and perceiving the environment, did not generate (with a few exceptions, e.g. mobile robots) much funding, and so this did not attract much research interest. However, recent programs on advanced robotics promoted by different agencies along with some decline in funding for industrial robotics have encouraged a shift of attention by many investigators towards those problems (including tactile sensing) associated with robots working in unstructured environments.

Even within the framework of the present interest for sophisticated robotic systems (terms like *mechanical redundancy*, *sensor-based control*, and *sensory fusion* are now popular keywords), it may still be instructive to present some considerations on tactile sensing in structured environments, an issue strongly debated until recently, with the purpose of deriving useful indications to direct future research efforts.

There are a number of arguments that one might raise concerning the usefulness of tactile sensing in industrial robotics. Those arguments are supported by a premise (that, unfortunately, is explicit only rarely), that the industrial robot is primarily a sophisticated, multi-degree-of-freedom, numerically controlled machine tool, and that, in most current industrial processes, used only as such. Many robotics researchers complain about the lack of intelligence in present industrial robots, and most industrial users claim to feel sympathetic with them. But, in practice, they continue to propose figures of merit to evaluate the performance of industrial robots that are the same used for conventional machine tools or machining centers, i.e. positioning accuracy and repeatability, speed of operation, payload, and so forth. Some of these figures of merit recently have become so extreme that, for instance, a new ADEPT robot has a position accuracy of about 25 microns and can operate at a maximum speed of about 10m/s! In fact, at least in a rigidly structured environment as the factory floor is, or can be quite easily arranged to be, a robot having a stiff mechanical structure, a primitive end effector, and very simple sensors (if any), can execute a number of useful tasks. Thanks to the flexibility provided by both its articulated mechanical structure and computer control, for certain tasks these current industrial robots can be economically competitive as compared with dedicated machines.

An intelligent robot capable of sensing environmental conditions and of dexterous manipulation could execute many more useful tasks than those carried out by present robots, for instance in the field of industrial assembly. The key problem is that sensing and manipulating requires (and will probably always require) the use of quite expensive hardware components, signal preprocessing and sensor-based control techniques which will inevitably complicate and slow down robot operation. Note that this holds for any type of sensing, including vision, proximity and force/torque (all of which are much more developed and understood than touch, and are more popular in industrial robotics). In fact, eventhough tactile sensing may have distinctive features and elective application domains (Harmon 1982), the key point is that the very same concept of perceiving through sensing (a lengthy and/or expensive process) is often incompatible with many types of industrial operations.

This observation leads to a dilemma which is still very common in industrial automation: when should one chose to incorporate as many sensors as possible in the robot system and to correspondently reduce the degree of "structure" of the working environment (a solution intrinsically flexible but also expensive), and when should one chose to neglect sensors and to increase the structure of the field of operation (thus keeping costs low, at the expense of flexibility)?

This problem is obviously not consistent with our definition of advanced robots, currently attributed only to those robots aimed at operating in unstructured environments. If we limit our discussion to the field of industrial robotics, and, in this context, to the problem of the appropriate use of touch, we observe that the long lasting debate between the "progressives", *i.e.* the supporters of a highly sensorized industrial robot (the same who, however, are forced to propose very demanding specifications, primarily in terms of cost and robustness, for the sensory systems to be used in those robots), and the "conservatives", who support the concept of an industrial robot which does not incorporate any but very simple sensors, can be perhaps resolved by the suggestion that there are two essentially distinct application domains for industrial robots (this for the sake of clarity; there are obviously intermediate solutions):

A. A class of tasks for which high speed and absolute position accuracy have a clear priority (as, for instance, in very large scale production). These tasks can be carried out most effectively by position controlled, poorly or totally non-sensorized manipulators operating in highly structured environments. The wide success of the SCARA robot (simple, stiff, precise, extremely fast and usually, lacking any type of external sensor) versus the anthropomorphic robot in assembly operations of the above class, confirms this observation. For these types of applications, sophisticated sensory and manipulatory skills might even be detrimental to the achievement of the required robot performances in terms of speed of operation. In fact, since an industrial robot for large scale production should be extremely fast, it may be convenient to put as much "structure" as possible within (*i.e.* dedicated end effectors) and outside (*i.e.* dedicated jigs) the robot manipulator, avoiding the use of sophisticated grippers and sensors. This solution is an excellent compromise between the often contrasting requirements for precision (in pick-and-place, in assembly, etc.) and speed of operation, since all available computing resources are committed only to optimal position control of the robot, and neither to the processing of sensory data and/or to the control of articulated end effectors.

B. A class of tasks which require more delicate operations, based on sensing local geometic features (e.g. curvatures) and on accurately controlling contact forces. A typical example is the assembly of parts for small batch, handicraft-like production. It is in these types of industrial applications, and perhaps only in these, that the complication and costs (including computational) inherently associated with sensing and analyzing environmental conditions can be justified.

Hence, if we agree on the hypothesis that the use of a truly intelligent (*i.e.* highly sensorized) robot may not be justified in conventional industrial applications, the domain of tactile sensing, intended as the process of detecting and interpreting the whole collection of information deriving from the contact between the robot end effector and the external environment (not just pressure distribution and the kinematic state of the contact, but also perceptual qualities like "is-fuzzy", "is-smooth", "is-warm", "is-oily", etc.) is an often delicate and slow process, then the use of touch might be justified only in those applications in which the *quality* rather than just the speed of operation is major concern. These considerations seem to apply equally well to define the domain of application of other artificial sensory systems (including vision), and perhaps, more in general, even of articulated end effectors (hands, feet).

Taking a skilled human operator (for instance a craftsman who fabricates with great care and patience a few, or even a single, object, investing all his sensory-motor capabilities, and creativeness, in this task) as a model for the robot in which a tactile sensing system would be truly justified, also provides intrinsic value to the anthropomorphic approach to the modelling of such artificial sensory system (even if non-anthropomorphic technical solutions can be adopted for the specific design of tactile sensors). Accordingly, this type of robot, that I shall call the *craftsman robot*, will also be equipped, in general, with some sort of dexterous hands, the only type of end effector capable not only of providing the robot with the necessary manipulation capabilities, but aso of truly implementing the very same notion of tactile sensing. Note that the concept of the *craftsman robot* is perfectly consistent with that of the advanced robot for non-industrial applications. Thus, most of the following observations hold for every robotic system using touch for perceptual purposes.

Based on these premises, let us proceed further in the attempt of defining more clearly what tactile sensing is, or should be, by referring to some of the many ideas and practical results that a number of investigators have reported so far (see, for instance, Cutkosky 1985), as well as to the experience that we have accumulated in our laboratory in the definition of a simple robotic system for the investigation of active touch and tactile perception (Dario and Buttazzo 1987).

I propose to adopt the term *contact sensing* to indicate generally the effects occurring when the robot end effector touches an object. Contact sensing includes force sensing and tactile sensing. The distinction between these two sensing modes, defined, respectively, as the measurement of the global mechanical effects of contact, and as the detection of the wide range of local parameters (physical and chemical) affected by contact, is sometimes rather subtle. Let us consider, for instance, the effects originated by the contact between the robot end effector and an object. At a macroscopic level, contact forces and moments result from a distribution of stresses produced by the interpenetration of the two contacting bodies. However, this interaction may also produce other physical effects in the two contacting objects (for instance, alterations of the thermal and/or magnetic equilibrium), or chemical modifications.

A further distinction between force and tactile sensing, perhaps more subtle, but consistent with the system approach that is extremely useful to follow in the design of tactile sensors, refers to the level of the control architecture at which the signal is processed. According to the proposed distinction, force sensing relates to low level processing for controlling compliant motion (a process that, after a learning phase, might be fast and almost "unconscious"), whereas true tactile sensing implies high level perceptual processing which requires attention and thus, usually, slow purposive movements (there are exceptions, though such as in sensing qualitative properties of surface conditions, like texture).

Having pointed out that true tactile sensing should include the detection of a number of parameters other than just local contact forces, let us analyze in greater detail, although still qualitatively, the mechanical effects of contact. Consider, for the sake of simplicity, the case of a rigid object indenting the compliant covering of a fingertip of a robot hand. As observed by Fearing and Hollerbach (1985), a finger must have a compliant covering to take advantage of the increased prehension stability possible at corners as well as to facilitate, at a higher control level, distinguishing between object features. In fact, a compliant covering will provide contact areas larger than hard skin. For example, with very hard skin, an edge and a side have a contact with the finger at only a few points and so may be indistinguishable without finger motion.

The same physical effect originated by contact, i.e. the indentation of the compliant skin, could be analyzed and interpreted, for control purposes, in two conceptually different ways (Dario 1988): measuring the actual contact stresses inside the skin, is important for controlling fine manipulation tasks, while measuring the deflection profile of the skin is useful for recognizing geometrical object features directly. It is important to recognize the distinct meanings which can be assigned to the same type of tactile information, even though the two effects (displacement profile and contact stress distribution) are obviously related, and even though there may be practical reasons for which it is usually more convenient to determine the actual contact stress rather than the contact profile. In fact, the main problem in measuring the contact profile is the indirect relation between the shape of the object and the profile of deformation. Since the compliant skin material is not perfectly compressible, it tends to pile up outside the contact region in a manner difficult to predict analytically. The mathematical description of the problem is further complicated by the fact that superposition, in general, holds for forces but not for displacements (Fearing and Hollerbach 1985).

There are examples of both deflection sensors and contact stress (more commonly strain) sensors in the growing literature on tactile sensor technology (Ogorek 1985), (Dario and De Rossi 1985), (Dario et al. 1988). For almost all of these sensing devices, however, performance is unacceptably poor, especially in terms of linear response and hysteresis. Let's reconsider this statement, based on the opinions expressed in the introduction to this article, and also with regard to the above observations on the role of displacement and contact stress sensing.

Sensing and controlling forces and moments generated at the end effector is necessary in order to control compliant motion, a fundamental aspect of dexterous behavior. Forces and moments can be sensed at different locations on the robot manipulator. In general, the finer the manipulation task, the closer to the contact area the force sensing device should be placed. Obviously, there is a tradeoff between sensor sensitivity and sensor ruggedness: a sensitive sensor capable of detecting locally small distributed contact forces is likely to be very delicate. The same holds for humans: in fine manipulation tasks (for instance in the manipulation of delicate electronic components or of thin and flexible wires) force control is obtained mostly by using the information provided by the cutaneous receptors; the gross manipulation of heavy objects is controlled almost completely through force feedback from deep sensors (muscle spindle organs, Golgi tendon organs), while the hand is often protected by gloves which reduce tactile sensations severely.

For the purposes of this analysis, three different types of manipulation can be considered, each requiring different contact sensing modalities:

a. gross manipulation,

b. fine manipulation of "regular" objects,

c. fine manipulation of "delicate" objects.

Most operations classifiable as type (a) (for instance those in which the hand grasps firmly the object, without truly manipulating it) could be controlled just by using torque sensors located at the robot joints, and/or at the wrist.

The dexterous manipulation of objects having "regular" shape and weight, e.g. easily and securely graspable, can usually be managed by sensing and controlling finger joint torques and fingertip contact forces and moments. The concept of fingertip force sensing, lucidly proposed by Salisbury (1984), is extremely powerful, and has two attractive features: first, being conceptually equivalent to a miniature 6-axis force sensing wrist transferred to the base of each fingertip, a fingertip sensor is capable of accurately resolving all three components of the applied force and all three components of the applied moment; second, if some assumptions on contact conditions (namely, contact occurring at a single point with friction, shape of the fingertip contact surface known and convex, contact exerting a force directed into the surface) are assumed, a fingertip sensor is capable of reading the magnitude and line of application of the resultant contact force as well as the location of the contact. As the overall performance of this type of device can be excellent in terms of sensitivity, accuracy, linearity, negligible hysteresis, fast response, and even ruggedness and low cost (Bicchi and Dario 1988) it is reasonable to argue that, for the class of manipulation tasks we are considering (that probably include most of the cases of practical interest), the function of sensing forces could be conferred entirely to the fingertip sensor.

A very interesting consequence of this hypothesis is that a tactile sensor might not be designed for sensing distributed contact forces. In fact, if the resultant contact force can be measured through contact resolving fingertip sensors, the cutaneous sensors would only have the function of locally sensing all the other parameters affected by contact. Being free from the constraints imposed on their performance by the requirement of accurately and reliably controlling contact force, tactile sensors could be designed without worrying too much about linearity, hysteresis and perhaps even static force response (all drawbacks that, incidentally, most receptors in the human skin possess to some extent).

The third case we are considering, *i.e.* the fine manipulation of delicate objects, is most challenging. In this situation, in fact, the contact forces can be so small that the fingertip force sensor may not be able to sense them. In addition, the accurate execution of the desired manipulatory or exploratory task could require not only the control of the resultant force, but also the fine control of local contact forces. For example, incipient slip conditions are different whether the finger is touching a corner or an edge, or whether or not it is manipulating a tiny wire that deeply indents the fingertip covering; in fact, the local distribution of contact forces varies, even if the resultant contact force may be the same. For these types of *extreme* applications, monitoring and controlling the distribution of contact stresses at the compliant covering of the fingertip, including both normal and tangential stresses, is important.

In conclusion, I will try to identify a few specific research areas that, according to the above observations, seem to be critical for the development of more effective robotic tactile sensing systems. I think that such areas are the following:

1. **Contact Problems.** The approach proposed by Fearing and Hollerbach (1985) had the merit of pointing out the importance of this study to guide the design of effective tactile sensors. The analysis of the local contact conditions is also a fundamental source of information for the interpretation of tactile signals both for control purposes (Fearing 1987) and for perceptual tasks. At the moment, however, the models proposed for describing and investigating the elastic contact problem at the interface between a compliant tactile sensor and an object are oversimplified. Further research efforts, also addressing other parameters affected by the contact conditions, such as temperature and humidity, are definitely required.

2. **Tactile Sensor Technology.** A number of different technologies have already been explored for the design of sensors capable of detecting distributed forces. For a detailed overview of tactile sensors see (Dario et al, 1988) in which tactile sensors were classified, based on the transducing technology they incorporate, e.g. piezoresistive, magnetic and electromagnetic, capacitive, electro-optic and ferroelectric polymer sensors. Representative of the technological state of the art and achievable performance by each class of sensor are the devices described in (Van Brussel and Belien 1987) for peizoresistive sensors, (Vranish 1986) for

magnetic sensors, (Siegel et al. 1986) for capacitive sensors, (Begey 1988) for electro-optic sensors, and (Fiorillo et al. 1987) for ferroelectric polymer sensors.

Some new ideas also seem particularly attractive for the development of tactile sensors. Worthy of mention are the photoelastic sensor proposed in (Cameron et al. 1988). This sensor has the very interesting capabilities of intrinsically enhancing edge detection and of exploiting motion for better sensing. Also, the polymer sensor proposed in (De Rossi et al. 1988), and the field effect sensor described in (Jacobsen et al. 1987) are of interest.

It is important to point out that only sensing the distribution of the normal component of contact forces may not be sufficient for the level of performance expected by robots in the future. In the execution of some fine manipulative tasks, for example, surface sensors capable of detecting small shear forces would considerably simplify the problem of sensing incipient slip, even in such difficult situations as manipulating tiny or slippery objects. In those cases, the sensitivity of a resultant force/torque sensor may not be high enough, whereas the spatial response of strain sensors distributed underneath compliant skin would be severely decreased by the same elastic medium. Furthermore, distributed multi-component strain sensors would also simplify the general solution of the mathematical problem of distinguishing indented object features (e.g. a vertex from a side). There are an increasing number of proposed multi-component strain sensors (Hackwood et al. 1983), (King and White 1985), (Jacobsen et al. 1987), (Femi et al. 1987), (Yao et al. 1987).

In my opinion, however, it is also important to note that simplified solutions should also be explored in order to encourage the practical use of tactile sensors, an invaluable source of feedback for designers of tactile sensing systems. For example, some of the suggestions given in this article may encourage reconsidering solutions worthy of further investigation (for instance, even the discredited conductive elastomer might deserve consideration). More emphasis should be given to the study of such problems as sensing multiple variables (for instance temperature and contact forces, or proximity and contact forces), sensor conformability to the shape of the fingers of the dexterous end effector, wiring, sensor calibration and signal processing. My feeling is that the pioneer phase of tactile sensing technology is over, and that any sensor that is unable to be arranged in dense arrays and incorporated in curved structures, and which would not allow massive signal preprocessing, should not deserve serious consideration any longer.

3. **System Design.** As already suggested, a tactile sensor cannot be designed appropriately without carefully considering the overall features of the manipulatory system with which it is expected to operate. An excellent example of this approach is represented by the work of Jacobsen and coworkers in the design of the Utah/MIT hand (Jacobsen et al. 1984). System design is particularly necessary in the case of the conceptual architecture outlined in this article, whose implementation requires the integration of multiple sensory data. For example, it is intuitive that various force sensors incorporated in a dexterous end effector (at the joints, at the fingertip base, and at the fingerpad) should measure different force ranges, which will partly overlap. For instance, depending on the overall scale of the application, the fingertip sensor could be dimensioned in order to detect resultant contact forces in the range 0.1-10 Nt, while the fingertip tactile sensors could measure distributed contact forces in the range 0.01 - 1 Nt. Investigating how the controller could manage the transition from very delicate operations (the domain of cutaneous sensors) to ordinary manipulation (when the cutaneous sensors saturate) is an intriguing problem.

Another problem that could be investigated in the same framework is how to deal with friction forces at the fingertip. In fact, those forces could also be measured, although in different ranges of intensity, both by the contact resolving fingertip sensor and by the tactile sensors. An example of real "sensory fusion" between force/torque and tactile sensors has been presented in (Bicchi et al. 1988). In the proposed application, sensory data, rather than being simply redundant, are integrated in order to control object slippage in a sensorized gripper. This would be impossible with any of the two types of sensors considered alone. Further investigating methods for organizing force and tactile data to obtain the desired sensory-motor behavior is a fundamental aspect to the application of artificial intelligence to robotics.

4. **Interpretation of Tactile Data.** This problem would be directly affected by the proposed "loosening" of the specifications on tactile sensors. However, the interpretation of the data provided by the tactile sensors would be facilitated by the previously proposed, more accurate and extensive calibration procedures that should become a standard requirement in the process of development of every new tactile sensor.

Three fundamental classes of problems are open in this field. The first problem relates to the interpretation of true tactile data, in order to extract as much information as possible from the local contact. A second problem is the geometrical interpretation of tactile data (Gaston and Lozano-Perez 1985). This study can be extended in order, for instance, to integrate local contact features with sparse tactile data, as proposed in (Browse 1987). A third fundamental problem concerns the very essence of tactile sensing, *i.e.* active touch.

A true step ahead is needed towards implementing robotic haptic perception. The idea of analyzing static tactile images (which implies the concept of understanding the environment through "camera shots") should be substituted by that of dynamic tactile images (tactile sensing as a sequence of shots, *i.e.* a movie; see, for instance, the paper by Cameron et al. (1988)). The analysis of techniques for building models of the environment by means of robotic tactile exploratory procedures which replicate those suggested by psychological studies on human behavior, has been addressed with particular attention by the GRASP laboratory of the University of Pennsylvania (Stansfield 1986).

A very recent and interesting approach to the analysis of tactile data involves the use of neural network techniques (Patti 1988). This approach has the potential to lead to future designs of smart tactile sensors incorporating not only the sensing sites but also the preprocessing unit, a research line whose interest is confirmed by the work being carried out in the field of retinal sensors (Chiang 1988).

The study of robotic tactile sensing, although quite young, has already produced a number of interesting concepts and accomplished significant results. Of course, as in many other fields of robotics research, much remains to be done.

An appropriate conclusion for this synopsis is a further remark concerning the fundamental importance of the ability to manipulate objects through tactile sensing. Even if there are fully respectable robot systems (for instance almost the entire class of mobile robots) which demonstrate their intelligence through sensate behaviors not involving touch, I think that the implementation of the full concept of a robot requires the development of artificial tactile sensing capabilities. I would even say that only those robots capable of extensively manipulating the outer world (and definitely not most of the present "puritan" robots, which can just "contemplate" some features of their environment, being almost "scared" to touch objects using something more than the equivalent of a pair of pincers) really to be called a *robot*.

References

Bicchi, A. and Dario P., 1987. Intrinsic Tactile Sensing for Robots. Robotics Research of the 4th ISRR, Santa Cruz, Aug. 1987, MIT Press.

Browse, R., 1987. Feature-based Tactile Object Recognition. IEEE PAMI.

Dario, P. and DeRossi, D., 1985. Tactile Sensors and the Gripping Challenge. IEEE Spectrum 22 (8): 46-52.

Dario, P., 1987. Contact Sensing for Robot Active Touch. Proc. SDF Symp. on Robotics Research, Santa Cruz, Aug. 1987, MIT Press.

Dario, P., Bergamasco, M. and Fiorillo, A., 1987. Force and Tactile Sensing for Robots. In: Sensors and Sensory Systems for Advanced Robots, Dario P., Ed., NATO ASI Series, Springer-Verlag, Heidelberg.

Dario, P. and Buttazzo, G., 1987. An Anthropomorphic Robot Finger for Investigating Artificial Tactile Perception, Int. J. Robotics Res.

Fearing, R.S., and Hollerbach, J.M., 1985. Basic Solid Mechanics for Tactile Sensing. Int. J. Robotics Res.

Gaston, P.C. and Lozano-Perez, T., 1984. Tactile Recognition and Localization Using Object Models: The Case of Polyhedra on a Plane. IEEE PAMI, 6(3): 257-265.

Jacobsen, S. et al., 1984. The Utah/MIT Hand: Work in Progress. Int. J. Robotics Research, 4(3), pp. 21-50.

Ogorek, M., 1985. Tactile Sensors. Manufacturing Engineering, 94 (2): 69-77.

Salisbury, K. 1984. Interpretation of Contact Geometries from Force Measurements. In: Robotics Research. Brady M. and Paul, R., Eds., MIT Press, pp. 567-577

Stansfield, S.A., 1986. Primitives, Features, and Exploratory Procedures: Building a Robot Tactile Perception System. Proc. IEEE Int. Conf. on Robotics and Automation, San Francisco, pp. 1274-1279.

Synopsis of Recent Progress on Camera Calibration for 3D Machine Vision

Roger Y. Tsai

IBM T. J. Watson Research Center
Yorktown Heights, NY 10598

1.1 Introduction

Camera calibration is the problem of determining the elements that govern the relationship or transformation between the 2D image that a camera perceives and the 3D information of the imaged object. Simply speaking, there are four main reasons why camera calibration is important:

- Only when the camera is properly calibrated can its 2-dimensional image coordinates be translated into real-world locations or constraints of locations for the objects in its field of view.

- In model based 3D Vision, it is important to model and predict the performance or accuracy capability of any vision algorithm, in order to plan the proper strategy for sensing. Without a solid understanding of camera calibration, the chain is broken and sensing strategy cannot be formed or realized.

- Part of the task of camera calibration is camera location determination relative to the calibration plate. This task can be applied to the general task of model based object location determination or 3D object tracking.

- Camera calibration is critical for vision-based robot calibration, robot hand-to-hand calibration and other vision related geometric calibration (see Tsai and Lenz, 1987B, 1988,and Lenz and Tsai, 1988).

There are two kinds of parameters that define this 2D/3D relationship:

- Intrinsic
  These are the parameters that characterize the inherent properties of the camera and optics:
    - Image Center (C_x, C_y)
    - Image X and Y scale factors
    - Lens principal distance ("effective focal length")
    - Lens distortion coefficients

Note that focal length and the image scales cannot be uniquely determined at the same time, although the aspect ratio is unique. You can increase the focal length and decrease the image scale without influencing the transformation between 2D and 3D. It is inherent in any imaging system but will not influence the goodness of fit between the model and the world. However, for solid state, discrete array camera (DAC), since the pixel spacing in the sensor plane can be known very accurately from camera manufacturers' data sheet, this ambiguity is not present. In spite of this, the x scale factor still needs to be estimated in any case since raster scan converts image signal into analog waveform and then sampled by the frame grabber. The conversion always entails uncertainty and needs to be calibrated.

- Extrinsic
  These are the parameters indicating the position and orientation of the camera with respect to the world coordinate system:

  - Translation (T_x, T_y, T_z)
  - Rotation about X, Y, Z axes

In this synopsis, we will first make a brief survey of all the camera calibration techniques that came into existence before 1985. Then, we will make some critical comparison among four post-85 techniques. The reasons for a break at 85 are twofold. First is that, simply speaking, prior to 1985, most of the camera calibration techniques are polarized between approaches closely related to classical Photogrammetry approach where accuracy is emphasized, and the approaches geared for automation, where speed and autonomy are emphasized. The former will be called category I and the latter category II. After 1985, the techniques are more versatile and tend to combine strengths from both categories. It is natural to discuss them separately. The second reason is that we are making some critical comparisons among selected recent advancement, and there must be a break for distinguishing "recent" from the older ones. We choose 1985 for this break.

We use three criteria for comparison of the post-85 techniques. They are speed, accuracy and autonomy. Some people might wonder why should speed be important for calibration. Well, if it can be done fast with less cost or effort, why not. Actually, the more important reason is that for automation environment, the camera may be mounted on the robot hand, which may bounce around wildly. The calibration parameters may change frequently and require fast recalibration. Also, in automation environment, the camera parameters might have to be changed purposely to meet changing tasks requirements. Furthermore, part of the calibration results are position and orientation of the world coordinate system relative to the camera, which can be applied to model based object location determination. Obviously, it pays very well to do it fast, even in real time.

Summary of Pre-85 Calibration Techniques

For analysis purposes, it is convenient to classify the techniques into the following three groups:

Category I: Techniques involving full scale nonlinear optimization

The transformation between the 3D object coordinates and 2D image coordinate is a nonlinear function of all the calibration parameters. Category I is the classical approach and it attempts to do nonlinear optimization to obtain the best estimate of the calibration parameters by minimizing the residual error of the above nonlinear equations. The main advantage is that no approximation is involved and the camera model can be quite elaborate. The problem is the same as that associated with any full scale nonlinear optimization, which is that it is very computation intensive, and needs a good initial guess. The latter of course makes automation difficult. Also, it needs a good nonlinear optimization software package, which may be cumbersome to setup for a mini or personal computer. The DLT (Direct Linear Transformation) is also included in this category if lens distortion is considered (see Abdel-Aziz and Karara, 1971, 1974; Karara,1979). DLT normally should belong to category II below. But when lens distortion is included, full scale nonlinear optimization again needs to be done. Typical examples for category I are: Faig, 1975; Abdel-Aziz and Karara, 1971, 1974; Karara, 1979; Brown, 1971; Sobel, 1974; Gennery, 1979; Malhotra, 1971; Wong, 1975; Okamoto, 1981, 1984.

Category II: Techniques involving computing perspective transformation matrix first using linear equation solving

Although the equations characterizing the transformation from 3D world coordinate to 2D image coordinate are nonlinear function of the camera internal parameters (intrinsic) and external orientation and position parameters (extrinsic), it is possible to linearize the problem by ignoring lens distortion and treating the coefficients of the 3 × 4 perspective transformation matrix as unknowns (see Duda and Hart, 1973, for a definition of perspective transformation matrix). These coefficients are functions of the camera parameters. Given the 3D world coordinates of a number of points and the corresponding 2D image coordinates, the coefficients in the perspective transformation matrix can be solved by least square solution of an overdetermined systems of linear equations. Given the perspective transformation matrix, the camera model parameters can then be computed if needed. The advantage obviously is that nonlinear search is avoided and automation is possible since no initial guess needs to be made. But there are several problems:

- Many investigators have found that ignoring lens distortion is unacceptable when doing 3D measurement (e.g., Itoh, Miyanchi and Ozwa, 1984, Luh and Klassen, 1985; Faugeras and Toscani, 1987).

- The resultant rotation matrix is usually not orthonormal. Recently, Grosky and Tamburino tried to correct this problem, and is reviewed after this pre-85 analysis.

- Although the equations are linearized, the number of unknowns for the linear equations is usually greater than necessary. For example, the two stage technique (Tsai, 1986; Tsai and Lenz, 1988) requires solving linear equation of five unknowns is stage 1 and linear equations in three unknowns in stage 2 only, while category II techniques generally need to solve linear equations with more than 11 unknowns.

- The calibration points cannot be coplanar. Allowing coplanar calibration points greatly facilitates the process of producing highly accurate calibration points as well as making the illumination and feature extraction problem much easier. Due to the artificial linearization of the unknowns in category II techniques, coplanar points cause a singularity problem. Recently, Grosky and Tamburino tried to resolve this issue, and will be reviewed below.

A partial list of publications related to category II techniques are: Yakimovsky and Cunningham, 1978; Sutherland, 1974; Strat, 1984; Ganapathy, 1984; Abdel-Aziz and Karara, 1971, 1974; Karara, 1979; Hall, Tio, McPherson and Sadjadi, 1982.

Category III: Two Plane Method

The two-plane methods model the transformation from the image coordinates to the 3D coordinates on several calibration planes to be linear, and once each individual linear transformation (one to each plane) is estimated, the rest of the 3D points within these calibration planes are interpolated. The advantage is the same as that for category II techniques, i.e., the computation is linear. But all the problems associated with category II come in here too. Particular, the number of unknowns is at least 24 (12 for each plane), much larger than the degrees of freedom, and larger than necessary for accurate and fast computation. The two-plane method developed by Martins, Birk and Kelly (1981) theoretically can be applied in general without having any restrictions on the extrinsic camera parameters. However, for the experimental results they reported, the relative orientation between the camera coordinate system and the object world coordinate system was assumed to be known (no relative rotation). In such case, the average error is about 4 mil with a distance of 25 inches. This is comparable to the accuracy obtained using Category II and the Two-Stage Calibration Technique (Tsai, 1986). A general calibration using the two-plane technique was proposed by Isaguirre, Pu and Summers (1985). Full scale nonlinear optimization is needed. No experimental results were reported. A partial list of publications related to category II techniques are: Martins, Birk and Kelley, 1981; Isaguirre, Pu and Summers, 1985.

Critical Comparison of Several Selective post-85 Techniques

Four methods are selected for comparisons. Brief introductions are first given, followed by critical comparisons. Then, in the conclusion, there will be some final recommendations.

Brief Introduction of four selected techniques

Method A: Faugeras and Toscani, 1987

This method basically falls into category II for the pre-85 survey, so all the introduction there apply here too. Since category II methods inherently ignore lens distortion, so does method I, except that after the linear coefficients are estimated, the lens distortion is corrected at a separate step. Furthermore, in order to estimate a measure of uncertainty of the calibration results, Kalman filtering is applied, assuming zero mean independent Gaussian noise. Also, in order to make the homogeneous transformation matrix separable into a product of two matrices, one for intrinsic, and one for extrinsic, they propose an alternative means which uses the actual intrinsic and extrinsic camera parameters as unknowns and minimize a nonlinear function in the formulation of extended Kalman filtering. The residual error is minimized by iterative linear approximation.

Method B: Grosky and Tamburino, 1987

This method again falls into Category II in the pre-85 survey, so all descriptions applies to this well, except that during the transformation from the homogenous linear coefficients to the actual intrinsic and extrinsic parameters, the orthonormality of the rotation matrix is preserved. Also, with a slightly different formulation, the calibration points can be coplanar. These are the two main features of this method. Depending on how many camera parameter are unknown, the formulation for the final solution is different. Each case needs to be worked out analytically and separately. Usually, in addition to solving linear equations with 11 unknowns, a polynomial equation of single unknown and of third or fourth order needs to be solved. As characteristic of the category II methods, lens distortion is not considered.

Method C: Tsai, 1987; Tsai and Lenz, 1987A, 1988, Lenz and Tsai, 1987, Tsai, 1988

This method avoids large scale nonlinear optimization while at the same time maintaining exact modelling without making any approximations, by decoupling the calibration parameters into two groups; each group can be solved easily and rapidly. A physical constraint is sought for that is only dependent by a subset of the calibration parameters and independent of the others. If such constraint can be found, then equations can be setup that is only a function of the former subset of parameters, not the latter. Furthermore, this equation should be easy to solve. It turns out that the best candidate for such physical constraint is the Radial Alignment Constraint (see Tsai, 1987). It is quite a simple one, yet it is capable of decoupling the parameters. The resulting equations can be used to solve for most of the extrinsic parameters by computing solutions of a linear matrix equation with five unknowns (if calibration points are coplanar) or seven unknowns (if calibration points are noncoplanar), plus some tricks. The solution of the linear equation is not exactly camera parameters, but there is a one-to-one correspondence between the two. The nice thing is, there exists a trick to convert one to the other efficiently. The rest of the parameters, mostly comprising intrinsic parameters, can be solved trivially. One other goal is to allow the calibration points to reside on a single plane. In turns out that the above mentioned approach allows this to happen.

Method D: Goshtasby, 1987

This method is for calibrating lens distortion. Lens distortion is formulated in terms of Bezier patches. The x distortion and y distortion are treated separately. A square grid is put in front of the camera. Assuming that the plane holding the grid is parallel to the image plane, and the x and y axes of the grid is aligned with the x and y axes of the image plane,and the patches are used as a transformation functions that map the deformed image into the ideal image. Iterative approach is employed to obtain the control vertices of the Bezier patch that best fit the transformation that carries the distorted image sample points to the ideal image sample points.

Critical Comparisons:

- Speed

    We discuss the speed of two separate tasks involved in a normal camera calibration:

    - Speed of image feature extraction

        Although image feature extraction is only a preprocessing step for camera calibration, it is influenced in part by certain generic nature of the calibration itself. This generic nature is primarily whether the calibration points can be coplanar or not. If they can be coplanar, then the calibration plate can be a glass plate with many circles or disks printed on it (using photographic process, for instance), and the illumination can be back lighting. It is hard to overemphasize how much contribution this single factor has on the speed, robustness and ease of feature extraction. Methods B and C require only single plane calibration points, and it is reported in the publication for Method C that one is able to extract image features of 36 points with an accuracy of 1/30 pixel within 65 milliseconds, including the frame grabbing time, using minicomputer and off-the-shelf general purpose image processing hardware. Although the same feature extraction algorithm can be implemented for Method A with equal speed, Method A requires multiple plane calibration, making it a necessity to move the calibration target or the camera to a different location with highly accurate moving mechanism, and perform another feature extraction.[1] One of course will start with a static set of calibration target points that are not coplanar. But this would make it vastly more difficult to produce a calibration target with more high accuracy while at the same time easy for the camera to extract image features. Illumination may easily cause problems frequently. Another issue is about calibration point patterns (note that this has nothing to do with the calibration algorithm itself). In the paper for Method A, it is mentioned that rectangular grids are preferred rather than circles since they think the perspective distortion would cause a mismatch between the 3D circle center and the centroid of the distorted ellipse. Through some analysis, we have found that such deviation is extremely small for a normal calibration setup (easily within 1/30 pixel). Using the circles makes the feature extraction a lot easier and faster.

- Speed of camera calibration itself

  If the basic approach in Method A (rather than extended Kalman filtering in Method A) is considered, then Method A and B should be similar in speed, since they both solve linear least square problems with about 11 unknowns. Method C however, only requires solving linear equations of five and two unknowns, and can be done in 25 milliseconds using a normal minicomputer. Note that this includes both intrinsic and extrinsic calibration. If extended Kalman filtering of Method A is used, then it is equivalent to full scale nonlinear optimization, and should be more time consuming. Also, for Method A, the distortion correction requires solving for linear affine transformation coefficients for about a hundred regions (eight coefficients for each region). It is not clear how efficient it would be. So far as distortion estimation is concerned, Method D seems to be similar to Method A and less efficient than Method C.

---

[1] Furthermore, if the calibration glass plate is sitting on top of the backlighting illuminator, which in turn has to sit on the translation device, the parallelism between the top and bottom surface of the illuminator will influence the accuracy significantly.

- Accuracy

    Since Method B does not calibrate distortion, its accuracy is naturally lower in general comparing with Method A and C. Also, experimental results on the accuracy for Method B is not yet available. In principle, Method A and C should be similar in accuracy unless tangential lens distortion is big. In all the lenses we have tested, tangential lens distortion is less than 1/30 of a pixel. This is also supported by the Manual of Photogrammetry, 1980, which states that the tangential distortion is only of historical importance, and is much less than the radial distortion. The publication for Method A listed above however indicated that for a set of experiments being performed, Method A is better than Method C (still within the same order of magnitude). But they indicated that it is probably due to the experiment setup since calibration plane was almost parallel to the image plane when using Method C. When using single plane calibration in Method C, it is necessary to tilt the calibration plane at least 30 degrees to make it work. Another reason could be that during those tests, the new center and scale calibration schemes developed by Lenz and Tsai, 1987, were not done (these were developed after the original paper for Method C were published). Still another reason could be that the tangential distortion is too big for the lens used during the tests and Method C suffers some error. There is one reason why we think Method A "might" suffer some error too due to the separation of lens distortion correction from the rest of the calibration. Method A first assumes there is no distortion. Based on that, the whole calibration is done. Then, the distortion is to be estimated. When distortion is significant, this will create a bias on the solution for the rest of the parameters, which are estimated before the correction of the distortion. We do not know how great this bias is, so this concern may not be important, at least not for all cases.

    One attractiveness of Method A is that Kalman filtering is employed, and therefore, if one is interested in obtaining an statistical measure of the uncertainty as the calibration point comes in one by one, Method A suits the purpose well.

    When multiplane calibration is used for Method C, a certain set of the parameters characterizing the motion of the xyz translation stage carrying the calibration plate can be modelled into the equation to be solved for so that the xyz stage need not be very accurate to give highly accurate results.

    For correcting distortion, Method D requires precise alignment of the calibration plate and the image plane. Unless that alignment is achieved, the accuracy should be less than Method A and C.

- Autonomy

There are a few factors to be discussed. The first is the dependence on *a priori* knowledge or information. If the basic approach of Method A (as opposed to extended Kalman filtering of Method A) is considered, which does not allow the independence of intrinsic parameters with the world coordinate system, then both Method A and C do not need a *a priori* information. Although Method C makes use of the manufacturer supplied information on the pixel distance in y direction, it does not harm automation in any sense for the reason explained in the paragraph following the definition of intrinsic parameters in the Introduction. Furthermore, the manufacturer's information on the the pixel distance on the actual DAC sensor plane is highly accurate, due to the manufacturing process of the DAC sensor.

Conclusion

The importance of camera calibration may be self evident by observing the wealth of work poured into this area in the past decade. The importance of camera calibration is not just for making accurate 3D measurement, but also for helping 3D model based vision system to model the performance or capability of any particular sensing strategy. The purpose of this synopsis is to review the key features of the existing techniques, and make critical comparisons, so that the community may benefit from it when comes to choosing a particular calibration approach. From the analysis above, the following is a brief summary:

From a user's point of view,

1. if you are most concerned with the simplicity of the setup and the speed and accuracy of feature extraction, use a technique that allows single plane calibration. This eliminates the need for a very expensive and accurate xyz translation stage for moving the calibration plate or camera for creating non-coplanar calibration points. Note that although one can create non-coplanar calibration target in a static setup, it makes the construction of the setup, image feature extraction, as well as illumination a lot more difficult. Use Method B or C if single plane calibration is desired.

2. if you are most concerned about accuracy, then use Method A or C. If you have reasons to believe that "tangential" lens distortion is abnormally large, use Method A or D. We think that such occasion rarely occurs. Even if it does occur, some of the tangential distortion people experienced may not be as significant as it appears to be since they estimate the distortion after the rest of the calibration is done, and some bias is already introduced.

3. if you are most concerned with speed, then use Method C. It only takes 25 milliseconds on a minicomputer (not including image feature extraction). Comparing with all calibration methods that does not ignore

lens distortion, this is at least two or three (even four) orders of magnitude faster. Even considering those methods that ignore lens distortion, Method C is still many times faster.

4. if you are most concerned with autonomy, use Method A, B or C.
5. if you have an application where the calibration points must come in sequentially, and you are interested in how each calibration point influences the statistical uncertainty of the results as the calibration points come in one by one, use the Kalman filtering approach in Method A.
6. if you are in a situation where a large portion of the calibration parameters are known already, then Method B is intended for it, although Method A and C suit the purpose equally well.

Each method has its own unique merit, and the above comparisons and recommendation should help the user make proper decision depending on his or her own needs.

References

1. Abdel-Aziz, Y.I. and Karara, H.M., 1971, Direct Linear Transformation into Object Space Coordinates in Close-Range Photogrammetry, *Symposium on Close-Range Photogrammetry,* University of Illinois at Urbana-Champaign, Urbana, Illinois, January 26-29, 1-18.
2. Abdel-Aziz, Y.I. and Karara, H.M., 1974, *Photogrammetric Potential of Non-Metric Cameras,* Civil Engineering Studies, Photogrammetry Series No. 36, University of Illinois at Urbana-Champaign, Urbana, Illinois, March..
3. Brown, Duane C., 1971, Close-Range Camera Calibration, *Photogrammetric Engineering,* Vol. 37, No. 8, 855-866.
4. Cohen, R.R. and Feigenbaum, E.A., editors, 1982, *The Handbook of Artificial Intelligence,* Vol. III, Heuris Tech Press, William Kaufmann, Inc.
5. Dainis, A. and Juberts, M., 1985, Accurate Remote Measurement of Robot Trajectory Motion, *Proceedings of Int. Conf. on Robotics and Automation,* 92-99.
6. Duda, R.O. and Hart, P.E., 1973, *Pattern Recognition and Scene Analysis,* New York, Wiley.
7. Faig, W., 1975, Calibration of Close-Range Photogrammetry Systems: Mathematical Formulation, *Photogrammetric Engineering and Remote Sensing,* Vol. 41, No. 12, 1479-1486.
8. Faugeras, O. and Toscani, G., 1987, Camera Calibration for 3D Computer Vision, *Proceedings of the International Workshop on Industrial Applications of Machine Vision and Machine Intelligence,* Seiken Symposium, Tokyo, Japan, February 2-5.

9. Ganapaphy, S., 1984, Decomposition of Transformation Matrices for Robot Vision, *Proceedings of Int. Conf. on Robotics and Automation,* 130-139.

10. Gennery, D.B., 1979, Stereo-Camera Calibration, *Proceedings Image Understanding Workshop,* November, 101-108.

11. Goshtasby, A., 1987, Correction of Image Deformation from Lens Distortion, Technical Report, Department of Computer Science, University of Kentucky, Lexington, Kentucky 40506-0027.

12. Grosky, W. and Tamburino, L., 1987, A Unified Approach to the Linear Camera Calibration Problem, Technical Report, Department of EE and CS, University of Michigan, Ann Arbor, Michigan.

13. Hall, E.L., Tio, M.B.K., McPherson, C.A. and Sadjadi, F.A.,1982, Curved Surface Measurement and Recognition for Robot Vision, Conference Record, *IEEE Workshop on Industrial Applications of Machine Vision,* May 3-5.

14. Itoh, H., Miyauchi, A. and Ozawa, S., 1984, Distance Measuring Method using only Simple Vision Constructed for Moving Robots, *7th Int. Conf. on Pattern Recognition,* Montreal, Canada, July 30-August 2, Vol. 1, p. 192.

15. Isaguirre, A., Pu, P. and Summers, J., 1985, A New Development in Camera Calibration: Calibrating A Pair of Mobile Cameras, *Proceedings of Int. Conf. on Robotics and Automation,* 74-79.

16. Karara, H.M., editor, 1979, *Handbook of Non-Topographic Photogrammetry,* American Society of Photogrammetry.

17. Lenz, R. and Tsai, R., 1987, Techniques for Calibration of the Scale Factor and Image Center for High Accuracy 3D Machine Vision Metrology, *Proceedings of IEEE International Conference on Robotics and Automation,* Raleigh, NC, Also to appear in IEEE Trans. on PAMI.

18. Lenz, R. and Tsai, R., 1988, Calibrating a Cartesian Robot with Eye-on-Hand Configuration Independent of Eye-to-Hand Relationship, *Proceedings of IEEE International Conference on CVPR,* Ann Arbor, MI. Also to appear in IEEE Trans. on PAMI.

19. Lowe, D.G., Solving for the Parameters of Object Models from Image Descriptions, *Proceedings Image Understanding Workshop,* April 1980, 121-127.

20. Luh, J.Y. and Klaasen, J.A., 1985, A Three-Dimensional Vision by Off-Shelf System with Multi-Cameras, *IEEE Transactions on Pattern Analysis and Machine Intelligence,* Vol. PAMI-7, No. 1, January, 35-45.

21. Malhotra, 1971, A Computer Program for the Calibration of Close-Range Cameras, *Proceedings of Symposium on Cloase Range Photogrammetric Systems,* Urbana, Ill.

22. *Manual of Photogrammetry,* 1980, fourth edition, American Society of Photogrammetry.
23. Martins, H.A., Birk, J.R. and Kelley, R.B., 1981, Camera Models Based on Data from Two Calibration Planes, *Computer Graphics and Image Processing,* 17, 173-180.
24. Moravec, H., 1981, *Robot Rover Visual Navigation,* UMI Research Press.
25. Okamoto, A., 1981, Orientation and Construction of Models, Part I: The Orientation Problem in Close-Range Photogrammetry, *Photogrammetric Engineering and Remote Sensing,* Vol. 47, No. 10, 1437-1454.
26. Okamoto, A., 1984, The Model Construction Problem Using the Collinearity Condition, *Photogrammetric Engineering and Remote Sensing,* Vol. L, No. 6, 705-711.
27. Sobel, I., 1974, On Calibrating Computer Controlled Cameras for Perceiving 3-D Scenes, *Artificial Intelligence,* 5, 185-198.
28. Strat, T.M., 1984, Recovering the Camera Parameters from a Transformation Matrix, *Proceedings: DARPA Image Understanding Workshop,* Oct., pp. 264-271.
29. Sutherland, I., 1974, Three-Dimensional Data Input by Tablet, *Proceedings of the IEEE,* Vol. 62, No. 4, April, 453-461.
30. Tsai, R., 1987, A Versatile Camera Calibration Technique for High Accuracy 3D Machine Vision Metrology using Off-the-Shelf TV Cameras and Lenses, IEEE Journal of Robotics and Automation, Vol. RA-3, No. 4, August. A preliminary version appeared in 1986 IEEE International Conference on Computer Vision and Pattern Recognition, Miami, Florida, June 22-26.
31. Tsai, R., 1988, Review of RAC-based Camera Calibration, to appear in *Vision,* MVA/SME's Quarterly on Vision Technology, November.
32. Tsai, R., and Lenz, R., 1987A, Review of the Two-Stage Camera Calibration Technique plus some New Implementation Tips and New Techniques for Center and Scale Calibration, *Second Topical Meeting on Machine Vision,* Optical Society of America, Lake Tahoe, March 18-20.
33. Tsai, R. Y., and Lenz, R., 1987B, A New Technique for Fully Autonomous and Efficient 3D Robotics Hand-Eye Calibration, *4th International Symposium on Robotics Research,* Santa Cruz, CA, August 9-14. Also to appear in IEEE Journal of Robotics and Automation.
34. Tsai, R. Y., and Lenz, R., 1988, Real Time Versatile Robotics Hand/Eye Calibration using 3D Machine Vision, *International Conference on Robotics and Automation,* Philadelphia, PA, April 24-29.
35. Wong, K.W., 1975, Mathematical Formulation and Digital Analysis in Close-Range Photogrammetry, *Photogrammetric Engineering and Remote Sensing,* Vol. 41, No. 11, 1355-1373.

36. Yakimovsky, Y. and Cunningham, R., 1978, A System for Extracting Three-Dimensional Measurements from a Stereo Pair of TV Cameras, *Computer Graphics and Image Processing,* 7, 195-210.

4D-Dynamic Scene Analysis with Integral Spatio-Temporal Models[1]

E. D. Dickmanns

Reviewed by:
H. Harlyn Baker
Artificial Intelligence Center, SRI International
333 Ravenswood Avenue, Menlo Park, CA 94025.

Most researchers in computer vision have had relatively little exposure to modern control theory. Having backgrounds in perhaps mathematics, computer science or electrical engineering, our approaches have been based primarily on theory from these disciplines. While enabling us to solve substantial problems in the processing of visual information, we have often failed to look at some of the more quantitive issues behind approximation. Indeed, much of this is quite understandable when one considers the amount of effort required to bring a system to performance. The issues of inherent robustness and the precision of results can easily become secondary to the mere accomplishment of specifying a process that can take raw imagery and produce results that are judged qualitatively satisfactory. There has not been a universal neglect of accuracy and robustness concerns, in the sense of being able to speak quantitatively about the accuracy of visual measurements, but only seldom and only recently has the quantifying influence of estimation theory begun to appear in vision work. We can learn much from control and estimation theory, and their application in fields such as photogrammetry and remote sensing, where robustness and precision are terms much more tightly defined than in the generic use to which we often put them. This paper is an excellent example of what people in these fields can offer us (see also Förstner (1987) for a detailed treatment of the matter).

This paper deals with the very topical issue of autonomous real-time visual control of a vehicle, and describes application of the approach the author's research team has developed to three vision-based navigation tasks: a van entering and continuing driving on a track at 60 km/h, a planar docking maneuver, and an aircraft landing using simulated data. In all three cases real-time performance has been demonstrated. They pose the task as one of optimal control: an analytic model is

---
[1] R. Bolles and B. Roth (editors), Robotics Research, Fourth International Symposium, August 1987, MIT Press, 1988, 311–318.

defined of the expected state space (vehicle, pathway etc.); the parameters for these state variables and their visual observables are related through linearized differential equations; operating over time with implicit high redundancy, the observations and differential equations form an overdetermined system, and this is solved, for the various parameters, by sequential maximum likelihood estimators (see Gelb (1974)). The author's point out that sequential estimation is an integrative process and has none of the noise amplification problems characteristic of difference methods. Broida and Chellappa (1986), and Gennery (1982) have also developed intriguing motion estimation procedures using principles from estimation theory, and these are well worth reading, but what makes the Dickmanns work stand out for me is its thorough demonstration of performance in a variety of applications, all carried out in real time on a parallel architecture of their design. In a more recent report, Dickmans (1988) describes the open-road control of a Mercedes van driving at 60 mph.

Their computational approach involves the use of template-based convolution of expected scene patterns (sides of the roadway, various views of docking target features, the runway from various aspects, etc.) to drive a battery of linear filters maintaining maximum likelihood estimates of the relevant scene variables and their parameters. For the roadway task the scene variables are those describing the road curvatures and those describing the attitude and velocity of the vehicle with respect to the road; in the 2-D docking task the variables are those which specify the position and attitude of the incoming vehicle with respect to the target; in the aircraft landing simulation, 12 state variables define the 6 degrees of freedom of aircraft motion and four controls are defined for elevator, aileron, rudder and throttle.

The prediction component of the sequential filters allows selective and focussed application of computing resources – in their words, processing "only the task-significant regions of a scene". The estimators direct the positioning of the feature templates in each new image, select the best responses, update the system parameters and their covariances, and determine control adjustments to be used in maintaining or attaining the desired goal. Occlusions are predicted, erroneous estimates are rejected, and new features to be tracked are introduced as required. Only the most current image in the sequence is needed for the estimation – the history is encapsulated in the state vector. In a companion paper, Mysliwetz and Dickmanns (1986) explain more of the initiation procedure for the road tracking mechanism.

The group has designed and implemented their processing in a distributed arrangement of dedicated 8086 microprocessors. It is structured as an MIMD multiprocessor. Some of the processors handle low-level image or actuator control operations (windowing, edge-tracking via convolution, steering, adjusting camera attitudes, etc.), others integrate the information across local windows and carry out the sequential parameter estimation. It is striking that the design of their algorithm enables such immediate implementation in a parallel or distributed processor architecture. This structuring is not so obvious in other autonomous navigation efforts where more involved local processing and aggregation is required, for example the Autonomous Land Vehicle projects of Martin Marietta or FMC (Kuan, Phipps and Hsueh, (1987)). This, along with the parallel nature of the filters, provides the key to their real-time performance.

The vision used is fairly rudimentary, but not particularly more so than that being used in projects with similar goals of autonomous navigation. The use of templates for tracking presumes detailed models of anticipated objects. What this work demonstrates is that these can be generic while remaining effective over a wide range of conditions – they use road-edge templates for their vehicle control, and only with standing water completely obscuring the track and shoulder did the first demonstration system require manual intervention. Of course the coupled premise is that those features for which the templates were designed are present in the scene, and the templates are sufficiently selective that there won't be many other features which they just happen to fit. Generalizations from the use of template matching in other areas of perception might cast well-founded doubt on their appropriateness here, and question extending the approach to more complex tasks (for example, where vehicle paths are difficult if not impossible to define by simple operators, or where paths are defined by derived characteristics such as local planarity). Increased capability beyond template matching might be demanded as the generality of the task and environment increase, yet this is not obvious: A critical point to bear in mind about the spatiotemporal approach is that massive redundancy and the rapid flow of imagery mean that ambiguity will be rare, and, if it exists, it is unlikely to persist over time. The local ambiguity of monocular processing may never be an issue in dynamic scene analysis, where simple techniques exploiting time for consistency may hold the real answers.

This work demonstrates that, when operating over time against a dynamic environment, simple techniques in visual processing can achieve impressive performance.

Higher-level issues that have occupied a principal spotlight in much of traditional robot vision work – recognition, inspection, hypothesis generation and verification, obstacle avoidance, planning, et cetera – have not been addressed here, although some of these have been mentioned as current considerations. In fact it is not clear that optimal estimation will have much of a role in these areas, but by providing more precise and reliable information to the various tasks, it will surely improve the capabilities of their later processing.

Others researchers, for example Waxman (1986) and Wells (1987), have demonstrated or proposed approaches similar to this – collecting local observations (*edges of a road*), aggregating them to more global constructs having semantic significance (*road, doorway*), and then using these descriptions and their parameters to specify and initiate some action (*turn, stop, etc*). What makes this work stand out is their use of an analytic model of the dynamics of the process, their unified incorporation of this in a sequential spatiotemporal computation, and their development of a real-time autonomous navigation control system to drive and be driven by the data acquired from the scene. All in all a remarkable achievement.

REFERENCES

Broida, T.J., and R.Chellappa. 1986. Kinematics and Structure of a Rigid Object from a Sequence of Noisy Images, *Proceedings, IEEE Workshop on Motion: Representation and Analysis,* Kiawah Island Resort, Charleston, South Carolina, 95–100.

Dickmanns, E.D. 1988. An Integrated Approach to Feature Based Dynamic Vision, *Proceedings of the Conference on Computer Vision and Pattern Recognition,* IEEE Computer Society, Ann Arbor, Michigan, 820–825.

Förstner, W. 1987. Reliability Analysis of Parameter Estimation in Linear Models with Applications to Mensuration Problems in Computer Vision, *Computer Vision, Graphics, and Image Processing,* **40**, 273–310.

Gelb, A. (editor). 1974. **Applied Optimal Estimation,** written by the technical staff, The Analytic Sciences Corporation, MIT Press, Cambridge, Massachusetts.

Gennery, D.B. 1982. Tracking Known Three-Dimensional Objects, *Proceedings of AAAI-82,* Carnegie-Mellon University, Pittsburgh, 13–17.

Kuan, D., G. Phipps, and A. Hsueh. 1987. Autonomous Land Vehicle Road Following, *IEEE International Conference on Computer Vision,* London, 557–566.

Mysliwetz B., and E. D. Dickmanns. 1986. A Vision System with Active Gaze Control for Real-time Interpretation of Well Structured Dynamic Scenes, *Conference on Intelligent Autonomous Systems,* Amsterdam, 522–532.

Waxman, A.M., J. Le Moigne, L. S. Davis, E. Liang, and T. Siddalingaiah. 1986. A Visual Navigation System, *IEEE International Conference on Robotics and Automation,* San Francisco, 1600–1606.

Wells, W.W. III. 1987. Visual Estimation of 3-D Line Segments From Motion – A Mobile Robot Vision System, *Proceedings of AAAI-87,* Seattle, Washington, 772–776.

An $n \log n$ Algorithm for Determining the Congruity of Polyhedra[1]

Kokichi Sugihara

Reviewed by

Jean-Daniel Boissonnat
INRIA
Route des Lucioles
06565 Valbonne, France

This paper describes an algorithm for determining whether two polyhedra, say P_1 and P_2, are congruent, i.e. are identical under translation and rotation. This problem can be viewed as a restricted instance of the scene analysis problem. Though the field is very active, only a very few theoretical results are known. Most proposed solutions in the literature are based on graph search techniques, and heuristics are used to make them running fast. This is a successful approach and systems have been described which can analyze quite complex scenes. However their efficiency cannot be analyzed precisely and may change dramatically under circumstances. This motivates further investigations to give more precise complexity results and to exhibit special cases of interest which admit exact solutions.

Sugihara attacks the problem along those lines and solves a non trivial though rather simple case. When complete descriptions of P_1 and P_2 are available, he shows that an exact solution can be found in $O(n \log n)$ time if n is the number of edges of the polyhedron. Moreover, he shows that the problem of partial congruity (only a connected part of the object is observed) can be solved in $O(n^2)$ time.

The paper is a clever application of an $O(n \log n)$ algorithm due to Hopcroft and Tarjan [1] which solves the graph isomorphism problem for triply-connected planar graphs. The main point of the paper is to show that a similar algorithm can determine whether two polyhedra are congruent , even though the graph composed of the vertices and the edges of a polyhedron is, in general, not planar nor triply-connected.

---
[1] Journal of Computer and System Sciences 29, 36-47 (1984)

The algorithm is quite simple and seems to be efficient and easy to code though no experimental results are given in the paper.

While the paper is generally concerned with an exact representation of polyhedra, the practical situation where data contains errors is also considered. It is shown that, when errors in the data are sufficiently small so that the abstract graph composed of the vertices and the edges of the polyhedron is not modified, a very similar algorithm with the same time complexity can been used.

In the second part of the paper, Sugihara presents an algorithm that judges whether a connected part Q of a polyhedron is congruent with some part P of another polyhedron. This is the usual situation in Computer Vision where only a view of a 3-dimensional object is seen. The analogous problem for graphs, the so called subgraph isomorphism, is known to be NP-complete. Nevertheless, Sugihara proves that this is not the case for polyhedra: this is essentially due to the fact that visibility constraints can be introduced here. The complexity of his algorithm is $O(nm)$ where n and m are respectively the number of edges of P and Q.

This theoretical result may be useful in several practical situations and moreover is an important contribution to a better understanding of the complexity of the scene analysis problem. Other results mixing complexity theory, practical problems and actual implementation would be welcome.

References

[1] HOPCROFT J.E. , TARJAN R.E., A $V \log V$ algorithm for isomorphism of triconnected planar graphs, J. Comput. System Sci. 7 (1973), 323-331.

Generating an Interpretation Tree from a CAD Model for 3D-Object Recognition in Bin-Picking Tasks[1]

Katsushi Ikeuchi

Reviewed by
Robert C. Bolles
SRI International, EK290
333 Ravenswood Avenue
Menlo Park, CA 94025

This paper describes an approach to automating the construction of computer vision programs. Such a capability is needed because an expert may spend months writing a special-purpose program for a moderately complex vision task. For machine vision to reach its full potential, techniques must be developed to simplify the programming. Whether such techniques are interactive or automatic, their goal is to reduce the cost of applying machine vision systems and, in so doing, make them more useful.

In this paper Ikeuchi describes a complete vision-programming methodology that, starting with a CAD model of an object, produces an interpretation tree program to recognize it. He illustrates his approach by applying it to a moderately complex object that has cylindrical and planar faces. His experimental system consists of a trinocular photometric stereo system, which gathers surface orientation and depth information, and a set of techniques, such as Extended Gaussian Image (EGI) analysis, which can estimate an object's position and orientation from this low-level information.

Ikeuchi assumes that the object-to-sensor projection can be modeled (locally) by an orthographic projection, which is appropriate for a large class of object recognition tasks. Similar assumptions have been recently made by several researchers, including Thompson (1987), Huttenlocker (1988), and Lamden (1988). Such assumptions make it possible to partition the six degrees of freedom associated with an object into two orientation angles that define the "viewing direction" of the object relative to the sensor, one rotation about the viewing direction, and three position parameters. As Ikeuchi points out in the paper, the two viewing-direction parameters are the ones that

---
[1] International Journal of Computer Vision, Vol.1, No.2, pp.145–165 (1987).

primarly determine the appearance of an object to a sensor. The rotation about the viewing direction only rotates the object's projection in the image plane. Once the three orientation angles have been computed, the three position parameters can be computed directly from the range data produced by Ikeuchi's photometric stereo system. His location strategy, therefore, is a three-step process: estimate the viewing direction, compute the rotation about that vector, and, finally, calculate the position parameters.

The first step is the most difficult because the appearance of an object can change dramatically over the range of viewing directions. Ikeuchi's approach is to reduce this complicated task to a tree of simpler ones. He therefore (1) forms groups of viewing directions within which the appearance of the object varies by just a small amount, (2) constructs a classification procedure that identifies the group to which an unknown view belongs, and (3) constructs a special-purpose location technique for each group. The vision program resulting from this approach is an interpretation tree in which the first few operations identify the group, while the remaining operations compute the object's six degrees of freedom.

Ikeuchi calls the groups of viewing directions "attitude groups." He forms them by sampling the range of viewing directions densely and then partitioning the set of sampled views into groups in which the same key surfaces are visible. The choice of surfaces (as opposed to edges or colored regions) for defining the groups was strongly influenced by the set of techniques he had available for gathering data, classifying views, and estimating parameters.

In summary, the basic steps in Ikeuchi's approach are as follows:

Attitude Group Formation and Analysis

1. Generate "all" views of the object by producing views from a large number of positions.

2. Form attitude groups of similar views by identifying those that contain the same key object features.

3. Select a representative attitude for each attitude group.

4. For each representative attitude, build a "work model" that describes six properties of the group; these include the shape of the primary object face, the location of key edges, and the EGI.

Interpretation Tree Construction

5. Build an interpretation tree to identify an unknown view's attitude group.

6. Extend the tree to include techniques for computing the viewing directions for each attitude group.

7. Extend the tree to include techniques for computing the rotation about the viewing direction.

The idea of partitioning the viewing directions into attitude groups is applicable to a wide range of recognition and location tasks, including those that involve recognition of multiple objects. The idea is closely related to concepts developed by other researchers, including "poses" by Birk et al. (1981), "characteristic views" of Chakravarty (1982), and the "locus of camera positions" used by Goad (1985). All of these concepts are designed to identify classes of viewing directions in which the key visible features were relatively stable. One goal of future research is to find ways of tailoring this basic strategy to a specific task, given a description of it. As Ikeuchi points out in his concluding remarks, some tasks may require additional control structures, such as backtracking, to make them computationally effective.

In his analysis of useful object features, Ikeuchi is careful to distinguish detectable surface regions from analytically defined surfaces. To make this distinction, he employs models of his sensor and feature-detection techniques to predict the detectable portions of the object surfaces. Although difficult, this step is crucial to the success of an automatic programming system because the effectiveness of a vision program depends on the detailed behaviorial aspects of the sensors and feature detectors. Despite the importance of this correlation, very little effort has been devoted to characterizing the behavior of sensors and processing techniques. Interest in this area of research is increasing, however, as its significance becomes clearer (e.g., see Sugihara (1979), Bolles (1982), Ponce and Chelberg (1987), Binford (1988), and Ikeuchi and Kanade (1988)). This work is challenging because the description of a technique's behavior needs to be understandable to a planning program, comprise several levels of abstraction, and span a number of factors, including both structural and statistical properties. However, a technique needs to be characterized only once, after which it can be added to the vision programming system's list of tools.

Some of the key steps in Ikeuchi's vision programming system were done interactively. For example, he chose the "eight identifying faces" (from the list of all possible object features) to be used in forming attitude groups (Step 2 above). He also helped select the techniques for accomplishing the classifications specified in the interpretation tree (Step 5 above). Automation of these tasks is a desirable goal of future research. It should be noted that the presence of more detectable features on an object often simplifies the recognition task because there is then a higher probability of having a unique feature for each attitude group. This somewhat counterintuitive phenomenon heightens the importance of the procedure (automatic or interactive) that finds these distinguishing features. And, since people quickly get tired of analyzing long lists of features, this is an excellent prospect for automation.

Another issue to be explored in the formation of vision programs is computational efficiency. Experienced vision programmers know that there are a number of data structures and search techniques that can be employed to reduce the execution times of some algorithms substantially. These include such things as limiting the search for a feature to small, high-probability regions, hash-coding the detected features, constructing tables of relationships (Grimson and Lozano-Perez (1984)), or compiling the interpretation tree into assembly code (Goad (1985)). Even though adding this level of analysis to the vision programming system would obviously result in more complexity it is nevertheless crucial to the automation of a large class of tasks.

In summary, Ikeuchi's paper is important because it describes a complete vision programming methodology, demonstrates its ability to generate programs for a system that analyzes depth and edge data, and suggests a number of promising areas for future work.

References

Binford, T.O., "Generic Surface Interpretation Observability Model," in the *Proceedings of the Fourth International Symposium on Robotics Research,* The MIT Press, Cambridge, Massachusetts, pp.265–272 (1988).

Birk, J., R. Kelley, and H. Martins, "An Orienting Robot for Feeding Workpieces Stored in Bins," *IEEE Transactions on Systems, Man, and Cybernetics,* Vol. SMC-11, No. 2, pp.151–160 (1981).

Bolles, R.C. and R.A. Cain, "Recognizing and Locating Partially Visible Objects: The Local-Feature-Focus Method," *International Journal of Robotics Research,* Vol. 1, No. 3, pp.57–82 (1982).

Brooks, R.A., "Symbolic Reasoning Among 3-D Models and 2-D Images," *Artificial Intelligence*, Vol. 17, pp.285–348 (1981).

Chakravarty, I., *The Use of Characteristic Views as a Basis for Recognition of Three-Dimensional Objects,"* Ph.D. thesis, Rensselaer Polytechnic Institute (1982).

Goad, C., "Fast 3D Model-Based Vision," in *From Pixels to Predicates,* edited by Alex P. Pentland, Ablex Publishing Company, Norwood, New Jersey, pp.371–391 (1985).

Grimson, W.E.L. and T. Lozano-Perez, "Model-Based Recognition and Localization from Sparse Range or Tactile Data," *International Journal of Robotics Research,* Vol. 3, No. 3, pp.3–35 (1984).

Huttenlocher, D.P., "Recognizing Solid Objects by Alignment," in the *Proceedings of the Image Understanding Workshop,* Cambridge, Massachusetts, pp.1114–1124 (1988).

Ikeuchi, K. and T. Kanade, "Modeling Sensor Performance for Model-Based Vision," in the *Proceedings of the Fourth International Symposium on Robotics Research,* The MIT Press, Cambridge, Massachusetts, pp.255–263 (1988).

Lamden, Y., J.T. Schwartz, and H.J. Wolfson, "On Recognition of 3-D Objects from 2-D Images," *Proceedings of the IEEE Conference on Robotics and Automation,* Philadelphia, Pennsylvania, pp.1407–1413 (1988).

Ponce, J. and D. Chelberg, "Finding the Limbs and Cusps of Generalized Cylinders," *International Journal of Computer Vision,* Vol. 1, No. 3, pp.195–210 (1987).

Sugihara, K., "Automatic Construction of Junction Dictionaries and Their Exploitation for Analysis for Range Data," in the *Proceedings of the 6th International Joint Conference on Artificial Intelligence,* pp.859–864 (1979).

Thompson, D. and J.L. Mundy, "Three-Dimensional Model Matching From an Unconstrained Viewpoint," in the *Proceedings of the IEEE Conference on Robotics and Automation,* pp.208–220 (1987).

Interpretation of Contact Geometries From Force Measurements*
J. K. Salisbury

Reviewed by

Mark R. Cutkosky
Mechanical Engineering Dept.
Stanford University
Stanford, CA 94305-3030

Salisbury's paper on "Interpretation of Contact Geometries From Force Measurements" presents a novel approach to obtaining tactile information for the control of a dextrous robotic hand. In particular, the paper provides several examples of how multiple measurements with just a few sensors can be used to determine the contact forces, contact type and contact location when a finger is pressed against an object. While the specific methods in the examples may be difficult to generalize to arbitrary combinations and arrangements of sensors, the paper is important because it represents one of the first efforts to use active sensing strategies to build up a more complete picture of the finger/object contact. The paper also makes it very clear that the ability to pinpoint the location of the contact on a fingertip does not depend on the number of sensing elements.

A considerable body of literature now exists on the kinematics, dynamics and control of robot hands, much of which makes it clear that tactile sensing is essential for manipulation. In reviewing "Interpretation of Contact Geometries From Force Measurements" we might therefore begin by asking what kinds of tactile information are needed. The necessary information falls basically into two categories:

1) *Information used directly for controlling the manipulation process:* Most importantly, this information includes forces and torques at the contact points and the locations of contact points on the fingers. Both of these quantities are essential for maintaining a stable grasp and for imparting desired forces and motions to the object.

2) *Information used to identify properties or features of the object:* This information includes the shapes of the contact patches and the patterns of pressure and shear tractions over the contacts, and is

---

* M. Brady and R. P. Paul (editors), "Robotics Research, First International Symposium," MIT Press, 1984. Also included in M. T. Mason and J. K. Salisbury, Jr., "Robot Hands and the Mechanics of Manipulation," MIT Press, 1984.

useful for estimating the object surface texture and for finding features such as edges, grooves or corners. The same information is also useful in planning manipulation strategies since it indicates whether the contact should be treated as a point contact, an area contact, and so forth.

An obvious approach to obtaining both kinds of information is to cover the fingertip with an array of sensing elements, each of which can measure local pressure or shear tractions. The contact forces and moments, and the centroid of the contact area, are then obtained by summing and taking moments of the signals from each element. Array sensors are especially suited for identifying object features, and much of the work on interpreting the output from tactile sensors has focused on recognition problems. By contrast, Salisbury's approach is especially suited to providing forces and torques for use in manipulation.

Many kinds of tactile arrays have been built and recent developments are summarized in [Fearing and Hollerbach 1985; Harmon 1982; Jacobsen et al. 1987]. While the robustness, accuracy and data rates from such devices is steadily improving, the criticisms that Salisbury leveled in 1984 remain largely true: Design tradeoffs among size, robustness and cost usually result in individual sensing elements that are less linear, more hysteretic and noisier than "classic" sensors for instrumentation. In addition, a large array of sensors typically requires either a large number of connecting wires (the routing of such wires has been acknowledged as one of the most difficult issues in hand design [Jacobsen et al. 1987]) or multiplexing schemes with attendant time delays. The overall result is that it is difficult to provide accurate measurements of the contact forces at hand servo rates. In the long run, the use of custom VLSI hardware and better software promises to dramatically increase the rates at which force information can be obtained, but one is nonetheless motivated to consider other sources of force/torque information. Two possibilities come to mind: joint torques and dedicated force/torque sensors. Most dextrous robotic hands are provided with joint torque or tendon tension sensors, but for improved sensitivity and higher bandwidth a small force/torque sensor in the fingertip is attractive. This state of affairs led Salisbury to consider a miniature version of the strain-gage wrists used in many robotic applications and to place it in the fingertip. In contrast with tactile arrays, the sensor can provide linear, low-noise measurements of forces and torques *directly*. But what about finding the location of the contact point? If the same device could also be used to locate the contact area on the fingertip and to provide some indication of the contact type (e.g. point, line, area) then the tactile array and its many wires might be dispensed with entirely.

At first glance, it seems unlikely that a single force sensor could provide the contact force, orientation and point of application with just one or two samples. But Salisbury gives several examples in which a hemispherical fingertip with six or three sensors, or a 6-axis sensor in a plane, can determine the locations and forces associated with point, line and planar (area) contacts. Although Salisbury uses Plücker coordinates and the twist/wrench formalism introduced in his earlier papers, the derivations are easily viewed in terms of vector algebra. For example, suppose we have a hemispherical fingertip of radius r, as shown in Figure 1, and we want to find the point of application of a contact force, **f**, upon the outer surface. The three translational components from a 6-axis force/torque sensor will give us **f** and the three moment components will give us **t**, the vector of torques produced by **f**. We need to find the contact location, which we represent by a vector **r**. We know that **f**, **t** and **r** are related by **t** = **r** × **f**. If we consider the lever arm associated with **r** and call it **d**, then since **d**, **t**, **f** are mutually orthogonal, **d**/|**d**| = **f**/|**f**| × **t**/|**t**|. Where |**d**| = |**f**|/|**t**| and |**f**| and |**t**| are the magnitudes of the force and torque vectors produced by the 6-axis sensor. Finally, **r** is obtained by intersecting the line of action (the line parallel to **f** and intersecting the tip of **d**) with the surface of the fingertip. For a fingertip with a known shape, two such points exist, of which only one will permit a positive contact force. Even with just the three moment measurements (i.e. with a three strain-gage device calibrated so as to measure torques about its origin) **r** and **f** can be determined with two successive force measurements. Again, since **t** is perpendicular to **f** and **r**, then for two different forces f_1 and f_2, applied at the same contact, the direction of **r** is given by $t_1 \times t_2$.

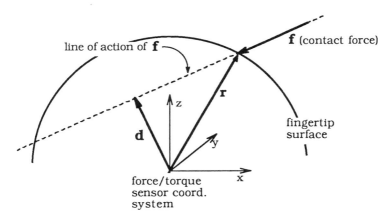

Figure 1: Using the measured force and torque to determine the line of action for a contact force.

Similar reasoning permits Salisbury to determine the force components and contact locations of point, line and planar contacts using a sensor located in the plane. As with the three-axis sensor in a fingertip, a single measurement is not enough to pin down all the unknown quantities. For planar contact, the convex hull of the contact area can be built up as a combination of line contacts.

The theme behind these examples is not one of gradually building up a picture of the contact through statistical or adaptive "calibration" procedures involving many samples, but rather of using analytic solutions to obtain all the necessary information from a minimum number of independent samples. However, in one section, Salisbury does discuss statistical methods. In analyzing the accuracy of locating a line contact using a planar sensor, he shows that the measurement is a function of sensor noise and sensor location with respect to the contact. Not surprisingly, the measurement is most accurate when the sensor is closest to the contact. Going a step further, Salisbury shows that by looking at the scatter of data points for a large number of such measurements, one can conclude whether the contact is a line, a point or a planar area.

Underlying Issues

The examples presented in this paper make it clear that the resolution with which one can locate the contact area is a function not of the number of sensing elements, but rather of sensor accuracy and of the accuracy of information about the fingertip geometry and position. One can draw a loose analogy with vision systems in which imaging arrays are compared to structured light or range-finding devices that gradually construct an image through multiple measurements. In fact, a finger equipped with a 6-axis force sensor was later used by Brock and Chiu [1986] to build up surface profiles of objects through repeated contacts. The point here is that while a single force/torque sensor, like a ranging device in vision, is slower than an array for such tasks as recognizing features on an object, it is much faster for obtaining contact forces because it obtains them directly -- just as a ranging devices are faster than imaging devices for obtaining depth information.

The distinction between sensor spacing and spatial resolution has been amplified in more recent work on tactile sensing. For example, Fearing and Hollerbach [1985] showed that spatial anti-aliasing limitations are a function of sensor spacing, sensor depth and the elastic properties of the fingertip material. With sensors somewhat below the surface of the skin, one can locate point contacts with the sub-element accuracy. Salisbury's solution, a single sensor located in a rigid material so that it is affected by forces acting anywhere over the fingertip[1], represents one extreme. Early array sensors, with sensing elements right at the surface, represent the other extreme since the sensors "see" only forces directly above them and therefore can only resolve point contacts to the nearest cell.

How can the examples given in this paper be generalized to different kinds of contacts? For example, what about soft fingers? Although the soft-finger kinematic case (in which three force components and one moment along the contact normal are transmitted) is not treated in the examples, it seems likely that a solution similar to those presented for point contact on a hemisphere or line contact in a plane would work. The deformation of the fingertip poses a more difficult problem since the deformed fingertip will no longer be a hemisphere. However, the error

---

[1]This is roughly equivalent to an array consisting of a single sensor at infinite depth below the surface of a rigid fingertip material.

involved in ignoring the change in fingertip geometry is probably a second order effect. Another possibility would be to estimate the deformed shape of the fingertip surface as a function of the contact force. But to some extent, this would be inconsistent with the approach in the paper, which uses rigid-body geometry throughout. The rigid-body assumption is also the fundamental reason that the planar sensor cannot find holes or grooves in the surface of the object, but only the convex hull of the contact area. Without modeling deformations, we cannot expect to resolve geometric features that are smaller than the scale of the fingertip itself (although large scale geometries such as those measured by Brock and Chiu [1985] can be obtained with multiple samples) and cannot produce information that depends on pressure distributions over the contact area. As Fearing and Hollerbach [1985] point out, with a complete elastic model of the fingertip and/or the object, a small number of sensors embedded in the fingertip can indeed identify small features. The real tradeoff then, is between the number of sensors (and attendant complexity) and the amount of information that can be obtained from a single sample.

None of this detracts, however, from what is likely to be the most lasting contribution of "Interpretation of Contact Geometries From Force Measurements," which is the demonstration of some active sensing strategies in which various forces are exerted by the fingers to disambiguate the contact type, location and forces. Admittedly, with more sensors (perhaps even sensors of multiple types, including force, vibration and pressure arrays) and with more complete models of the fingertip behavior (e.g. elastic or viscoelastic deformations) one can produce a more complete picture of the grasp. But for a given combination of sensors, one can always do even better with active sensing. This is perhaps obvious from looking at how people actively explore the objects that they touch and hold; and yet, examples of active sensing are rare in robotics.

References

Brock, D. and Chiu, S., "Environment Perception of an Articulated Robot Hand Using Contact Sensors," PED Vol. 15, *ASME Winter Annual Meeting*, November, 1985, Miami, pp. 89-96.

Fearing, R. S. and Hollerbach, J. M., "Basic Solid Mechanics for Tactile Sensing," *International Journal of Robotics Research*, Vol. 4, No. 3, 1985, pp. 40-54.

Harmon, L. D., "Automated Tactile Sensing," *International Journal of Robotics Research*, Vol 1, No. 2, 1982, pp. 3-32.

Jacobsen, S. C., McCammon, I. D., Biggers, K. B. and Phillips, R. P., "Tactile Sensing System Design Issues in Machine Manipulation," *Proc. 1987 IEEE International Conference on Robotics and Automation*, March, 1987, Raleigh, pp. 2087-2096.

On Optimally Combining Pieces of Information with Application to Estimating 3-D Complex-Object Position from Range Data[1]

Ruud M. Bolle and David B. Cooper

Reviewed by

Hugh F. Durrant-Whyte
Department of Engineering Science
Oxford University
19 Parks Road
Oxford OX1 3PJ
England

This paper addresses the problem of estimating the location of known 3-D objects from uncertain range data. The objects considered are modeled by collections of planar or quadratic surface and boundary patches. The goal is to estimate the transform describing object location by comparison of prior and observed object patches.

The authors approach differs from that presented in (Grimson and Lozano-Perez 84), in using a Bayesian, maximum-likelihood estimation scheme, and extends the work in (Faugeras and Herbert 86), by providing a richer geometric structure. Each geometric primitive is described by a location vector and normalized orientation vector.

Errors in range data are modeled by a one-dimensional, zero-mean Gaussian noise component, defined perpendicular to the observed surface. A key element of this formulation is the exploitation of the assumption the range observations are all statistically independent. This means that the location likelihood can be considered as the

---

[1] *IEEE Trans. on Pattern Analysis and Machine Intelligence*, vol. 8, no. 5, pp. 619-638, 1986 (September).

product of contributions from each range observation, which in turn reduces the global location estimation problem to one of combining locally optimal parameter estimates.

The algorithm for estimating object location consists of three phases: First the range data is divided up into sets associated with primitive surface and boundary regions, using the Mahalanobis distance as a matching metric. Second, the range data in each set is combined to provide a local, partial, maximum likelihood estimate (m.l.e.) of object location. In this phase, the parameters of interest are the differences between prior and observed surface or boundary descriptions. Finally, the contributions from each local location estimate are combined to provide a global m.l.e of object location.

The authors have developed a method of using local observations of geometric surface primitives to provide global estimates of object location. Although the method is complete in itself, it raises three important questions about the general problem of object identification, and in particular the application of statistical techniques to localization. The first, and most common question, concerns the form of noise model employed. The authors of this paper justify their assumption of Gaussianity by appealing to the central limit theorem, though further assuming that this process has significant components only in the sensed direction. The Gaussian assumption is universally employed in signal processing, generally for reasons of expediency rather than correctness. Noise models, particularly for CCD arrays, are unlikely to be Gaussian, (only simulations were demonstrated in this paper), and although they may be approximated as such, future work should consider the ramifications of this model. The assumption that noise is only significant perpendicular to observed surfaces provides a computationally efficient solution to the m.l.e problem. However, it should be observed (Ayache and Faugeras 87) that this immediately precludes the *use* of direction-dependent noise to successively estimate location from different viewpoints.

A second, more subtle point, is raised by the authors regarding minimal parameterizations of observed geometric primitives. The representations used in this paper are over-parameterized, resulting in constraints between components of feature descriptions. Statistically, only unconstrained parameterizations will provide true m.l. estimates of feature descriptions. This problem is deftly side-stepped by the authors using two mechanisms; defining only scalar noise components on surface patches, and by using a small-perturbation argument to demonstrate that the constraint does not greatly affect the resulting m.l.e. If the ideas in this paper are to be utilized in situations where large errors are experienced or prior information is unavailable, either unconstrained representations must be used (Durrant-Whyte 88), or techniques for explicitly evaluating the effects of constraints should be developed. This problem has received considerable attention in the past (Kendall and Moran 63; Davidson 68).

A final point concerns the use of constraints between different geometric surface and boundary primitives, the power of which is adequately demonstrated in (Grimson and Lozano-Perez 84). The authors make no provision for such constraints, especially in cases where they may be used to reduced the process of recognition and matching. An argument often voiced by computer scientists against statisticians is the latters neglect of environment structure in favor of homogeneity. It would seem that the techniques developed by the authors could benefit from this argument and encorporate, statistically, the constraints used in more domain-specific searching and matching tasks.

The use of statistical techniques in complex robotics and geometric recognition problems is, so far, underdeveloped. This paper is an important step in remedying this state of affairs.

References

N. Ayache and O. Faugeras, 1987. Building, registrating, and fusing noisy visual maps. In *Int. Conf. Computer Vision*, pages 73–79, London, UK.

R. Davidson, 1968. *Some Arithmetic and Geometry in Probability Theory*. PhD thesis, University of Cambridge, U.K.

H.F. Durrant-Whyte, 1988. Uncertain geometry in robotics. *IEEE J. Robotics and Automation*, 4(1):23–31.

O.D. Faugeras and M. Herbert, Fall 1986. The representation, recognition and locating of 3-d objects. *Int. J. Robotics Research*, 5(3):27–52.

W.E.L. Grimson and T. Lozano-Perez, Fall 1984. Model-based recognition and localization from sparse range or tactile data. *Int. J. Robotics Research*, 5(3):3–34.

M.G. Kendall and P.A.P. Moran, 1963. *Geometric Probability*. Griffin Academic Press.

Epipolar-Plane Image Analysis: An Approach to Determining Structure from Motion[1]

R. C. Bolles, H. H. Baker and D. H. Marimont

Reviewed by

W. Eric L. Grimson
Artificial Intelligence Laboratory
Massachusetts Institute of Technology
545 Technology Square
Cambridge, MA 02139

This paper addresses the problem of building a three-dimensional description of a static scene from a sequence of images. Of course, this is a general and widespread problem in vision research, and has been the focus of considerable research. Of the variety of possible approaches to the problem, this paper focuses on the particular case of computing structure from a sequence of images obtained from a moving camera.

In its general form, this structure-from-motion problem can be defined as follows. A model of the image formation process is assumed to be known, allowing one to predict the image of a scene feature from the camera's position and orientation. As the camera moves through a static environment, the image formation model will in most cases predict a moving image feature. The structure-from-motion problem attempts to invert this process, by finding a set of moving image features and estimating the motion of the camera and the 3D structure of the scene features that gave rise to the image features. Most of the work in this area has considered the problem in two stages: estimating the image features' motions, which usually requires identifying feature points in successive images corresponding to the same scene feature; and then estimating the camera's motion and the corresponding scene structure.

The authors use two key observations to reduce this problem. The first idea is to assume that the camera motion parameters will be supplied by an independent process, such as an inertial-guidance system. This implies that the relationship between the coordinate systems of successive images is known, and this allows the authors to use a technique

---

[1] *International Journal of Computer Vision* **1**(1):7–55, 1987.

common in stereo vision analysis to significantly simplify the problem of estimating the motion of an image feature. In particular, if the relationship between two camera positions is known, then for any image feature in one view, the image feature of the other view corresponding to the same scene feature must lie along a known line in the second view. This **epipolar constraint** reduces the search for corresponding feature positions, and hence the computation of feature motion, from a two-dimensional to a one-dimensional search.

The second key observation is to take successive images with very small separations between the camera positions. This observation again utilizes a known effect from stereo vision, that the difficulty of matching image features between views increases with the separation of the lens centers, but at the same time the accuracy of the reconstructed scene features improves with this separation. Earlier work by Moravec [1] circumvented this trade-off by taking a sequence of nine views at low separation. Across the full range of images, features had a large baseline and led to accurate reconstruction, but tracking of features between frames was simplified by the small separation between successive pairs of frames. The authors push this idea considerably further, using hundreds of frames acquired by very small displacements of the camera.

The advantages of these two observations are very strikingly demonstrated by the following simple case. Suppose the camera moves in a straight line orthogonal to the its optic axis, and a sequence of images are acquired. These images can be thought of as forming a block of data, with the x and y axes of the image forming two of the dimensions, and the time t forming the third dimension. In this simple case, the epipolar lines are horizontal, so that we can slice this block of data into a set of x–t planes, or **epipolar images (EPIs)**. Any scene point that projects into an image point on this epipolar plane in the first image will remain in this plane (unless it passes from the view of the camera). Hence, the determination of feature motion (i.e. tracking an image feature spatially and temporally) is reduced from a 3D problem to a set of 2D problems. Secondly, the very short spacing between images implies that an image feature moves only a small amount between time frames, and thus in the epipolar image, moves only a small amount between lines in the image. This has the very nice effect of reducing the problem of matching features in spatial images to one of finding feature paths in spatiotemporal images. In particular, one can apply edge detection techniques to the epipolar images to extract the paths over time of features in the image. Related versions of this idea of an epipolar image have been used in [2, 3].

By analyzing EPIs, the authors can deduce the paths of features over time, but they also must relate those feature paths to the 3D structure of the corresponding scene features. In the simple case of a camera moving linearly in a direction orthogonal to its optic axis, the epipolars are horizontal lines in the image, and the feature paths are all straight lines. Some simple trigonometry shows that the slope of the line and the speed of the camera can be used to determine the three-dimensional position of the scene point. If the camera is oriented at some other fixed angle relative to the path, the epipolar lines are no longer horizontal, but easily can be determined from the camera geometry. Using this, one can again extract EPI's, where now the feature paths are hyperbolae. The authors develop a simple technique for linearizing these images, by projecting them onto a virtual image plane oriented with normal orthogonal to the line of travel. The projection of the hyperbolic paths into this plane yield straight lines, that can be analyzed using the earlier method to determine scene structure. The same technique holds even when the orientation of the camera varies with time.

As a result of this process, the estimation of 3D structure can be solved by applying edge detection techniques to a sequence of epipolar images, extracting linear segments by using split-and-merge methods on the resulting edge points, and using properties of the lines to determine 3D structure. In principle, only the 3D positions of scene points corresponding to edge points in the EPI's can be deduced. By analyzing the T junctions in the edge map, however, one can determine which objects occlude other objects. This can be used to sweep out a map of the free space around the system, since there can be no obstacles along the line of sight between the camera and the observed point. This allows one to deduce some information about the structure of the scene at locations other than those giving rise to image edges.

The authors report on an implementation of this approach, and a variety of empirical tests of its performance, with impressive results. They indicate experiments on synthetic data and on natural sequences of both indoor and outdoor scenes.

Finally, the authors also establish a basis for generalizing the method to situations other than linear camera motions and stationary point objects. They use projective duality, a technique with a long history in vision [4, 5], to transform more general problems into versions in which the developed epipolar techniques can be applied. This allows them to deal with a camera moving in a planar but non-linear path, with curved objects, and with moving objects. These results have been further developed in a recent generalization [6].

The problem of building a three-dimensional description of a static scene from a sequence of images is a central on in computer vision. This paper makes a strong contribution towards the solution of that problem. Through the concept of epipolar images, the authors provide an elegant solution to the ubiquitous problem of tracking features across images. This solution appears to have considerable advantages over other techniques in the field, especially with respect to problems of noise corruption in the data, and of handling false matches in tracking features. The experimental results are quite impressive, and provide solid evidence of the utility of the method on real data, something that many other methods in structure-from-motion have not clearly demonstrated. At this point, the computational efficiency of the current scheme is not clear, but that remains an issue for additional work, as do the generalizations and extensions to the method suggested by the authors.

References

[1] H. P. Moravec, "Visual mapping by a robot rover," in *Proceedings of the International Joint Conference on Artificial Intelligence*, Tokyo, 1979, pp. 598–600.

[2] E. H. Adelson and J. R. Bergen, "Spatiotemporal energy modesl for the perception of motion," *Journal of the Optical Society of America* **A 2**, pp. 284–299, 1985.

[3] N. J. Bridwell and T. S. Huang, "A discrete spatial representation for lateral motion stereo," *Computer Vision, Graphics, and Image Processing* **21**, pp. 33–57, 1983.

[4] D. A. Huffman, "A duality concept for the analysis of polyhedral scenes," *Machine Intelligence* **6**, pp. 295–324, 1971.

[5] A. K. Mackworth, "Interpreting pictures of polyhedral scenes," *Artificial Intelligence* **4**, pp. 121–137, 1977.

[6] H. H. Baker and R. C. Bolles, "Generalizing epipolar-plane image analysis on the spatiotemporal surface," *Computer Vision and Pattern Recognition Conference*, Ann Arbor, MI, 1988.

Machine Interpretation of Line Drawings[1]
Kokichi Sugihara

Reviewed by

Deepak Kapur
State University of New York at Albany
Department of Computer Science
Albany, NY 12222

and

Joseph L. Mundy
General Electric Company
Corporate Research and Development
Schenectady, NY 12345

One of the central problems in computer vision is the recovery of three-dimensional object descriptions from two-dimensional projections of the objects onto an image viewplane. Major steps towards understanding the constraints imposed by such projections were made by Huffman, Clowes and later extended by Waltz in the 1970's. These papers examined the relationship between the configuration of polyhedral surface faces and the appearance of the edges and vertices between these faces from all possible general viewpoints. The edges between faces are labeled according to convexity, concavity or occlusion. The analysis of the behavior of these labels is largely topological and the process can be viewed as a propagation of constraints over a graph formed by the edge and vertex connections. A major observation by Waltz was that, in many cases, the constraints among labels admit only a small number of consistent solutions. This result generated a great deal of interest in constraint propagation and its application in domains other than polyhedral projections.

This monograph introduces an important new source of constraints on the image projection of polyhedra. These constraints take the form of linear algebraic equations and inequalities that express the coplanarity

---
[1] Published by the MIT Press, 1986, 233 pages.

of vertices and edges that lie in the same face and the constraints imposed by image projection itself. The approach uses image labeling to identify coplanar vertex sets and to identify depth inequalities which correspond to occlusion of one face by another. The three-dimensional configuration of the object can be recovered from a single projection, provided that some assumptions are made about reflectance or texture properties.

Perhaps the most important contribution of the monograph is that the formal algebraic properties of polyhedral projections considered as line drawings are clearly developed and the criteria for a projection to correspond to the image of a polyhedron are precisely given. The monograph is well written, quite precise and easy reading. It is an elegant account of solutions to a clearly stated problem, which is often difficult to find in the artificial intelligence (AI) literature. We now give a chapter by chapter treatment.

Chapter 1 is a brief introduction to the problem, an overview of the literature on the subject and the rest of the monograph. A short history of the research on this problem is given. The author compared his approach with others reported in the literature; the comparison, though adequate, is far from being very comprehensive.

Chapter 2 discusses labeling as a mechanism to rule out a large class of spatial interpretations of line drawings. The presentation is nice and well structured as it clearly states assumptions made in the approach for labeling. Constraint propagation which plays a crucial role in determining labels is briefly discussed. Its significance in arriving at all possible labelings of line drawings could have been emphasized more. The chapter is lacking in the discussion of the literature on constraint propagation algorithms including the work of Mackworth and Freuder (1985).

Any one familiar with a labeling algorithm and related relaxation techniques would have experienced the combinatorial explosion of the search space. The problem of deciding whether a line drawing can be labeled is known to be NP-complete (see Kirousis and Papadimitriou (1984)). However, by exploiting the full vertex-edge-face topology, this search space can be drastically reduced for many examples as demonstrated by the recent work of Nguyen (1987). For instance, if one considers a line drawing as complex as an image of a model Jeep, a polyhedral object with 122 vertices, 167 edges, and 47 faces, there are nearly 10^{100} possible labels but there are only 96 globally consistent labelings which can be compactly represented and easily computed (Nguyen, 1987). Malik's thesis (1985) which extends the results of Huffman, Clowes,

Waltz, and Kanade, to curved objects is also worth mentioning.

Chapter 3 introduces the author's main contribution on this subject, namely the concept of planar-panel scenes for arriving at a necessary and sufficient conditions for the two-dimensional image of a three-dimensional scene to be labelable (Theorems 3.1, 3.2 and 3.3). The derivation of linear equations as well as linear inequalities from a consistent labeling is described well. The problem of determining whether a line drawing represents a polyhedral scene is shown to be equivalent to the problem of finding a solution to a system of linear equations and inequalities obtained from the line drawing and its labeling. The treatment is very elegant.

Chapter 4 is the discussion of the generalization of the theory in Chapter 3 to "hidden-part-drawn" pictures. Some additional issues and assumptions to generalize the conditions are addressed. Except for a few additional details, the discussion is similar to that in Chapter 3.

Chapter 5 is a discussion about recovering the three-dimensional structure from the algebraic characterization of pictures. The treatment is formulated in terms of degrees of freedom inherent in a picture, namely what depth values uniquely determine other depth values. The theory of matroids is related to finding independent subsets of depth values which can be arbitrarily assigned values and which determine the shape of the three-dimensional scene. This is followed by a detailed discussion of pictures with 4 degrees of freedom and axonometric projections in which axes are assumed to be parallel to most faces and lines in a three-dimensional scene. The approach here is based on the system of equations obtained in Chapters 3 and 4, and thus driven only by topological constraints of coplanarity of vertices.

Chapters 6, 7 and 8 are attempts by the author to use this algebraic approach to deal with errors in positions of vertices in the picture. Due to digitization errors as well as errors in low level segmentation to determine lines and vertices, the system of equations and inequalities obtained from line drawings is almost always inconsistent. As stated by the author, because of the superstrictness of mathematical characterization of the problem, it is necessary to introduce some additional mechanism to provide human flexibility. This is done by identifying subsets of independent equations within a given system, leaving aside redundant information.

Sugihara seemed to take the view that as long as a maximally consistent subset of equations representing some constraints in a line drawing (called *generically reconstructible incidence substructures*) can be identified, that portion of the line drawing can be viewed to be correct. The author relied on certain vertices in a line drawing being in a generic position so that even if they are moved slightly, the topological structure of the corresponding object does not change. The rest of the constraints imposed by the line drawing can be viewed as redundant and possibly incorrect. The author went a step further by suggesting that incorrect portion of the line drawing can in fact be corrected using what is identified as correct portion insofar as vertex positions are assumed in error. Of course, the suggested correction can be deemed acceptable only if it meets certain error criterion proposed by the user. An algorithm for judging the correctness of labeled pictures is proposed in Figure 7.3 on page 141.

It is not quite clear from the monograph how one goes about finding a possibly correct subportion of the line drawing as there are quite a many of them. Furthermore, there is little discussion of any criteria that can be used to decide what vertex positions are likely to be correct (or more correct) than other vertex positions. We believe that in the absence of any information about scene and camera characteristics, it is not practical to make any assumption of correctness of positions of some subset of vertices over other vertex positions; all vertex positions are equally likely to be in error.

A more serious issue is the assumption that for a real image, a system of equations can be obtained from its line drawing. This, first of all requires a very good segmentation method of real images. Subsequently, an algorithm is needed for labeling of line drawings obtained from real images especially when in real images, it is often the case that some of the vertices are missing or misplaced, line segments are broken, and faces are not closed. The author has not discussed these problematic aspects related to real data in much detail. He seemed to suggest relying on the user to separate interesting portion of a segmented image from the background noise as well as for providing missing edges.

Chapters 9 and 10 discuss sources of additional information to fix the depth values in an independent subset. The author discussed light intensity and texture as two such sources of additional information. He proposed a least square estimate to completely recover a unique shape from a line drawing. Since it is quite difficult to obtain a proper line drawing from a real image which can be labeled, this step of using intensity and texture information can become extremely error-prone.

This essentially concludes the author's approach for recovering three-dimensional shape from two-dimensional images. As the author himself said in Chapter 1, "Chapter 11 is rather a digression from the main story of the book." This chapter is a discussion of the gradient space approach which is the dual of the labeling approach. The material is not as well discussed. Sugihara's lack of enthusiasm about this approach which has been suggested to overcome some of the limitations of the labeling technique, is quite evident from the discussion. We would have liked to see this material integrated with the rest of the monograph.

Overall, the monograph is excellent in presenting an approach to solving an idealized AI problem. Its practical relevance needs to be explored further. We recommend it to those interested in understanding the theoretical foundations of an AI problem.

References

Kirousis, L, and Papadimitriou, C. 1984. The Complexity of Recognizing Polyhedral Scenes. Technical Report, Dept. of Computer Science, Stanford University.

Mackworth, A.K. and Freuder, E.C. 1985. The Complexity of Some Polynomial Network Consistency Algorithms for Constraint Satisfaction Problems. *Artificial Intelligence,* Vol. 25, 65-74.

Malik, J. 1985. Interpreting Line Drawings of Curved Objects. Ph.D. Thesis, Dept. of Computer Science, Stanford University.

Nguyen, V. 1987 (August). Exploiting 2D Topology in Labeling Polyhedral Images. Proc. *IJCAI-87,* Milan, Rome.

Nguyen, V. 1987. A Parallel Algorithm for Labeling Polyhedral Images. Unpublished Manuscript, G.E. R&D Center, Schenectady, NY.

TINA: The Sheffield AIVRU vision system*

J. Porrill, S.B. Pollard, T.P. Pridmore, J.B. Bowen, J.E.W. Mayhew, and J.P. Frisby

Reviewed by

David G. Lowe
Lab. for Computational Vision
Computer Science Department
University of British Columbia
Vancouver, B.C., Canada   V6T 1W5

Most researchers in computer vision devote themselves to a single problem at a time, such as shape reconstruction, object modeling, or scene analysis. While this is often necessary in order to focus a research effort, it has the unfortunate result that the goals of the individual projects are often poorly defined with respect to the role that they might play in a complete vision system. As a result, it could be argued that much current research in computer vision is of little relevance for improving the performance of practical robot vision, due to a mismatch between the goals defined by the individual projects and the actual capabilities that the surrounding system components would have. One solution might be to define standardized interfaces between various components, so that individual research efforts can be tested in the context of full systems. Unfortunately, the computer vision field seems to be still at too early·a stage of development to allow for widespread agreement upon such standards.

The Sheffield AIVRU vision project is one of best examples of an alternative approach, in which all levels of the vision problem are tackled at once. In this case, they have developed components for edge segmentation, stereo matching, curve fitting, model-acquisition, and recognition. The numerous design decisions that must be made for each component are justified according to their impact on the performance of the system as a whole rather than in terms of some artificially-defined metric. As a result, many of the techniques developed during

---

* *Proc. of IJCAI-87,* Milan, Italy (August 1987), pp. 1138–1144.

this systems-oriented approach are quite different from the results of more theoretical approaches. These differences include a much greater emphasis on noise tolerance and robustness, incorporation of methods for calibration and quantitative analysis, less attention paid to special cases that are unlikely to arise in practice, and a more realistic analysis of efficiency trade-offs.

The Sheffield team consists of a half-dozen individual researchers who have been unusually productive and successful at integrating their work over the past few years. At its current stage, the project has already produced one of the most general vision systems for robotics applications, in which the three-dimensional locations and orientations of known objects can be determined from stereo images. Although not yet operating at a speed that would make it practical for industrial tasks, the system has been demonstrated for use in robot acquisition of randomly oriented parts. In addition, their work has made substantial contributions to each of the subfields which make up the system, and has shown the successes and limitations of many existing approaches when applied to realistic imaging data. Some of these results will be briefly reviewed in the following paragraphs.

Two of the principles involved in the project, Mayhew and Frisby, are well known for their earlier research in the psychophysics of human stereo vision [2]. Therefore, it is no surprise that the stereo matching component of the Sheffield system is based on a model of human perception that goes well beyond most previous approaches to stereo in the computer vision community. One of the most interesting aspects of this approach is the use of a "gradient disparity constraint" to solve the central problem of identifying corresponding features in the two stereo images. This constraint simply limits the rate of change in disparity between matching features across the image, and it is based on solid psychophysical measurements of human abilities to fuse stereograms. This one constraint replaces a number of looser constraints that have typically been applied to the stereo problem, and it results in a simple algorithm with high performance. This algorithm, dubbed "PMF," was the topic of a PhD thesis by Pollard, who remains a member of the Sheffield group, and has been described in detail in another publication [3]. The Sheffield group has applied it to the output of the Marr-Hildreth and Canny edge detection algorithms for a wide range of images with very successful results.

The depth measurements provided by stereo vision are quite noisy as they are based on small disparity differences between the two images, and depth information is entirely lacking for horizontal edges. Therefore, the Sheffield group has devoted considerable effort to methods for smoothing, segmenting, and fitting 3-D lines and circular arcs to the raw stereo data [5]. An important aspect of this data fitting is to use estimates of the variance in each measurement to weight the data points during the fitting process, so that the most reliable measurements can be propagated into regions lacking good data. One particularly general way to do this is through the application of Kalman filtering, which is likely to have applications throughout computer vision. Kalman filtering uses a weighted least-squares formulation to provide an efficient method for incrementally updating the estimated mean and standard deviation of desired parameters as each new measurement is obtained. The application of Kalman filtering to problems in computer vision has also been studied extensively by the group at INRIA in France [1]. This is exactly the kind of general method for integrating noisy measurements that is needed to make computer vision a practical real-world technology, and its development should be followed with interest by anyone working in the field.

Almost all work in model-based recognition can be cast in a hypothesize and test framework, and the high-level recognition component of the Sheffield system is no exception. Potential matches are identified by examining each line segment or circular arc that is longer than a certain length, matching it against each potential model feature, and then attempting to form maximal cliques of nearby features that are consistent with pairwise constraints that have been precomputed from the model [4]. The final estimate for location and orientation is provided by a least-squares fit. For the most part, the Sheffield group has selected these techniques from among the best of previous approaches to model-based recognition and combined them into a new arrangement. An interesting aspect of this is to look at how the various methods perform in comparison to one another after they have been reimplemented by a new group. There is much scope here for further experimentation and systematic testing.

It is still somewhat early to predict the final effect that this research is likely to have on the development of reliable, low-cost, real-time robot vision. While there is an ongoing debate regarding the best form of sensing for robot vision, ranging from the use of single two-dimensional images to the development of scanning laser depth sensors, there is likely to always be a useful role for stereo vision as a low-cost flexible method for deriving depth measurements. The various methods that have been developed to combine noisy data measurements are certain to play an important role in any practical application of computer vision. But most importantly, the experimental work of implementing, combining and testing the many components in such a system provides a great deal of information for enabling researchers to judge the most promising directions for the field. While it requires considerable time and effort to read the many papers giving the details of the Sheffield system, anyone undertaking this will be rewarded with many insights into a large body of carefully chosen techniques for the construction of functioning computer vision systems.

References

[1] Ayache, Nicholas, and Olivier D. Faugeras, "Building, registrating, and fusing noisy visual maps," *Proc. of First International Conference on Computer Vision,* London, England (June 1987), 73–82.

[2] Mayhew, John E.W., and John P. Frisby, "Psychophysical and computational studies towards a theory of human stereopsis," *Artificial Intelligence,* **17** (1981), 349–385.

[3] Pollard, Stephen, John E.W. Mayhew, and John P. Frisby, "PMF: A stereo correspondence algorithm using a disparity gradient limit," *Perception,* **14** (1985), 449–470.

[4] Pollard, S.B., J. Porrill, J. Mayhew, and J. Frisby, "Matching geometrical descriptions in three-space," *Image and Vision Computing,* **5** (1987), 73–78.

[5] Pridmore, A.P., J. Porrill, and J. Mayhew, "Segmentation and description of binocularly viewed contours," *Image and Vision Computing,* **5** (1987), 132–138.

Three-Dimensional Model Matching from an Unconstrained Viewpoint[1]

D. W. Thompson and J. L. Mundy

Reviewed by

David H. Marimont
Electronic Documents Laboratory
Xerox Palo Alto Research Center
Palo Alto, CA 94304

In this paper, the authors present a promising new technique to estimate the location of a previously-identified instance of a polyhedral model from a single, possibly cluttered image. They present some new theory, explain its relevance to the problem, describe an algorithm based on the technique and some interesting aspects of the algorithm's implementation in an experimental system, and review the system's performance on synthetic and real images. While their claims for the technique's domain of applicability may be somewhat exaggerated, the technique itself appears to be a valuable contribution to the literature in its area.

The technique is based on a new feature called a "vertex pair," which can either be on the model or in the image. A vertex pair on the model consists of two vertices, one of which is defined by two edges that pass through it. A vertex pair in the image is the perspective projection of the vertex pair on an instance of the model. The authors point out that because vertices in the image can be estimated by intersecting nearby edge fragments, vertex pairs in images can be extracted efficiently and reliably.

A single view of a vertex pair suffices to estimate its location (position and orientation) relative to the camera. By approximating perspective projection with orthogonal projection followed by scaling by the distance of the polyhedron's centroid from the camera, the authors

---

[1] *Proc.* IEEE Conference on Robotics and Automation, Raleigh, North Carolina, pp. 208–220, 1987.

show that this location can be computed quite easily from an image vertex pair.

To take advantage of these simple features and estimation procedures, the authors suggest computing an estimate of the object's location from every possible match of model to image vertex pairs rather than establishing correspondences as a first step. This computation is feasible because of the typical complexity of polyhedral models and the efficiency with which the computation of location can be implemented as a table look-up.

The six-dimensional estimate of location resulting from each possible match is used to increment the appropriate bin in a (six-dimensional) histogram; the authors argue (based more on intuition and experience than theory) that if reasonable care is taken to cope with the effects of noise in the image and quantization in the histogram, a peak can detected in the neighborhood of the bin that corresponds to the object's location. They describe a clustering technique that divides the problem into smaller, more manageable problems in lower-dimensional subspaces of the full six-dimensional space.

The authors have implemented this algorithm and tested it on synthetic and natural images. They describe the more interesting implementation issues in some detail. The performance of the system seems satisfactory, even on fairly complex natural images.

The advantages of the authors' approach include the simplicity of the image and model features used, the location estimates' apparent insensitivity to noise and partial occlusion, the possibility of efficient implementation in parallel hardware, and the elimination of the need to establish correspondences between features. This last advantage is particularly important, because establishing feature correspondences is typically difficult and expensive in problems of recognition and location estimation.

The authors' claim that their technique can be used for recognition does not seem well-founded. Recognition usually involves choosing the appropriate model, but they assume that the appropriate model is given. The authors do not discuss using their technique to select the appropriate model for an image from a set of possibilities. A related issue is how location estimation can take advantage of computations involved in recognition: if the choice of model was based on feature correspondences, avoiding their use in location estimation is less of an advantage.

What happens when many vertex pairs in the image do not lie on the modelled object, which can happen in cluttered scenes, is not discussed very clearly in the paper. In the examples with natural image, the image vertex pairs do seem confined to the object of interest. Perhaps the effect is just more background noise in the histogram.

These are minor failings in light of the overall quality of the paper. The authors deal with an important problem, formulate a theory with novel analytical and computational components, and obtain results suggesting that their approach warrants further exploration.

III

Kinematics, Dynamics, and Design

The very nature of robotics implies that mechanisms of some kind must be designed and constructed. It is clear that this physical nature of a robot dictates that the field of mechanics will always play a central role in robotics research.

Kinematics is the study of geometry and motion (i.e., position, velocity, acceleration, and all higher derivatives) without regard for the forces that cause the motion. Such studies generally make use of a set of generalized coordinates which describe the position and motion of the device with respect to certain angles or lengths measured between elements of the machine. These coordinates are often called *joint variables*, and the set of all values these coordinates can take on is called the *joint space* of the device. The difficulties encountered when such parameterizations are used to describe task specifics soon lead one to consider a description of the geometry and the motion of the device in terms of external *operational* coordinates.

In one way or another, every kinematic problem is essentially one of describing joint space or operational space, and generally such problems are concerned with making relations between the two. Perhaps the most basic kinematic problem is that of the map from joint space to operational space and vice versa in a static situation. Particularly in the case of redundant manipulators, this problem continues to recieve the attention of researchers. One basic form of path planning is concerned with the mapping of a spatially continuous set of points (i.e., a path) from one space to the other. Another consideration is the local sensitivity of the map between joint space and operational space, which captures whether a given design is "well-conditioned" or "dextrous" in a kinematic sense. The accuracy characteristics of this mapping as a function of errors in various parameters of the system are also of interest. Another problem is that of workspace characterization: the representation of the entire joint space of a device in terms of operational space. Kinematics relates strongly to the planning problem in that any geometric aspects of such planning are largely kinematic in nature. This includes the use in planning systems of geometric uncertainty and tolerance propagation. The field of kinematics is fundamental to the science of robotics, as it provides a foundation for the investigation of an abundance of design and control issues.

Dynamics is the study of the forces required to cause motion. Clearly, actuator design and mechanism design rely on dynamic considerations, and the design of control systems would not be possible without such considerations. From a practical point of view, the computational burden of making dynamic calculations for multi-body problems in robotics has resulted in a great deal of work on the complexity of these computational algorithms. This same complexity has led to the development of symbolic manipulation systems that can generate the dynamic equations of motion for a system given its parametric description. Many of the problems considered from a kinematic viewpoint can be more fully explored by taking the additional complexity of dynamics into account. Examples of problems that have been studied at the "kinematic level" but which should more properly be considered to be dynamic problems include path planning, local properties of manipulator workspace, redundancy, accuracy, and geometric reasoning in planning systems.

Design of robotic mechanisms is a problem characterized by a complex blend of analysis, intuition, and experience. By nature design centers around the problem of carefully making choices between conflicting alternatives. Any one design might be optimized for any of several criteria. Design for an intended *function* should be the goal, and is achieved by the simultaneous consideration of many technical questions. Design of the actuators themselves is an important problem in robotics that has recieved some attention recently. Kinematic and static force analysis techniques have been suggested that allow the designer to quantify such diverse aspects of a potential design as workspace size and quality, load capability, and fine motion and fine force control capabilities. An important area of current research is the attempt to consider dynamics and control interactions at design time – i.e., to design for control. Other issues involve paarallel mechanisms, optimal gearing and transmission systems, flexible manipulators, and how to best use redundant degrees of freedom in a design.

In part III we have two annotated bibliographies (one by Hollerbach and one by McCarthy), two survey articles (one by Youcef-Toumi and one by Kumar and Waldron), and eight reviews of selected articles in the area of kinematics, dynamics, and design. Below, we give a brief synopsis of of all these pieces, arranged in an order suggested by the preceeding introduction.

McCarthy presents an annotated bibliography on robot kinematics, workspace analysis, and path planning. The references he has compiled cover the fundamentals of kinematics. There are many important references concerned with various representations of spatial kinematics, as well as references that address most of the problems of kinematics mentioned above.

Renaud reviews a paper in which Borrel and Liegois view the workspace of a manipulator as the union of connected subsets of joint space called *aspects*. These aspects aid in understanding the workspace of manipulators and multiple kinematic solutions.

Hollerbach's extensive article covers a large amount of work on the kinematic calibration problem. This is an extremely important topic, as it concerns the practical application of robotics to industrial and other domains. It is interesting that most of the reported work has been performed by universities and research laboratories and not by industrial robot manufacturers!

Featherstone reviews a paper in which Paul and Zhang deal with the efficient computation of kinematics for a class of manipulators. Walker reviews a paper by Featherstone that is concerned with the computational complexity of robot dynamics algorithms. Orin reviews a paper in which Lee and Chang present an algorithm that allows the inverse dynamic computation to be performed in parallel by several processors in an efficient manner. These three papers are representative of the large amount of attention that has been given to the problem of the efficiency of various kinematic and dynamic computational algorithms during the last several years.

Mason reviews a paper by Lötstedt on contact and friction problems in rigid-body mechanics. An interesting approach to automatic planning research has been to investigate the physics that underlies the seemingly simple manual operations that humans perform practically without thought. These investigations will someday allow planning systems to contain a rudimentary understanding of the mechanics of manipulation.

Kumar and Waldron present an extensive review of artificial legged locomotion systems, with particular emphasis on statically stable walking vehicles. The main thrust of the article is on motion planning, control, and coordination. Through the included historical review of past designs, as well as the authors' statement of the state of the art, one becomes aware of the abundance of design issues to be considered in the design of such vehicles.

Youcef-Toumi surveys the design and control of direct-drive robots. This class of robots, which use little or no gearing, represents one direction that robot design is taking. Youcef-Toumi reviews designs that have acheived very high speeds and acceleration capabilities, and whose structures possess remarkably high natural frequencies.

Inoue reviews a paper by Hirose, Ikei, and Ishii on their development of a "holonic" manipulator. This device has sensors and computational elements built into the structure which communicate over a computer network internal to the manipulator. It represents an integrated approach to design and a new direction in which manipulator design may proceed in the future.

Nakamura reviews a paper on multifingered hands by Kerr and Roth. The paper is representative of recent research that aims to quantify the dexterity of multi-fingered hands for fine positioning and fine force application. Also considered are internal forces used in grasping, and other properties of hands.

Scheinman reviews a book by Rivin on the mechanical design of manipulators. Rivin presents the many conflicting requirements that a mechanical designer must consider when faced with the design of a robot manipulator.

A Survey of Kinematic Calibration

John M. Hollerbach
Artificial Intelligence Laboratory
Massachusetts Institute of Technology
Cambridge, MA 02139

Abstract

The literature on the kinematic calibration of manipulators is surveyed, according to five categories. (1) Parametric estimation of endpoint location involves finding the parameters relating link coordinate systems. (2) Non-parametric estimation of endpoint location involves table lookup schemes that directly map joint angles to endpoint location. (3) Robot repeatability involves how precisely the robot can achieve a particular endpoint location under repeated trials. (4) Robot registration involves determining the relative location of the robot and objects in the environment. (5) Advanced instrumentation for endpoint tracking is reviewed.

1 Introduction

Recently, there has been an explosion in the literature on the topic of kinematic calibration, reflecting its current importance and development in robotics. Kinematic calibration concerns the determination of an accurate relationship between joint angles and endpoint positions. Kinematic calibration encompasses primarily geometrical effects, such as variations in link lengths, joint axis orientation, and base location, but also non-geometrical effects, such as gear eccentricity, backlash, and joint compliance. To provide bounds on this topic, kinematic calibration is restricted to a static analysis of endpoint position, leaving such time-varying or dynamic effects as servo and trajectory errors, link flexibility during motion, and endpoint vibration to other topics. Nevertheless, kinematic calibration does include such static effects as joint compliance due to gravity loading.

As often stated in the literature, one major need for kinematic calibration is to correct the kinematic model of the robot after its manufacture. Variations in the kinematic model arise from imprecisions in the manufacturing process, but improving the precision in the manufacturing process is costly. Thus every robot needs to be calibrated after it is built, and the corrected kinematic parameters incorporated into the software for accurate positioning. For the PUMA 560 robot, for example, endpoint positioning accuracies before such calibration are only on the order of 1 cm, whereas after kinematic calibration some of the papers that follow report accuracies of 0.2-0.3 mm. This is a drastic improvement of almost two orders of magnitude.

Determining the parameters of a kinematic model is only one manifestation of the more general issue of kinematic calibration. Under kinematic calibration is also included robot registration, namely locating the robot with respect to the environment or vice versa, and determining the tool transform of grasped objects or hand sensors. Determining robot repeatability is also an important aspect of kinematic calibration. Other issues include the instrumentation employed, the choice of coordinate system, the modeling of the nongeometric parameters, and the system identification method. A final issue is the inverse kinematics for a calibrated manipulator. While the manipulator may have been designed for its inverse kinematics to be analytically solvable, the calibration may result in a non-analytically solvable manipulator.

For convenience of review, the papers are placed in one of the following categories:

- parametric estimation of endpoint location

- nonparametric estimation of endpoint location

- robot repeatability

- robot registration

- advanced instrumentation for endpoint tracking

These categories are closely related, and it is easily possible, for example, to apply a parametric kinematic modeling effort to a repeatability effort. Nevertheless, papers have been categorized on the basis of their primary emphasis. Another review and categorization of the literature may be found in Roth, Mooring, and Ravani (1986, 1987).

2 Parametric Estimation of Endpoint Location

There have been many papers on estimating the parameters of a robot model to improve robot positioning. In the most general instance, the model contains both geometric and non-geometric parameters. Most often these papers have dealt with just the geometric parameters: the relative location of link coordinate origins. A few papers have also dealt with non-geometric factors, which are particularly important for geared robots, such as backlash, gear eccentricity, joint compliance, and base motion. To provide some order, these papers will be discussed based on their coordinate system representations. More detailed comparisons of some of these representations may be found in Everett, Driels, and Mooring (1987) and Ziegert and Datseris (1988).

2.1 Four-Parameter Representations

The attraction of four-parameter representations is that this is the minimal parameter set required to locate link coordinate origins and define transformations between neighboring links. The majority of papers have assumed the Denavit-Hartenberg (1955) representation, which labels links with four parameters: the link lengths a_i, joint offsets s_i, joint angle offsets $\delta\theta_i$, and skew angles α_i (Figure 1).

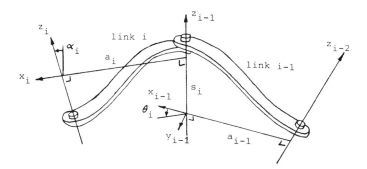

Figure 1: Denavit-Hartenberg link labeling.

For clarity of exposition, below is presented a typical procedure for kinematic calibration, based on iteration of the linearized kinematic equations. For a hypothetical rotary manipulator, the forward kinematics derives the endpoint position and orientation **x** as a function of the joint angles $\boldsymbol{\theta}$ and the Denavit-Hartenberg parameters $\boldsymbol{\alpha}$, **a**, and **s**:

$$\mathbf{x} = \mathbf{f}(\boldsymbol{\theta}, \boldsymbol{\alpha}, \mathbf{a}, \mathbf{s}) \tag{1}$$

The form of this function **f** can be found in many publications, and is not the main issue here other than its nonlinearity. Taking the first difference of (1),

$$\Delta \mathbf{x} = \frac{\partial \mathbf{f}}{\partial \boldsymbol{\theta}} \Delta \boldsymbol{\theta} + \frac{\partial \mathbf{f}}{\partial \boldsymbol{\alpha}} \Delta \boldsymbol{\alpha} + \frac{\partial \mathbf{f}}{\partial \mathbf{a}} \Delta \mathbf{a} + \frac{\partial \mathbf{f}}{\partial \mathbf{s}} \Delta \mathbf{s} \tag{2}$$

The term $\partial \mathbf{f}/\partial \boldsymbol{\theta}$ is just the ordinary manipulator Jacobian matrix **J**. The other partial derivatives of **f** with respect to the remaining three Denavit-Hartenberg parameters are also different Jacobian matrices. The difference $\Delta \mathbf{x}$ may be interpreted as the error in position and orientation of the endpoint, obtained by subtracting the endpoint position and orientation computed from the current kinematic model from the measured endpoint position and orientation. The differences $\Delta \boldsymbol{\theta}$, $\Delta \boldsymbol{\alpha}$, $\Delta \mathbf{a}$, and $\Delta \mathbf{s}$ may be viewed as the corrections to the kinematic parameters. The difference $\Delta \boldsymbol{\theta}$ is interpreted as the joint offset, having to do between the difference in the zero angle as determined from a joint encoder versus the zero angle as defined by the Denavit-Hartenberg link labeling.

Equation 2 can be written more compactly as

$$\Delta \mathbf{x} = \begin{bmatrix} \frac{\partial \mathbf{f}}{\partial \boldsymbol{\theta}} & \frac{\partial \mathbf{f}}{\partial \boldsymbol{\alpha}} & \frac{\partial \mathbf{f}}{\partial \mathbf{a}} & \frac{\partial \mathbf{f}}{\partial \mathbf{s}} \end{bmatrix} \begin{bmatrix} \Delta \boldsymbol{\theta} \\ \Delta \boldsymbol{\alpha} \\ \Delta \mathbf{a} \\ \Delta \mathbf{s} \end{bmatrix} \tag{3}$$

$$= \mathbf{C} \Delta \boldsymbol{\phi} \tag{4}$$

Calibration proceeds by positioning the manipulator in many points of the workspace and combining the error vectors $\Delta \mathbf{x}_i$ and Jacobians into a single equation:

$$\begin{bmatrix} \Delta x_1 \\ \Delta x_2 \\ \vdots \\ \Delta x_m \end{bmatrix} = \begin{bmatrix} C_1 \\ C_2 \\ \vdots \\ C_m \end{bmatrix} \Delta \phi \qquad (5)$$

or more compactly,

$$\mathbf{b} = \mathbf{D}\Delta\phi \qquad (6)$$

The least squares solution for $\Delta\phi$ is the pseudoinverse of $\mathbf{D}$:

$$\Delta\phi = (\mathbf{D}^T\mathbf{D})^{-1}\mathbf{D}^T\mathbf{b} \qquad (7)$$

The updated parameter values ϕ' are then obtained by

$$\phi' = \phi + \Delta\phi \qquad (8)$$

Since this is a nonlinear estimation problem, this procedure is iterated until the variations $\Delta\phi$ approach zero and the parameters ϕ have converged to some stable values. At each iteration, the Jacobians are evaluated with the current parameters.

There are a number of issues that arise when executing this procedure, which are handled differently in the various references. One issue has to do with the computational efficiency of deriving the Jacobian matrices: most references employ differential four-by-four homogeneous matrices, while others employ essentially screw theory. Potentially the most serious issue is that $\mathbf{D}^T\mathbf{D}$ may not be invertible, which may arise either from the data not being "persistently exciting" or from singularities in the representation. The data will not be persistently exciting if not enough poses are chosen that allow a parameter variation to be manifested as a measurable endpoint error. Singularities in the Denavit-Hartenberg representation happen when neighboring joint axes are nearly parallel. The Denavit-Hartenberg parameters are ill-conditioned because the common normal varies wildly with small changes in axis orientation. Hence a number of papers have modified this representation for the parallel-axis case.

2.1.1 Unmodified Denavit-Hartenberg Representation

A number of papers have proposed the estimation procedure outlined above for the Denavit-Hartenberg parameters. This includes Wu (1983, 1984), Ahmad (1985, 1988), Payannet, Aldon, and Liegeois (1985), Zhen (1985), Ibarra and Perriera (1986), Khalil and Gautier (1986), Knapczyk and Morecki (1987), and Kirchner, Gurumoorthy, and Prinz (1987). Kirchner, Gurumoorthy, and Prinz (1987) showed the form of the linearized equations to second order, but in their recursive estimation scheme they adopted just the normal first-order linearized equations. Ibarra and Perriera (1986) attempted to circumvent the singularity problem of the Denavit-Hartenberg representation by assuming that parallel axes are indeed parallel, but then slightly misaligned parallel axes can never be calibrated.

Most of these papers presented just the mathematics without simulation or experimental results, but three papers presented such results. Payannet, Aldon, and Liegeois (1985) calibrated an MA23 six degree-of-freedom manipulator, using a three-dimensional position measurements system comprising three wound cables attached to potentiometers. Knapczyk and Morecki (1987) presented simulations for an RPA-80 industrial robot. Ibarra and Perriera (1986) simulated a PUMA 560 robot.

Most of these papers also dealt only with geometric effects, except for Ahmad (1985, 1988), who modeled in addition the nongeometric effects from backlash, gear eccentricity, and joint compliance.

2.1.2 Modified Denavit-Hartenberg Representation

Hayati (1983) pointed out the singularity of the Denavit-Hartenberg representation, and presented a modified representation for the parallel axis case. The distal origin was restricted to lie in the previous origin's XY plane, and an angular alignment parameter β_i was substituted for the Denavit-Hartenberg parameter d_i (Figure 2). Hayati noted that this modification will in turn have a problem when neighboring joint axes are nearly perpendicular. Thus a hybrid representation is used depending on whether adjacent joint axes are parallel or not, although in each case there are only four kinematic parameters, the minimum number. Hayati suggested an estimation procedure by minimizing the mean-square error from a number of measurements, although no simulations or experimental results were given, and geometric parameters only were considered.

Hayati and Mirmirani (1985) extended this work to use the normal iterative linearized estimation procedure, and presented simulation results for the PUMA 560 and Stanford arms. Hayati and Roston (1986) experimentally verified this procedure for a PUMA 260 robot. The calibration was achieved with a calibration block

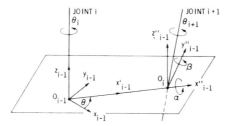

Figure 2: Hayati's modification to the Denavit-Hartenberg parameterization.

and precision points, and the accuracy was improved from 2-3 mm to 0.5 mm. A summary of these results was presented in Hayati, Tso, and Roston (1988).

A number of papers subsequently used Hayati's modified Denavit-Hartenberg representation. Sugimoto and Okada (1985) simulated a recursive least-squares calibration of geometric parameters based on the linearized equations.

Judd and Knasinski (1987) experimentally calibrated an Automatix AID-900 robot, by sighting an unspecified number of theodolites on three target points carried on the robot end effector tool plate. Non-geometric factors due to gear runout and orientation errors were modeled and estimated by single-joint motions; gravitational deflections were also modeled, and any unmodeled residual errors were approximated by the first Fourier terms. They concluded that joint angle offset accounted for 90% of the RMS error, link errors for 5%, and gearing errors for 0.5%.

Puskorius and Feldkamp (1987) experimentally calibrated a Merlin robot, by means of a stereo vision system placed on the end effector. The cameras were calibrated with a calibration block, and were used to observe a spherical target in the environment. Backlash and joint compliance were modeled and found to be significant. Overall accuracy was improved by a factor of three, limited by the repeatability of the robot being three times better than the vision system.

Hollerbach and Bennett (1987, 1988; also in An, Atkeson, and Hollerbach, 1988) investigated kinematic calibration of the geometric errors only for the MIT Serial Link Direct Drive Arm, which does not exhibit significant non-geometric errors since it does not have gears. A mixed Denavit-Hartenberg and Hayati parameterization was used, and identification was accomplished through iterative linearized least squares. Experimental results were given using a Watsmart System, which is a commercial 3-D motion tracking system based on lateral effect photodiode cameras similar to the Selspot II System (see Section 4). They pointed out the benefits in computational complexity through the use of screw theory for the linearization, rather than the usual use of differential homogeneous transformations. They also advocated the use of the Levenberg-Marquardt algorithm rather than straight iterative least squares, when the initial parameter value guesses are far from the final values. Thus Equation 7 becomes

$$\Delta \phi = (\mathbf{D}^T \mathbf{D} + k\mathbf{I})^{-1} \mathbf{D}^T \mathbf{b} \qquad (9)$$

where a small diagonal element $k\mathbf{I}$ makes this relation invertible. When k is large, this procedure just boils down to gradient descent. Although gradient descent is slower than straight use of the pseudoinverse, it is robust to the singularities. Hollerbach and Bennett often found it essential that this procedure be used. Some of the initial values of the parameters may be hard to know beforehand, such as the transformation between an external measuring system and the robot base and the tool transform, as well as the joint offsets. During iterative least squares, it is quite likely that an intermediate parameter set will drift into the singularity of the Denavit-Hartenberg representation and prevent a solution. Incorporating the Hayati representation doesn't solve the problem, since it too has a singularity when axes are nearly perpendicular.

2.2 Five-Parameter Representations

Hayati (1983) proposed a representation for prismatic joints that involved five parameters. The five-parameter representation of Hsu and Everett (1985) is equivalent to this, but is applied to rotary joints also. The difference from the four-parameter Hayati representation is that an additional offset d_i is allowed along the joint axis (Figure 3). The reason they give for this modification is to have a uniform representation, although it is not clear why this is important. Simulated results for a PUMA 600 robot were said to have been obtained but were not presented. As reviewed by (Everett, Driels, and Mooring, 1987), the independence and robustness of these parameters is not known. Ziegler and Datseris (1988) point out that this kind of five-parameter representation still has a sensitivity problem when neighboring coordinate origins are close

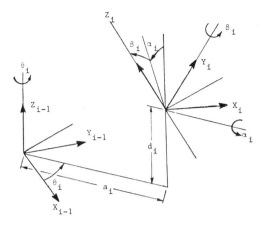

Figure 3: Five-parameter representation of Hsu and Everett (1985).

together: a small shift in the distal axis location can result in a large change in θ.

Veitschegger and Wu (1986) corrected Wu's earlier papers by adding Hayati's β parameter to the Denavit-Hartenberg parameter set, yielding five parameters. This is done by post-multiplying the regular transformation matrix by a rotation β about the y axis. They also simulated the importance of the second-order terms in the linearized kinematic equations in case of larger parameter errors for a PUMA 560 robot, and concluded that the first-order model is sufficiently accurate for most applications. More recently, Veitschegger and Wu (1987) experimentally calibrated the geometric parameters for a PUMA 560 robot, using a probe placed into precision points in a machined fixture. By considering the errors accounted for, they concluded that non-geometric errors such as joint deflection due to gravitational loading are less than 0.3 mm and are not significant.

2.3 Six-Parameter Representations

A number of papers have abandoned the Denavit-Hartenberg parameters entirely, and treated the general case of two coordinate systems related by six parameters. Three parameters are required for coordinate origin displacements, and three parameters for relative coordinate system orientation. Papers differ as to the representation of orientation, some using Euler angles and others using variations of the Euler vector, namely rotation about a vector. An advantage of a six-parameter representation is that higher-order kinematic pairs than just revolute joints can be handled (Everett, Driels, and Mooring, 1987), a disadvantage is that the constraints of lower-order kinematic pairs are not incorporated and the representations are redundant. Two of these six-coordinate representations, the S-Model of Stone (1987) and the C-Model of Chen and Chao (1986), are analyzed for completeness by Everett and Hsu (1988). Everett and Suryohadiprojo (1988) present a method for identifying the dependent parameters by examining the linear dependence of differential motions induced by small parameter changes. Other authors below have used similar procedures to eliminate the dependent parameters.

Mooring (1983) also pointed out the problems with the Denavit-Hartenberg representation about the same time as Hayati (1983), but proposed instead a six-parameter representation using Euler vectors. Simulations for a PUMA 600 indicated that a one-degree error in the skew angle caused endpoint errors of more than one inch. An estimation procedure was outlined but not implemented, in which single joints were moved to two positions, the position and orientation of the gripper measured somehow at these two positions, and the six parameters for each joint determined by direct solution. Mooring and Tang (1984) modified this procedure to use six-dimensional precision points in a fixture, avoiding the need to determine the gripper position and orientation by external sensing. An insight of this paper is to estimate orientation parameters before position parameters: while the former leads to a nonlinear estimation problem, the latter can then be treated as a linear estimation problem. For the orientation errors and parameters, the sum of squares of Euler angle differences is minimized by a finite-difference Levenberg-Marquardt algorithm, commonly employed for nonlinear estimation problems. The estimated orientation parameters are then incorporated to apply direct linear estimation to the position parameters. Simulation results for a PUMA 600 robot were given. Although the insight that the length parameters enter linearly is indeed important, their method is probably suboptimal because the effect of the angular parameter errors on position errors is not incorporated. There is clearly a tradeoff: throw out the coupling terms in the Jacobian and derive two simpler and separable least squares problems, or include the full Jacobian and potentially get better results. Simulations in fact show that the latter approach yields better results.[1]

One of the best works in kinematic calibration is that of Whitney, Lozinski, and Rourke (1984, 1986). Six-parameter coordinate systems with Euler angles were used, and both geometric and non-geometric parameters were modeled. The non-geometric parameters included gear transmission error, joint compliance, backlash, and base motion. Experimental results were obtained for a PUMA 560 robot using a theodolite and length standard, and a nonlinear estimation procedure based on pattern search. After calibration, their robot's accuracy was reduced from 5 mm to 0.3 mm. In contrast to the work of Veitschegger and Wu (1987), non-geometric factors were found to be as important as geometric factors. Whitney and Shamma (1986) emphasized the distinction between an internal kinematic model and an exact kinematic model, in commenting on terminology by Bazerghi, Goldenberg, and Apkarian (1984), who corrected an internal PUMA model but referred to it as an experimentally exact model.

Chen, Wang, and Yang (1984) employed six-parameter transformations with Euler angles. A recursive least squares procedure was applied to the linearized equations, necessitating a reduction of the dependent geometric parameters. Simulation results were given. Chen and Chao (1986, 1987) extended this work to include non-geometric models for gravitational joint deflection and backlash. Linearization was suggested as a means for identifying dependent parameters. Experimental results were given for a PUMA 760 robot, the measurements of an end effector target ball being made with three theodolites. Accuracy was improved by an order of magnitude: correcting the geometric parameters brought the accuracy down to 1 mm, adding corrections for joint 3 compliance and backlash to 0.5 mm, and adding to this joint 2 compliance to 0.27 mm. To avoid the problem of a redundant parameter set, Chao and Yang (1986) applied the nonlinear Levenberg-Marquardt procedure to the same data, and achieved identical results.

Stone, Sanderson, and Neuman (1986) employed a six-parameter coordinate representation as a preliminary step towards deducing the Denavit-Hartenberg parameters. Geometric parameters only were modeled. Their measurement apparatus consisted of an ultrasonic system: individual sparkers were attached to each link, and a square jig with four microphones served as pickup. The sonic digitizer was carefully calibrated for environmental effects such as temperature and humidity, which affect the speed of sound. The

---

[1] David Bennett, private communication.

identification procedure involved rotation of single axes at a time, and least-squares estimation of the plane and center of rotation. Experimental results on seven PUMA 560 robots indicated that accuracy could be improved to 0.2 mm. Independent verification of the calibration was performed with the Ranky (1984) repeatability jig mounted on a milling machine. The hardware and software system was described in more detail in Stone and Sanderson (1987). More recently, Stone and Sanderson (1988) provided a statistical analysis of the effect of measurement error on the calibrated parameters as well as on the endpoint position. This work is also summarized in (Stone, 1987).

Broderick and Cipra (1988) investigated just the geometric factors using a six-parameter representation based on the shape matrix. Their identification scheme also involved motion of single joints alone, moving distally to proximally, analogous to the scheme proposed by Stone (1987). Along the way, the shape matrix was recursively determined for each link. Given the few number of poses that were employed and the way in which the plane and center of rotation were determined, there seems to be an issue of robustness. Simulations were performed on a kinematic model of the PUMA 600.

The ability of six-parameter representations to model higher-order kinematic pairs, such as bent prismatic joints, was discussed by Driels and Pathre (1987a) and Everett, Driels, and Mooring (1987). Driels and Pathre (1987a) restricted their analysis to geometric parameters, applied to a PUMA 600 robot model, and a linearized estimation procedure was proposed. No results were given. Driels and Pathre (1987b) proposed a forward kinematic compensation method for parameter errors by utilizing Jacobians in terms of these parameters. One wonders why the forward kinematics could not have been solved more straightforwardly by just using the modified parameters in the regular evaluation. Different nonlinear functions for kinematic parameters were considered by Everett, Driels, and Mooring (1987), such as a quadratic function.

Chen and Chen (1987) investigated the correction of beam warping in prismatic joints. The beam warping was modeled by third-order polynomials, and identification was performed by recursive least squares of the linearized equations. Simulations for a three-coordinate measuring machine verified the feasibility of using ball bars, placed throughout the workspace, as length standards.

Vaishnav and Magrab (1987) actually employed a nine-parameter representation, where three additional rotation parameters allow skew coordinate systems to be represented. This enables wobble in the rotation axis to be modeled. A differential error analysis was coupled to a parameter reduction scheme based on a correlation matrix analysis, and a least squares estimation procedure derived the parameter errors from end effector six-dimensional errors. A simulation was performed for a six degree-of-freedom polar robot, to predict endpoint error based on statistical variations in the kinematic parameters, but no calibration results were given.

2.4 Closed-Loop Manipulators

So far, the discussion of parametric estimation has taken place in the context of open-loop kinematic chains. Closed-loop manipulators offer special problems, since extra constraint equations are required for the loops. Everett and Lin (1988) present a calibration procedure for manipulators with arbitrary numbers of loops, where the loop constraints are adjoined to a measurement-error objective function by lambda multipliers and the combined equations solved by nonlinear optimization. Simulation results are presented for a planar five-bar linkage arm.

Bennett and Hollerbach (1988) investigated the kinematic calibration of closed-loop manipulators comprised of a single chain. This single chain could be formed either from a redundant manipulator with fixed endpoint, or from two manipulators rigidly attached at their endpoints. Surprisingly, they showed that such manipulators could be calibrated without endpoint sensing, provided that certain identifiability conditions were satisfied. If the closed-chain mechanism has positive mobility, then consistency conditions during internal motion of the mechanism permit the mechanism to be identified, provided that one length parameter somewhere is known.

2.5 Inverse Kinematics

Even though a manipulator may have been designed so that the inverse kinematics is analytically solvable, kinematic calibration could lead to a non-analytically solvable manipulator. For example, a spherical wrist could be made not spherical. This situation has been recognized in a number of references, and a typical iterative solution to the inverse kinematics is outlined below.

Presumably the corrected kinematic structure would only differ slightly from the nominal analytically solvable kinematic structure. The nominal model $\mathbf{x} = \mathbf{f}_N(\boldsymbol{\theta})$ can then be used in conjunction with the calibrated model $\mathbf{x} = \mathbf{f}(\boldsymbol{\theta})$ for solution by iteration. Suppose

$\mathbf{x}_d$ is the desired endpoint position. Then the nominal joint angle solution $\boldsymbol{\theta}_N = \mathbf{f}_N^{-1}(\mathbf{x}_d)$ can be found since the manipulator was designed to be analytically solvable. The actual endpoint location that would result from the nominal joint angle solution is $\mathbf{x} = \mathbf{f}(\boldsymbol{\theta}_N)$, based on the calibrated model. The endpoint error $\Delta \mathbf{x} = \mathbf{x}_d - \mathbf{x}$ can be corrected to first order by

$$\Delta \boldsymbol{\theta} = \mathbf{J}^{-1}(\boldsymbol{\theta}_N)\Delta \mathbf{x} \tag{10}$$
$$\boldsymbol{\theta} = \boldsymbol{\theta}_N + \Delta \boldsymbol{\theta} \tag{11}$$

Equation (10) is valid because the error $\Delta \mathbf{x}$ is presumably small, and the Jacobian matrix $\mathbf{J}$ relates incremental endpoint motion to incremental joint motion. The updated joint angle estimate (11) can be iterated if the first correction $\Delta \boldsymbol{\theta}$ is not adequate, but presumably this would not be necessary if the nominal and calibrated models are very close. Thus this inverse kinematics position calculation should still be efficient.

The Jacobian matrix $\mathbf{J}$ in (10) is based on the calibrated model. Even though the inverse kinematics positions may not be analytically solvable, the inverse kinematics velocities for a square Jacobian matrix can always be found (except for singularities) by simple matrix inversion or Gaussian elimination. While mathematically it may be best to use the Jacobian $\mathbf{J}$ of the calibrated model, nevertheless it is computationally advantageous to use the nominal Jacobian $\mathbf{J}_N$. As with the inverse kinematics position calculation, the inverse kinematics velocity calculation is considerably simplified for kinematically simple manipulators as well. Highly efficient algorithms have been proposed that avoid matrix inversion or formation of the Jacobian matrix (e.g., Featherstone (1983)). Again, this scheme is only feasible if the nominal and calibrated models. are close, otherwise convergence cannot be guaranteed.

References that have proposed the procedure above include Payannet, Aldon, and Liegeois (1985), Sugimoto and Okada (1985), Hayati and Roston (1986), Kirchner, Gurumoorthy, and Prinz (1987), Broderick and Cipra (1988), and Wu, Ho, and Young (1988). Simulations of this procedure were presented by Hayati and Roston (1986) for the PUMA 560 when the wrist becomes non-spherical. Payannet, Aldon, and Liegeois (1985) generalize Equation 10 to handle redundant manipulators by substituting the pseudoinverse of $\mathbf{J}$.

So far, actual experimental results in kinematic calibration have shown that wrist offsets are not significant (Shamma and Whitney,

1987), leaving still a solvable manipulator from a standpoint of geometric link parameters. Nevertheless, nongeometric factors such as gear eccentricity, backlash, and joint compliance may also prevent the formulation of an analytic inverse. The procedure outlined above could still be useful in the latter case.

3 Non-Parametric Estimation of Endpoint Location

Parametric models of geometric or non-geometric factors cannot always be conveniently posed to account for all error sources. Some factors may be difficult to express analytically, and some error sources may display local variations. While parametric models are convenient because they are global and cover the whole workspace, they are not easily adapted for such arbitrary error sources. Tuning a parametric model for one part of the workspace will most likely degrade performance in another part of the workspace.

An alternative is to employ non-parametric methods to map locally the relation between endpoint positions and joint angles. A sufficient number of such relationships must be developed in a local area, and hence one disadvantage of these methods is simply the number of measurements required. Moreover, different local areas will require additional measurements, and a change in operating condition such as load may invalidate the whole local map.

Foulloy and Kelley (1984) applied local error corrections at particular manipulator configurations, through a residual matrix representing a differential linear correction. A residual matrix was obtained by inserting a calibration plate with four pins into a calibration cube with 25 holes, and applying a least squares estimation technique. Experimental results indicated that robot error was reduced by a factor of ten.

Shamma and Whitney (1987) distinguished between forward calibration, which determines the endpoint position from joint angles, and inverse calibration, which determines the joint angles from the endpoint position. Non-parametric representations therefore represent inverse calibration. Shamma and Whitney applied third-order trivariate polynomials as approximation functions to relate endpoint positions to joint angles. For a six degree-of-freedom robot such as the PUMA 560, the correction process was partitioned into two parts: first the wrist position was corrected, and then the wrist rotations were corrected. Data points were generated by Tchebychev spacing, and simulations indicated that accuracy was reduced to below 0.5 mm. Drawbacks to this scheme include limitations to one arm configuration and to one quadrant of the workspace.

Driels et al. (1988) discussed two table lookup schemes, one based on CMAC and another on a finite element method. The CMAC implementation quantizes endpoint position, and linearly combines quantizing functions that are addressed by the indexed endpoint cells. While generalization is achieved by neighboring cells sharing similar quantizing functions, Driels et al. mention that interpolation is not readily performed. A finite element method, originally presented in Everett and McCarroll (1986), was also discussed.

Atkeson et al. (1988; also in An, Atkeson, and Hollerbach, 1988) presented a learning scheme for implementing inverse calibration, based on a nominal model of the inverse kinematics and learning. This problem is related to solving for the inverse kinematics by iteration of the linearized equation (section 2.5). Suppose that a nominal forward kinematics model $\mathbf{x} = \mathbf{f}_N(\boldsymbol{\theta})$ and its inverse $\boldsymbol{\theta} = \mathbf{f}_N^{-1}(\mathbf{x})$ are available, but that the actual kinematics $\mathbf{x} = \mathbf{f}(\boldsymbol{\theta})$ is unknown. An iterative solution begins with an estimate of the joint angles through $\boldsymbol{\theta}_0 = \mathbf{f}_N^{-1}(\mathbf{x}_d)$, where $\mathbf{x}_d$ is the desired endpoint position. At iteration i, let $\mathbf{x}_i = \mathbf{f}(\boldsymbol{\theta}_i)$ be the actual endpoint position that results when the manipulator repositions itself corresponding to the current estimate of the joint angles $\boldsymbol{\theta}_i$. Then define the joint angle error as

$$\Delta \boldsymbol{\theta}_i = \boldsymbol{\theta}_0 - \mathbf{f}_N^{-1}(\mathbf{x}_i) \tag{12}$$

The joint angle solution at the next iteration is then updated as $\boldsymbol{\theta}_{i+1} = \boldsymbol{\theta}_i + \Delta \boldsymbol{\theta}_i$. The main difference with this approach over that in section 2.5 is the use of finite differences involving the nominal nonlinear inverse function, whereas the latter involves the use of the Jacobian matrix. The method of Atkeson et al. also relies on $\mathbf{f}_N$ being a reasonably good kinematics model, and they provided convergence conditions on how good the model has to be. This method really falls between parametric and nonparametric calibration, since a nominal kinematics model is used.

4 Robot Repeatability

The issue in robot repeatability is how precisely the robot can achieve a particular endpoint location under repeated trials. Repeatability sets the limits of accuracy, and is determined by a variety of factors such as sensor precision, servo controller, backlash,

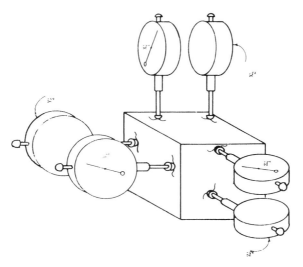

Figure 4: Gauge block position and orientation measured with six dial indicators.

and load. Moreover, the repeatability will be a strong function of the manipulator configuration.

Repeatability most often has been measured by placing an end-effector fixture into a gauge block. McEntire (1976) employed a gauge block with three faces forming a concave corner and six dial indicators, two per face (Figure 4). The end-effector fixture was a cube that was placed against the dial indicators, allowing the 6 parameters of position and orientation to be measured. Simulations for a typical robot were given. Experimental results were presented for a Fanuk robot. To automate the repeatability measurements, Warnecke and Brodbeck (1977) and Brodbeck and Schiele (1978) used three inductive sensors, one per face, to detect three position parameters only. Twenty industrial robots of different types were said to be tested with this apparatus.

Inagaki (1978) discussed the effect of servo stiffness and sensor resolution on positioning accuracy. A low servo stiffness may result in deadzones from static friction and steady-state disturbances, while a high servo stiffness may lead to oscillations. In a later paper, Inagaki (1978) distinguished positioning accuracy from orienting accuracy, and discussed several ways of measuring the latter. Dependence of accuracy on manipulator rigidity was mentioned, as well as backlash leading to the loss of repeatability. Accuracy results were given for 20 industrial robots.

Morgan (1980) investigated the relative accuracy of manipulators, using a concept called the Volumetric Accuracy Mapping. Given a reference corner on a cubic volume in the workspace, the accuracy of the robot in this volume relative to the reference corner was measured. A three-dimensional measurement system was described, comprising a clock gauge and two verniers. Fohanno (1982) investigated repeatability both during static positioning and during a trajectory. The static position measurements were made with an end effector cube mounted with unspecified sensors that measured points in a gauge block. The trajectory measurements were made with three guide wires attached to the end effector and wound on spring-loaded drums instrumented with potentiometers. Results were given for three robots.

Ranky (1984) also studied repeatability with a gauge block and dial indicators. His apparatus was almost identical to McEntire's, except that a total of nine dial indicators, three per face, were used. The extra three indicators, however, are redundant. Fenton, Davison, and Goldenberg (1984) measured the repeatability in both position and orientation of a PUMA 600 robot using six dial indicators in a gauge block and a special tool with three spherical tips connected normally by rods. The results indicated differences from the manufacturer's specifications along different axes.

Matthews and Hill (1985) employed five LVDTs in a gauge block, and devised an end-effector fixture with two large ball bearings to be placed against the LVDTs. Only five parameters of endpoint location can be determined this way, which was adequate in their case since their Microbot Alpha robot had five degrees of freedom. Experimental results were presented. More recently, Mooring and Pack (1986) used six LVDTs in the gauge block and three ball bearings in the endpoint fixture, to allow all six location parameters to be determined for a six degree-of-freedom robot. They reported that the position and orientation repeatability was roughly normally distributed for their robot, with standard deviations in position ranging from 1.5 to 3.2 mils and in orientation from 0.2 to 0.5 degrees.

Griebel and Higger (1985) devised a system for accuracy and repeatability measurements during linear trajectories. Five proximity inductance sensors were arranged orthogonally on a special tool, which also contained an optical sensor that measured trigger marks on a specially prepared test bar fixture. Accuracy and repeatability measurements were made along the linear trajectory defined by the bar.

On a more theoretical level, Waldron and Kumar (1979) and Kumar and Waldron (1981) computed the probability distribution of endpoint position errors based on a normal distribution of joint position errors, and simulated a six degree-of-freedom rotary robot. Kumar and Prakash (1983) extended these results to consider endpoint variations due to variations in the skew angles α_i of the Denavit-Hartenberg kinematic parameters, obtained from Equation (2) as

$$\Delta x = \frac{\partial f}{\partial \theta}\Delta \theta + \frac{\partial f}{\partial \alpha}\Delta \alpha \qquad (13)$$

They presented simulations for the rotary robot.

Wu (1983, 1984) examined endpoint variability by considering all four Denavit-Hartenberg parameters as random variables, normally distributed with zero means. The distribution of endpoint errors were evaluated at a nominal position, about which the kinematic equations were linearized. The distribution was Gaussian zero-mean but correlated. No results were given. Wu and Lee (1984, 1985) observed that the distribution of kinematic parameters is hard to know a priori, and proposed instead to measure the endpoint distributions directly by placing the arm in different measured locations and using maximum likelihood estimation. As stated, it does not seem that this procedure is valid, because placing the arm at different positions violates the linear analysis, since the kinematics are nonlinear. No results were given.

Benhabib, Fenton, and Goldenberg (1987) defined various measures of task-space tolerances for both position and orientation. An inverse error analysis was performed to relate bounds on joint tolerances given task-space tolerances. Simulations were performed for a Cartesian robot.

Menq and Borm (1988) examined the endpoint variations given a normal distribution of kinematic parameter errors, using a six-parameter representation of link coordinate origins (see Section 2.3). They showed that the envelope of these variations described an ellipsoid, since the endpoint distribution is correlated, as pointed out by Wu (1984). They also showed the dependence of this ellipsoid on the workspace location, and the use of the major axis of the ellipsoid to predict the most precise directions of motion.

Sturm and Lee (1988) presented a stochastic model for kinematic parameter errors, and predicted the tip positioning accuracy. Different probability distributions were chosen for different kinds of kinematic parameter errors. Theodolite measurements on a gantry robot of the endpoint error were used for comparison to the model.

Bhatti and Rao (1988) also presumed that both the Denavit-Hartenberg parameters and the joint angles have a normal distribution. They simulated and characterized as normal the distribution of endpoint positioning and orienting errors for the Stanford arm.

5 Robot Registration

Robot registration involves determining the relative location of the manipulator and objects in the environment, and has a number of manifestations. When an object is placed or identified in the environment, there is of course the problem of relating the object's coordinate system to the robot coordinate system. The robot itself may be placed on a mobile platform and moved from workstation to workstation; its location at each workstation must be determined. An even more demanding manifestation is tracking the manipulator endpoint when the base is moving or in case the manipulator has flexible links. The apparent position of the manipulator with respect to objects may drift with time, due either to electronic drift or environmental perturbations. Vision and tactile sensors also need to be located with regard to the robot reference frame. Finally, whenever a robot picks up an object, the appropriate tool transform for the object must be set up.

5.1 Calibration Fixtures

Approaches to robot registration have been as varied as the manifestations of robot registration. A number of approaches have employed calibration fixtures or mechanical measuring devices. Ho (1982) corrected for dimensional drift in an IBM RS-1 robot by having the robot touch three flat calibration posts on the robot table. The posts were adequate because vertical deflection and twist were not as significant as horizontal deviation for this robot.

Whitney and Junkel (1982) used Kalman filtering to calibrate the holes in a jig with an instrumented RCC, and gave simulations to show how dimensional variations can be tracked. A general discussion was presented on robot registration, where the applications have in common a contact operation that can be measured by the instrumented RCC.

Irwin (1986) automatically corrected for drifts in robot position, by inserting a special probe into champfered holes of a fixture with precision points. Experimental results were given for a Cincinnati T3 robot.

Ishii et al. (1987b) used a coordinate measuring device to register the base of a robot with respect to a global coordinate system. The joint offsets were also determined; the first three joints were identified separately from the wrist joints. Identification was accomplished by linearization and the use of a Newton's method.

Mooring and Pack (1988) calibrated just the joint offsets for a commercial robot, which changed due to dissassembly during normal maintenance. The wrist joint offsets were determined by attaching the wrist assembly to a special fixture and generating a number of positions. The three proximal joints were separately calibrated with a tooling ball attached to joint 3 and contacting a special fixture fixed in the robot's workspace.

5.2 Visual Sensing

The following papers employed some form of visual sensing to achieve robot registration. Foulloy and Kelley (1984) applied their non-parametric calibration procedure to determine the relationship between a camera coordinate system and the robot coordinate system, by visually observing the calibration cube with the camera.

El-Zorkany (1984) proposed the use of the Selspot II System, a commercially available opto-electronic three-dimensional movement monitoring system, for robot registration. Infrared light-emitting diodes are attached to the robot, and are tracked at high frequencies by special cameras with lateral-effect photodiode detectors. The endpoint tracking such a system is capable of has a number of uses. A robot program could be verified. Positioning errors, for example due to steady-state servo errors or to an inaccurate kinematic model, such as link deflection, could be corrected. No results were given.

El-Zorkany, Liscano, and Tondu (1985) proposed that a human hand-held target be used to register object features and to teach a trajectory. Experimental results on trajectory teaching were presented. Ganapathy (1986) also discussed robot teaching by tracking the human hand movement. A novel apparatus was proposed, consisting of one video camera tracking five planar points; experimental results were stated but not presented.

Agin (1985) determined the tool transform for a hand-held camera and laser striper, to be moved around a scene by the robot to obtain stereo information. The tool transform for the camera was obtained by observing an external spot in several positions, and then the tool transform of the laser was obtained by observation of the stripe by the camera. A manual procedure was also presented for correcting the joint offsets, through use of a level and displacement measurements. Experimental results for a PUMA 560 robot were discussed.

Lin, Ross, and Ziegler (1986) calibrated the tool transform of a grasped turbine blade, using a laser scanning system and triangulation and a CAD model of the part. The location of the robot relative to the laser system is assumed, and the tool is placed in several locations.

Ishii et al. (1987a) investigated robot teaching with a custom built tracking apparatus similar to the Selspot System. A 3-D teaching device was composed with four planar infrared LEDs, and the camera detector contained a lateral-effect photodiode. By taking advantage of constraint relations, it was possible to determine the position and orientation of the hand-held teaching device with just a single camera. Nevertheless, the accuracy attainable with this system (around 5 mm) seems to be somewhat worse than the two-camera commercial systems, and the sampling rates are much slower (113 msec to obtain 3-D information). The teaching device was used to define trajectories with the single camera. By mounting arrays of LEDs on the robot end effector, Ishii et al. also demonstrated endpoint tracking. Finally, they used a two-camera stereo system to define features in the environment by pointing with the hand-held device.

Shiu and Ahmad (1987) indicated how the tool transform of wrist-mounted sensor systems can be determined, such as cameras, range sensors, and tactile sensors, by adopting three poses of the robot arm relative to a calibration object. Results were stated but not presented. Chou and Kamel (1988) solved a similar problem, where just the rotational part of the tool transform was determined. Quaternions were employed as the orientation representation, and an analytical procedure was presented based on singular value decomposition for determining the orientation from two poses. Simulations were given.

Tsai and Lenz (1988a, 1988b) presented a procedure for determining the tool transform of an end-effector mounted CCD camera using a calibration block. The procedure is similar to that of Shiu and Ahmad, except that a number of poses are averaged and the number of unknown parameters is independent of the number of poses. Experimental results showed that the method was fast and accurate.

6 Advanced Instrumentation for Endpoint Tracking

Much of the instrumentation for kinematic calibration requires intensive human involvement and time. The eventual goal is to automatically calibrate a robot accurately and quickly, with minimal human involvement or special-purpose fixtures. Ideally, the endpoint measurements should be in real time, to allow movement of the robot base to be tracked or to compensate for unmodeled error sources. A more complete review of the measuring systems mentioned in this survey can be found in (Driels et al., 1988).

A number of papers have primarily emphasized improved instrumentation for kinematic calibration, without at the present time presenting calibration results. Gilby and Parker (1982) presented a laser scanning system for tracking the robot endpoint. A two-dimensional optical scanner external to the robot deflected a laser beam towards a retroreflector on the manipulator end effector, returning a parallel displaced beam whose displacement was measured by a photodiode quadrant detector. Tracking was accomplished by centering the reflected beam on the photodiode. No experimental results were presented.

A laser tracking interferometer system was presented by Lau, Hocken, and Haynes (1985). This time there was an active target mirror system on the end effector as well as a tracking mirror system. The resolution of the system was stated as 1 part in 100,000, or 12.5 microns. An initial position measurement was separately required. This apparatus was tested but not yet used to calibrate a robot. One disadvantage of this scheme is the complicated end effector; another is that the beam cannot be broken by an obstruction or tracking will be lost.

Dainis and Juberts (1985) modified a Selspot I System to obtain a camera detector resolution of 1/10,000 and an accuracy of 1/2000, obtained after 100 readings. The cameras were individually calibrated with a flat bed digital plotter, and their locations were determined by observation of four control points whose locations were specified in a reference coordinate system. Three-dimensional coordinates were obtained using direct linear transformations. This system was not actually applied to a robot calibration task. Along the same lines, Antonsson (1982) extensively calibrated a Selspot I System to bring the accuracy to the level of resolution, which was 1/4000.

A motion tracking system similar to the Selspot II System is the Watsmart System, produced by Northern Digital, which has the same accuracy and resolution. The Watsmart System has a real-time option that allows the pose tracking of two rigid objects at 100 Hz. Hollerbach and Bennett (1988), however, concluded that this kind of system is not currently accurate enough for kinematic calibration, especially when compared to the use of theodolites. The accuracy limitations derive primarily from the use of lateral effect photodiode detectors, because the resolution cannot be improved much beyond 1/4000 due to noise and fundamental reflection problems. Northern Digital has announced its next generation of motion tracking systems, the Optotrak System, whose resolution is up to 1/100,000 and which does not suffer from reflection problems. The Optotrak System operates with linear CCD arrays and cylindrical lenses.

Scheffer (1982) considered several possible measurement techniques, all under development, including a wire system attached to the end effector. The geometric and non-geometric factors in calibration were also thoroughly discussed.

Jarvis (1987), in considering calibration with theodolites, presented a procedure for determining the baseline, namely the relative transformation between two theodolites. This transformation was simply accomplished by observing points along an arbitrary line. Simulations were given. More recently, Jarvis (1988) completed the calibration of the theodolites themselves, each of which can be considered a two-link rotary manipulator. The theodolites were constructed from motorized rotary tables, under computer control. The cameras were simply lenses with quadrant photodetectors. The measurement system to calibrate the theodolites consisted of a Seiko robot executing a radial motion and an illuminated ball which could be tracked automatically by the theodolite system. The ball motion was tracked by a Hewlett Packard 5528A Laser Measurement System.

7 Conclusion

In reflecting on this literature, it is clear that many alternatives have been presented in terms of instrumentation, coordinate system representations, non-geometric factors, and system identification procedures. The area of kinematic calibration is still young, as attested to by the dates of the publications, and no consensus has yet developed among these alternatives. Experimental results are still the exception rather than the norm, and there are some disagreements among the few results that do exist.

One issue that should be settled in the future is the choice of coordinate system representation. One strong alternative seems to be the Hayati modification of the Denavit-Hartenberg representation. It is not clear at this point what advantages the six-parameter representations would have for modeling lower-order kinematic pairs, while they have the disadvantage of redundancy.

Another issue has to do with the system identification procedure. Many investigators linearized the kinematic equations and applied an iterative estimation procedure, while others applied other nonlinear estimation procedures. The relative advantages of the various nonlinear estimation methods have not been definitively settled. Some investigators have begun to consider the statistics of the identification process, but more work needs to be done, especially in regard to the effect of measurement noise and statistical parameter variations on the choice of an optimal estimation procedure.

The instrumentation for kinematic calibration is quite varied and for the most part laboratory built. Good commercial systems are beginning to emerge, some which in fact are already better than certain laboratory systems. One can expect in the future the continued emergence of commercial systems, which then become standard for kinematic calibration.

Finally, this survey reveals that there has been a great deal of duplication of published results. This duplication may be due partly to simultaneous activity at various laboratories, and hence may arise from research imperatives from a timely topic. It may also be due to the diverse nature of the publications that results have appeared in, especially conference proceedings which are not generally available. It is hoped that this survey has helped provide a historical account of the evolution of the field of kinematic calibration, and an overview of what exactly has been done and what remains to be done.

Acknowledgments

This report describes research done at the Artificial Intelligence Laboratory of the Massachusetts Institute of Technology. Support for the laboratory's artificial intelligence research is provided in part by the Office of Naval Research University Research Initiatives Contract N00014-86-K-0180 and the Defense Advanced Research Projects Agency under Office of Naval Research contracts N00014-85-K-0124. Support was also provided in part by an NSF Presidential Young Investigator Award. The author is grateful for comments on drafts by David Bennett.

References

Agin, G.J., 1985, "Calibration and use of a light stripe range sensor mounted on the hand of a robot," *IEEE Int. Conf. Robotics and Automation*, St. Louis, March 25-28, pp. 680-685.

Ahmad, S., 1985, "Second order nonlinear kinematic effects, and their compensation," *IEEE Int. Conf. Robotics and Automation*, St. Louis, March 25-28, pp. 307-314.

Ahmad, S., 1988, "Analysis of robot drive train errors, their static effects, and their compensation," *IEEE J. Robotics and Automation*, 4, pp. 117-128.

An, C.H., Atkeson, C.G., and Hollerbach, J.M., 1988, *Model-Based Control of a Robot Manipulator*, Cambridge, MA, MIT Press.

Antonsson, E. K., 1982, A three-dimensional kinematic acquisition and intersegmental dynamic analysis system for human motion, Ph. D. Thesis, Massachusetts Institute of Technology, Mechanical Engineering.

Atkeson, C.G., Aboaf, E.W., McIntyre, J., and Reinkensmeyer, D.J., 1988, "Model-based robot learning," *Robotics Research: The Fourth International Symposium*, edited by Bolles, R., and Roth, B., Cambridge, MA, MIT Press, pp. 103-110.

Bazerghi, A., Goldenberg, A.A., and Apkarian, J., 1984, "An exact kinematic model of PUMA 600 manipulator," *IEEE Trans. Systems, Man, Cybern.*, SMC-14, pp. 483-487.

Benhabib, B., Fenton, R.G., and Goldenberg, A.A., 1987, "Computer-aided joint error analysis of robots," *IEEE J. Robotics and Automation*, RA-3, pp. 317-322.

Bennett, D.J., and Hollerbach, J.M., 1988, "Self-calibration of single-loop, closed kinematic chains formed by dual or redundant manipulators," *Proc. 27th IEEE Conf. Decision and Control*, Austin, Tx, Dec 7-9.

Bhatti, P.K., and Rao, S.S., 1988, "Reliability analysis of robot manipulators," *ASME J. Mechanisms, Transmissions, and Automation in Design*, 110, pp. 175-181.

Brodbeck, B., and Schiele, G., 1978, "Measurements and analyses of geometrical quantities concerning industrial robots," *Proc. 8th Int. Symp. on Industrial Robots*, Stuttgart, May 30–June 1, pp. 255-268.

Broderick, P.L., and Cipra, R.J., 1988, "A method for determining and correcting robot position and orientation errors due to manufacturing," *J. Mechanisms, Transmissions, and Automation in Design*, 110, pp. 3-10.

Chao, L.M., and Yang, J.C.S., 1986, "Development and implementation of a kinematic parameter identification technique to improve the positioning accuracy of robots," *Robots 10 Conference Proceedings*, Chicago, April 20-24, pp. 11-69 – 11-81.

Chen, J., and Chao, L.M., 1986, "Positioning error analysis for robot manipulators with all rotary joints," *Proc. IEEE Int. Conf. Robotics and Automation*, San Francisco, April 7-10, pp. 1011-1016.

Chen, J., and Chao, L.M., 1987, "Positioning error analysis for robot manipulators with all rotary joints," *IEEE J. Robotics and Automation*, RA-3, pp. 539-545.

Chen, J., and Chen, Y.F., 1987, "Estimation of coordinate measuring machine error parameters," *Proc. IEEE Int. Conf. Robotics and Automation*, Raleigh, NC, March 31-April 3, pp. 196-201.

Chen, J., Wang, C.B., and Yang, J.C.S., 1984, "Robot positioning accuracy improvement through kinematic parameter identification," *Proc. 3rd Canadian CAD/CAM and Robotics Conf.*, Toronto, June 19-21, pp. 4.7-4.12.

Chou, J.C.K., and Kamel, M., 1988, "Quaternions approach to solve the kinematic equation of rotation $A_a A_x = A_x A_b$ of a sensor-mounted robotic manipulator," *Proc. IEEE Int. Conf. Robotics and Automation*, Philadelphia, April 24-29, pp. 656-662.

Dainis, A., and Juberts, M., 1985, "Accurate remote measurement of robot trajectory motion," *IEEE Int. Conf. Robotics and Automation*, St. Louis, March 25-28, pp. 92-99.

Denavit, J., and Hartenberg, R.S., 1955, "A kinematic notation for lower pair mechanisms based on matrices," *J. Applied Mechanics*, 22, pp. 215-221.

Driels, M., Mooring, B.W., Everett, L.J., and Roth, Z.S., 1988, "Fundamentals of Robot Calibration," *Lecture Notes for the Tutorial on Robot Calibration, 1988 IEEE Int. Conf. Robotics and Automation*, Philadelphia, April 25.

Driels, M.R., and Pathre, U.S., 1987a, "Generalized joint model for robot manipulator kinematic calibration and compensation," *J. Robotic Systems*, 4, pp. 77-114.

Driels, M.R., and Pathre, U.S., 1987b, "Robot manipulator kinematic compensation using a generalized Jacobian formulation," *J. Robotic Systems*, 4, pp. 259-280.

El-Zorkany, H.I., 1984, "Automatic location correction in off-line programming of industrial robots," *Proc. 14th Int. Symp. Industrial Robots*, Gotheburg, Sweden, Oct. 2-4, pp. 335-346.

El-Zorkany, H.I., Liscano, R., and Tondu, B., 1985, "A sensor-based approach for robot programming," *Int. Conf. Intelligent Robots and Computer Vision, SPIE Proc. Vol. 579*, Cambridge, Mass., Sept. 15-20, pp. 289-297.

Everett, L.J., Driels, M., Mooring, B.W., 1987, "Kinematic modelling for robot calibration," *Proc. IEEE Int. Conf. Robotics and Automation*, Raleigh, NC, March 31-April 3, pp. 183-189.

Everett, L.J., and Hsu, T.-W., 1988, "The theory of kinematic parameter identification for industrial robots," *Trans. ASME. J. Dy-*

namic Systs., Meas, and Control, 110 no. 1, pp. 96-99.

Everett, L.J., and Lin, C.Y., 1988, "Kinematic calibration of manipulators with closed loop actuated joints," *Proc. IEEE Int. Conf. Robotics and Automation*, Philadelphia, April 24-29, pp. 792-797.

Everett, L.J., and McCarroll, D.R., 1986, "Using finite element methods to approximate kinematic solutions for robot manipulators when closed form solutions are unattainable," *Proc. IEEE Int. Conf. Robotics and Automation*, San Francisco, April 7-10, pp. 1164-1167.

Everett, L.J., and Suryohadiprojo, A.H., 1988, "A study of kinematic models for forward calibration of manipulators," *Proc. IEEE Int. Conf. Robotics and Automation*, Philadelphia, April 24-29, pp. 798-800.

Featherstone, R., 1983, "Position and velocity transformations between robot end-effector coordinates and joint angles," *Int. J. Robotics Research*, 2 no. 2, pp. 35-45.

Fenton, R.G., Davison, T.C., and Goldenberg, A., 1984, "Accuracy and repeatability of robots and manipulators," *Proc. 3rd IASTED Int. Symp.: Advances in Robotics and Automation*, pp. 126-128.

Foulloy, L.P., and Kelley, R.B., 1984, "Improving the precision of a robot," *Proc. IEEE Int. Conf. Robotics*, Atlanta, March 13-15, pp. 62-67.

Ganapathy, S., 1986, "Teaching robots by hand movements," AT&T Bell Labs, Holmdel, N.J..

Gilby, J.H., and Parker, G.A., 1982, "Laser tracking system to measure robot arm performance," *Sensor Review*, 2 no. 4, pp. 180-184.

Griebel, U., and Higger, R.A., 1985, "A new method to determine accuracy and repeatability of robots," *Proc. IASTED Int. Symp. Robotics and Automation*, Lugano, June 24-26, pp. 253-257.

Hayati, S.A., 1983, "Robot arm geometric link parameter estimation," *Proc. 22nd IEEE Conf. Decision and Control*, San Antonio, Dec. 14-16, pp. 1477-1483.

Hayati, S.A., and Mirmirani, M., 1985, "Improving the absolute positioning accuracy of robot manipulators," *J. Robotic Systems*, 2, pp. 397-413.

Hayati, S.A., and Roston, G.P., 1986, "Inverse kinematic solution for near-simple robots and its application to robot calibration," *Recent Trends in Robotics: Modeling, Control, and Education*, edited by M. Jamshidi, J.Y.S. Luh, and M. Shahinpoor, Elsevier Science Publ. Co., pp. 41-50.

Hayati, S.A., Tso, K., and Roston, G.P., 1988, "Robot geometry calibration," *Proc. IEEE Int. Conf. Robotics and Automation*, Philadelphia, April 24-29, pp. 947-951.

Ho, C.Y., 1982, "Study of precision and calibration for IBM RS-1 robot system," *Assembly Automation*, 2 no. 4, pp. 100-104.

Hollerbach, J.M., and Bennett, D.J., 1987, "Automatic kinematic calibration using a motion tracking system," *Modeling and Control of Robotic Manipulators and Manufacturing Processes, DSC-Vol. 6, ASME Winter Annual Meeting*, Boston, MA, December 13-18, pp. 93-100.

Hollerbach, J.M., and Bennett, D.J., 1988, "Automatic kinematic calibration using a motion tracking system," *Robotics Research: the Fourth International Symposium*, edited by R. Bolles and B. Roth, Cambridge, MA, MIT Press, pp. 191-198.

Hsu, T.-W., and Everett, L.J., 1985, "Identification of the kinematic parameters of a robot manipulator for positional accuracy improvement," *Proc. 1985 Computers in Engineering Conference*, Boston, August, pp. 263-267.

Ibarra, R., and Perreira, N.D., 1986, "Determination of linkage parameter and pair variable errors in open chain kinematic linkages using a minimal set of pose measurement data," *J. Mechanisms, Transmissions, and Automation in Design*, 1986, pp. 159-166.

Inagaki, S., 1978, "Problems awaiting solutions of servomechanisms on industrial robots," *Proc. 8th Int. Symp. on Industrial Robots*, Stuttgart, May 30 – June 1, pp. 558-565.

Inagaki, S., 1979, "A discussion on positioning accuracy on industrial robots," *Proc. 9th Int. Symp. on Industrial Robots*, Washington, D.C., Mar. 13-15, pp. 679-690.

Irwin, C.T., 1986, "Automatic calibration system," *Robots 10 Conference Proceedings*, Chicago, April 20-24, pp. 7-43 – 7-57.

Ishii, M., Sakane, S., Kakikura, M., and Mikami, Y., 1987a, "A new approach to improve absolute positioning accuracy of robot manipulators," *J. Robotic Systems*, 4, pp. 145-156.

Ishii, M., Sakane, S., Mikami, Y., and Kakikura, M., 1987b, "A 3-D sensor system for teaching robot paths and environments," *Int. J. Robotics Research*, 6 no. 2, pp. 45-59.

Jarvis, J.F., 1987, "Microsurveying: towards robot accuracy," *Proc. IEEE Int. Conf. Robotics and Automation*, Raleigh, NC, March 31-April 3, pp. 1660-1665.

Jarvis, J.F., 1988, "Calibration of theodolites," *Proc. IEEE Int. Conf. Robotics and Automation*, Philadelphia, April 24-29, pp. 952-954.

Judd, R.P., and Knasinski, A.B., 1987, "A technique to calibrate industrial robots with experimental verification," *Proc. IEEE Int. Conf. Robotics and Automation*, Raleigh, NC, March 31-April 3, pp. 351-357.

Khalil, W., and Gautier, M., 1986, "Identification of geometric parameters of robots," *Robot Control (SYROCO '85): Proceedings of the 1st IFAC Symposium*, edited by L. Basanez, G. Ferrate, and G.N. Saridis, Oxford, Pergamon Press, pp. 27-30.

Kirchner, H.O.K., Gurumoorthy, B., and Prinz, F.B., 1987, "A perturbation approach to robot calibration," *Int. J. Robotics Research*, 6 no. 4, pp. 47-59.

Knapczyk, J., and Morecki, A., 1987, "Analysis of the positioning and orientation accuracy in 6R manipulators (direct task)," *RoManSy 6: Proceedings of the Sixth CISM-IFToMM Symposium on Theory and Practice of Robots and Manipulators*, edited by A. Morecki, G. Bianchi, and K. Kedzior, Cambridge, MA, MIT Press, pp. 90-98.

Kumar, A., and Prakash, S., 1983, "Analysis of mechanical errors in manipulators," *Proc. 6th World Congress on Theory of Machines and Mechanisms*, pp. 960-963.

Kumar, A., and Waldron, K.J., 1981, "Numerical plotting of surfaces of positioning accuracy of manipulators," *Mechanism and Machine Theory*, 16, pp. 361-368.

Lau, K., Hocken, R., and Haynes, L., 1985, "Robot performance measurements using automatic laser tracking techniques," *Robotics & Computer-Integrated Manufacturing*, 2, pp. 227-236.

Lin, W.-C., Ross, J.B., and Ziegler, M., 1986, "Semiautomatic calibration of robot manipulator for visual inspection task," *J. Robotic Systems*, 3, pp. 19-39.

Matthews, S.H., and Hill, J.W., 1985, "Repeatability test system for industrial robots," MS874-1045, Society for Manufacturing Engineers.

McEntire, R.H., 1976, "Three dimensional accuracy measurement methods for robots," *The Industrial Robot*, pp. 105-112.

Menq, C.H., and Borm, J.H., 1988, "Statistical measure and characterization of robot errors," *Proc. IEEE Int. Conf. Robotics and Automation*, Philadelphia, April 24-29, pp. 926-931.

Mooring, B.W., 1983, "The effect of joint axis misalignment on robot positioning accuracy," *Proc. Computers in Engineering Conf. and Exhibit*, pp. 151-155.

Mooring, B.W., and Pack, T.J., 1986, "Determination and specification of robot repeatability," *Proc. IEEE Int. Conf. Robotics and Automation*, San Francisco, April 7-10, pp. 1017-1023.

Mooring, B.W., and Pack, T.J., 1988, "Calibration procedure for an industrial robot," *Proc. IEEE Int. Conf. Robotics and Automation*, Philadelphia, April 24-29, pp. 786-791.

Mooring, B.W., and Tang, G.R., 1984, "An improved method for identifying the kinematic parameters in a six-axis robot," *ASME Proc. Int. Computers in Engineering Conf.*, Las Vegas, pp. 79-84.

Morgan, C., 1980, "The rationalization of robot testing," *Proc. 10th Int. Symp. on Industrial Robots*, Milan, pp. 399-406.

Northern Digital Inc., 1987, "Watsmart Product Literature," 403 Albert St, Waterloo, Ontario, Canada N2L 3V2.

Northern Digital Inc., 1988, "Optotrak Product Literature," 403 Albert St, Waterloo, Ontario, Canada N2L 3V2.

Payannet, D., Aldon, M.J., and Liegeois, A., 1985, "Identification and compensation of mechanical errors for industrial robots," *Proc. 15th Int. Symp. on Industrial Robots*, Tokyo, pp. 857-864.

Puskorius, G.V., and Feldkamp, L.A., 1987, "Global calibration of a robot/vision system," *Proc. IEEE Int. Conf. Robotics and Automation*, Raleigh, NC, March 31-April 3, pp. 190-195.

Ranky, P.G., 1984, "Test method and software for robot qualification," *The Industrial Robot*, pp. 111-115.

Roth, Z., Mooring, B.W., and Ravani, B., 1986, "Robot precision and calibration issues in electronic assembly," *Proc. IEEE Southcon, Program Session Record 21*, Orlando, pp. 1-13.

Roth, Z., Mooring, B.W., and Ravani, B., 1987, "An overview of robot calibration," *IEEE J. Robotics and Automation*, RA-3, pp. 377-386.

Scheffer, B., 1982, "Geometric control and calibration method of an industrial robot," *Proc. 12th Int. Symp. Industrial Robots*, Paris, June 9-11, pp. 331-340.

Selspot Systems Ltd., 1988, "Selspot II Product Literature," 1233 Chicago Rd., Troy, MI, 48083.

Shamma, J.S., and Whitney, D.E., 1987, "A method for inverse robot calibration," *ASME J. Dynamic Systems, Meas., Control*, 109, pp. 36-43.

Shiu, Y.C., and Ahmad, S., 1987, "Finding the mounting position of a sensor by solving homogeneous transform equation of the form AX=XB," *Proc. IEEE Int. Conf. Robotics and Automation*, Raleigh, NC, March 31-April 3, pp. 1666-1671.

Stone, H.W., 1987, *Kinematic Modeling, Identification, and Control of Robotic Manipulators*, Boston, Kluwer Academic Publ..

Stone, H.W., and Sanderson, A.C., 1987, "A prototype arm signature identification system," *Proc. IEEE Int. Conf. Robotics and Automation*, Raleigh, NC, March 31-April 3, pp. 175-182.

Stone, H.W., and Sanderson, A.C., 1988, "Statistical performance evaluation of the S-Model arm signature identification technique," *Proc. IEEE Int. Conf. Robotics and Automation*, Philadelphia, April 24-29, pp. 939-948.

Stone, H.W., Sanderson, A.C., and Neuman, C.P., 1986, "Arm signature identification," *Proc. IEEE Int. Conf. Robotics and Automation*, San Francisco, April 7-10, pp. 41-48.

Sturm, A.J., and Lee, M.Y., 1988, "A contribution to robot tooltip spatial accuracy identification," *Proc. IEEE Int. Conf. Robotics and Automation*, Philadelphia, April 24-29, pp. 1735-1738.

Sugimoto, K., and Okada, T., 1985, "Compensation of positioning errors caused by geometric deviations in robot system," *Robotics Research: The Second International Symposium*, edited by H. Hanafusa and H. Inoue, Cambridge, Mass., MIT Press, pp. 231-236.

Tsai, R.Y., and Lenz, R., 1988a, "A new technique for fully automated and efficient 3D robotics hand-eye calibration," *Robotics Research: The Fourth International Symposium*, edited by Bolles, R., and Roth, B., Cambridge, MA, MIT Press, pp. 287-298.

Tsai, R.Y., and Lenz, R., 1988b, "Real time versatile robotics hand/eye calibration using 3D machine vision," *Proc. IEEE Int. Conf. Robotics and Automation*, Philadelphia, April 24-29, pp. 554-561.

Vaishnav, R.N., and Magrab, E.B., 1987, "A general procedure to evaluate robot positioning errors," *Int. J. Robotics Research*, 6 no. 1, pp. 59-74.

Veitschegger, W.K., and Wu, C.-H., 1986, "Robot accuracy analysis based on kinematics," *IEEE J. Robotics and Automation*, RA-2, pp. 171-179.

Veitschegger, W.K., and Wu, C.-H., 1987, "A method for calibrating and compensating robot kinematic errors," *Proc. IEEE Int.*

Conf. Robotics and Automation, Raleigh, NC, March 31-April 3, pp. 39-44.

Waldron, K.J., and Kumar, A., 1979, "Development of a theory of errors for manipulators," *Proc. 5th World Congress on Theory of Machines and Mechanisms*, pp. 821-826.

Warnecke, H.J., and Brodbeck, B., 1977, "Test stand for industrial robots," *Proc. 7th Int. Symp. on Industrial Robots*, Tokyo, Oct. 19-21, pp. 443-451.

Whitney, D.E., and Junkel, E.F., 1982, "Applying stochastic control theory to robot sensing, teaching, and long-term control," *Proc. 12th Int. Symp. Industrial Robots*, Paris, June 9-11, pp. 445-457.

Whitney, D.E., Lozinski, C.A., and Rourke, J.M., 1984, "Industrial robot calibration method and results," *Proc. 2nd ASME Conf. on Computers in Engineering*, Las Vegas.

Whitney, D.E., Lozinski, C.A., and Rourke, J.M., 1986, "Industrial robot forward calibration method and results," *J. Dynamic Systems, Meas., Control*, 108, pp. 1-8.

Whitney, D.E., and Shamma, J.S., 1986, "Comments on "An exact kinematic model of the PUMA 600 manipulator"," *IEEE Trans. Systems, Man, Cybern.*, SMC-16, pp. 182-184.

Wu, C-H., 1983, "The kinematic error model for the design of robot manipulators," *Proc. American Control Conf.*, San Francisco, June 22-24, pp. 497-502.

Wu, C.-H., 1984, "A kinematic CAD tool for the design and control of a robot manipulator," *Int. J. Robotics Research*, 3 no. 1, pp. 58-67.

Wu, C.-H., Ho, J., and Young, K.Y., 1988, "Design of robot accuracy compensator after calibration," *Proc. IEEE Int. Conf. Robotics and Automation*, Philadelphia, April 24-29, pp. 780-785.

Wu, C.-H., and Lee, C.C., 1984, "On an accuracy problem of robot manipulators," *Proc. 23rd IEEE Conf. Decision and Control*, Las Vegas, Dec. 12-14, pp. 1636-1637.

Wu, C.-H., and Lee, C.C., 1985, "Estimation of the accuracy of a robot manipulator," *IEEE Trans. Automatic Control*, AC-30, pp. 304-306.

Zhen, H., 1985, "Error analysis of robot manipulators and error transmission functions," *Proc. 15th Int. Symp. on Industrial Robots*, Tokyo, pp. 873-878.

Ziegert, J., and Datseris, P., 1988, "Basic considerations for robot calibration," *Proc. IEEE Int. Conf. Robotics and Automation*, Philadelphia, April 24-29, pp. 932-938.

A REVIEW OF RESEARCH ON WALKING VEHICLES

Vijay R. Kumar
Department of Mechanical Engineering and Applied Mechanics
University of Pennsylvania
Philadelphia, PA 19104

Kenneth J. Waldron
Department of Mechanical Engineering
The Ohio State University
Columbus, OH 43210

ABSTRACT

Research into artificial legged locomotion systems with particular reference to statically stable walking vehicles is reviewed. The primary technical problem areas are briefly discussed and the principal references in the literature are cited.

1. INTRODUCTION

Mobility in Robots
Many robotic applications today require or would benefit from mobility. However, most present day robots are stationary or have, at most, limited motion along guideways in one or two directions and operate only on smooth level floors. This is because, until recently, it was not economical or practical to have mobile robots, particularly in situations involving uneven terrain. With recent advances in robotics and allied sciences (especially the development of relatively inexpensive digital computers), more generalized mobile robots have become feasible and very attractive prospects for appropriate applications. Researchers have realized the enormous potential of having mobility in robots performing repetitive tasks in industrial environments and for transportation in hostile surroundings and mobile robotics has become a very active area of research.

Legged Locomotion
Legged locomotion systems offer advantages over wheeled or tracked systems in appropriate situations. Legged systems can provide mobility in environments which are cluttered with obstacles where wheeled or tracked vehicles could not. They can usually provide superior mobility in difficult terrain or soil conditions. Further, such systems are also inherently omni-directional, whereas wheeled and tracked systems are directional, two degree of freedom systems. Finally, a passive suspension, as used in conventional vehicles is subject to unpredictable dynamics which compromise its performance as an instrument platform. Fully terrain adaptive legged vehicles (McGhee 1984) can completely isolate the vehicle

body from small wavelength terrain irregularities, thereby providing a stable and predictable instrument platform.

As will be seen in the paper, there are significant research issues which are important, or unique to the design and operation of legged locomotion systems. These issues have been the focus of a very active research effort in recent years (Hirose 1984; Mosher 1969;Okhotsimski et al 1977; Orin 1982; Raibert and Sutherland 1982; Russell 1983; Waldron and McGhee 1986).

Objective and Scope

In this paper, some of the recent advances in legged locomotion are described, and a critical assessment of the state of the art in walking vehicles is presented. Component design and scaling, prime-movers, actuators and transmissions, and computer control have been discussed exhaustively elsewhere (McGhee 1984; Waldron et al 1984). The main thrust of this paper is on motion planning, control and coordination. Also, for the most part, we limit our discussion to statically stable vehicles. An excellent analysis and review of dynamically stable vehicles is presented by Raibert (1985). Biped locomotion which is largely dynamically stable is also outside the scope of this paper. The interested reader can refer to (Kato et al 1983) and (Muira and Shiyoma 1985) as a starting point.

In this paper, we present a review of those walking machines that have been built so far, which are of most interest scientically, and the key research contributions that led to the design and fabrication of these systems in Section 2. The present state of the art is briefly discussed. Section 3 focusses on statically stable gaits. A brief history of research in this area is presented. The gait selection problem, practical issues in implementation, and automatically adapting the gait to the environment are of interest here. In Section 4, we discuss the kinematic and dynamic models that have been used for walking vehicles. From a kinematic point of view, the actuation scheme is a parallel one. Further, typically, such systems are redundant in actuation. The problem of allocating the load on the vehicle between the actuators (and feet) is analyzed and progress in this area is reviewed. In Section 5, we elaborate on the problem of motion planning for vehicles in unstructured environments, especially, uneven terrain. Finally, the research problems that need to be investigated and future challenges are described.

2. REVIEW OF WALKING ROBOTS

Walking machines possess an immense potential for rough-terrain locomotion, and several articulated legged vehicles have been built in the past two decades to demonstrate this potential (Hirose 1984; Mosher 1969; Okhotsimski et al 1977; Orin 1982; Raibert and Sutherland 1983; Russell 1983; Waldron and McGhee 1986). This area of science is relatively new and until now, most of the research has been geared to understanding the mechanics of locomotion and control and coordination of articulated limbs. Some of the significant projects in the area of legged locomotion, both past and current, are described in this section.

The first legged machine completely controlled by computers was built at the University of South Californa (McGhee and Frank 1968). The

Phoney Pony was a four legged machine which had two degrees of freedom in each leg which were coordinated by a computer. It was inspired by a machine built by Mosher (1969) at General Electric Corporation. The GE quadruped was a 3000 lb. machine in which the operator controlled the twelve degrees of freedom in the four legs by his hands and feet through a master-slave type valve controlled hydraulic servo system. No computer coordination was used. This proved to be extremely cumbersome and it demonstrated the necessity of automation in the coordination of the legs in a walking machine. In 1969, the largest off-road vehicle, *Big-Muskie*, a 15000 ton walking dragline was built for strip shining by Bucyrus-Erie Company (Sitek 1976). Work on analog computer controlled bipeds was pioneered in Yugoslavia (Vukobratovic, Frank and Juricic 1970) and in Japan (Kato *et al* 1983). At this stage, walking machines required a lot of operator participation but several proofs of the concept now existed.

In the USSR, significant theoretical work on locomotion and gaits was done by Bessonov and Umnov (1973). A few years later, Okhotsimski *et al* (1977) developed a six-legged walking vehicle. Unfortunately, very few details on this machine are available to us. It was thought that the number of legs reflected a tradeoff between stability and complexity and six was judged to be an optimal number. Although the machine was powered externally it was under complete computer control and was also equipped with a scanning range-finder. This machine was the first to demonstrate application of artificial intelligence concepts to legged locomotion systems. The legs were similar to those of insects and had a total of 18 degrees of freedom. The intelligence structure was hierarchical and the decision-making process was organized into a situation-action dictionary. Perception and motion planning algorithms were developed for locomotion in unstructured terrains. It seems to have been successful in adapting its gait to the surroundings to some extent. The problem of force allocation between the legs of the machine was also studied.

The *OSU Hexapod* was a 300 lb. six legged machine built at the Ohio State University (McGhee 1977; Orin 1982). It had "insect type" legs similar to its Russian counterpart and was driven by electric motors controlled by SCRs. It was powered externally and controlled through a PDP-11/70 computer. The Hexapod was equipped with force sensors at the feet for force control and gyroscopes for attitude control. Work on the design of walking machines was accompanied by research on intelligent control schemes and terrain-adaptive locomotion. Among several gait experiments, a *follow-the-leader* gait was implemented on the machine (Özgüner *et al* 1984). The operator used a hand-held laser to designate candidate footholds, which were used for the front legs if the computer found them acceptable. Two CID television cameras were used to detect the laser beam. The four legs behind the two fore legs stepped on the footprints selected for the fore legs.

The *PVII* was a small quadruped walking machine constructed at the Tokyo Institute of Technology (Hirose and Umetani 1980). It weighed only 10 kg. and had a 10 watt power consumption. It was powered externally and connected via an umblical cord to a minicomputer. A hierarchical control system was implemented whose functions included navigation, planning, gait control, posture regulation and the generation of commands for the servomechanisms. The machine demonstrated the ability to climb stairs and to avoid small obstacles through tactile sensing at the feet. More recently, another quadruped, *TITAN III* has been developed (Hirose *et al* 1985). It is a 320 watt machine which is also equipped with wheels as alternative locomotion elements. The emphasis in design has been on improving the efficiency of locomotion.

In 1983, Ivan Sutherland built the first six legged machine which was entirely self-contained and had an on-board computer and power supply at Carnegie Mellon University (Sutherland and Ullner 1984) . This also had a total of 18 degrees of freedom which were hydraulically actuated. The hydraulic circuits were designed to make the legs move "usefully" without being digitally controlled by a microprocessor. The operator controlled the speed, direction and attitude of the body. The questions of optimal parameters for gaits or foothold selection for navigation were not addressed. About the same time, also at CMU, Raibert built a one-legged hopping machine (Raibert and Sutherland 1983). It was the first successful statically unstable but dynamically stable machine. The project was a study in dynamic balancing and height and attitude control through hopping. Similar efforts later led to the design and fabrication of a four-legged machine based on the same principle (Raibert 1986; Raibert *et al* 1986).

The ODEX is a six-legged walking machine with an axisymmetric leg configuration built by Odetics Inc. in 1983 (Russell 1983). It had an unprecedented strength to weight ratio (5.6 when stationary and 2.3 while walking) and reasonable agility. It had eighteen degrees of freedom too but proved to be very easy to control. The mode of control was teleoperator-like and the vehicle received commands through a joystick via radio telemetry. The actions of the legs were coordinated by on-board computers but the operator used the camera system to view the surroundings and appropriately direct ODEX. ODEX uses a tripod gait used by many six-legged arthropods. A newer version of the machine is being evaluated for maintenance tasks in nuclear power plants.

Gradually, with an improvement in the understanding of the mechanics of locomotion, research on planning and coordination of the legs began to gain momentum. Kessis *et al* (1985) reported a four-level control architecture for an autonomous six-legged hexapod built at the University of Paris in 1980. The lowest level (leg level) involved the control of individual legs using a variable leg compliance to adapt to uneven terrain. The second level (gait level) generated gaits according to the commands from level three (plan interpreter). Level four (planner) coped with the perception and modeling of the universe and planned actions according to the situation encountered. The terrain model was built by a rule based production system and path planning used the A^* algorithm in a 3-dimensional environment. In Japan, adaptive gait control strategies to negotiate forbidden footholds have become the subject of active research (Hirose 1984). Similar efforts in the USA have lead to the development of algorithms for foothold selection and gait control for walking on unstructured terrain (McGhee and Iswandhi 1979; Kwak 1986; Kumar 1987; Qui and Song 1988).

Recently, a research effort at the Ohio State University has resulted in the successful design, fabrication and testing of a six-legged vehicle called the Adaptive Suspension Vehicle (ASV) (Waldron *et al* 1984a; Waldron *et al*

1986; Song and Waldron 1988). The ASV is designed to carry an operator but it may eventually operate as a completely autonomous unmanned system. It is a proof-of-concept prototype of a legged vehicle designed to operate in rough terrain that is not navigable by conventional vehicles (Waldron and McGhee 1986). It is 3.3 meters (10.9 feet) high and weighs about 3200 kg (7000 lb.). It presently operates in a supervisory control mode. To this end, it possesses over 100 sensors, 17 onboard single board computers and a 900 c.c. motorcycle engine rated at 50 kW (70 hp) continuous output. It has three actuators on each of the six legs thus providing a total of 18 degrees of freedom. The 18 degrees of freedom are hydraulically actuated through a hydrostatic configuration. The engine is coupled through a clutch to a 0.25 kW/hr flywheel which smooths out the fluctuating power requirements of the pumps.

The ASV senses over 100 control variables. The most important sensor for motion planning is an optical scanning rangefinder which is a phase modulated, continuous wave ranging system with a range of approximately 30 feet and a resolution of 6 inches at maximum range (Zuk and Dell'Eva 1983; Klein *et al* 1987). It has a field of view of 40 degrees on either side of the body longitudinal axis and from 15 to 75 degrees below the horizontal. A inertial sensor package consisting of a vertical gyroscope, rate gyroscopes for the pitch, roll and yaw axes, and three linear accelerometers provide information to determine body velocity and position. Leg position feedback is used from the legs in the support phase for the purpose of correcting for gyro and integration drift in the inertial reference system. Thus, there is considerable scope for sensor cross checking and error detection. The leg control system is based on a force control scheme, when the leg is on the ground, and on a position control scheme, when the leg is in transfer phase. Thus the position, velocity and pressure difference across each of the eighteen hydraulic actuators are monitored during operation.

The ASV, unlike its predecessors is completely computer controlled and mechanically autonomous. The operator performs the function of path selection and specifies the linear velocities of the vehicle in the fore-aft and lateral directions, and the yaw velocity. The roll and pitch rates and the velocity in the vertical direction are automatically regulated by the guidance system. In addition to path selection, the operator may also suppy important gait parameters such as stride length and duty factor, and set the mode of control (cruise/dash, terrain-following or large obstacle mode).

In conclusion, the walking machines that have been built so far, have been bulky, heavy, slow and awkward. Consequently, they have not been useful as vehicles or even prototypes. However, as laboratory vehicles, they have served as proofs of concept. In particular, they have demonstrated that active coordination of mechanical articulated limbs in order to drive a vehicle is a very practical idea. The need for intelligent control to minimize human participation in coordinating the multiple degrees of freedom has been underscored. These efforts have spurred research on biological systems to better understand the neural control of locomotion and has bridged the gap between work on animal walking and robot walking.

3. GAITS

Introduction
A vast variety of examples of legged locomotion can be found in nature. Several studies of animal gaits have been undertaken with a view to discovering strategies for control and coordination in locomotion. These studies have lead to the development of a mathematical theory for gait analysis. Very briefly, the history of research in this area, the mathematical models, and the application of the theory to gaits for walking robots is discussed here.

Gait Stability
Broadly speaking, animal locomotion can be classified into two categories. The first type is the one exhibited by insects. Insects are arthropods and have a hard exoskeletal system with jointed limbs. They use their legs as struts and levers and the legs must always support the body during walking, in addition to providing propulsion. In other words, the metachronal or sequential pattern of steps must ensure static stability. The vertical projection of the center of gravity must therefore always be within the support pattern (the two dimensional convex polygon formed by the contact points). This kind of locomotion has been described as *crawling* and the legs have to provide at least a tripod of support at all times.

Another kind of locomotion may be observed in humans, horses, dogs, cheetahs and kangaroos which have a more flexible structure. These animals require dynamic balance, which is a less stringent restriction on the posture and gait of the animal. The animal may not be in static equilibrium; to the contrary, there may be periods of time when none of the support legs are on the ground, as is observed in trotting horses, running humans and, of course, hopping kangaroos.

Until now, most efforts to build dynamically balanced robots have been confined to bipeds (Kato *et al* 1983; Muira and Shimoyama1985) and hopping machines (Raibert 1985). This is because the complexities of the locomotion system in biological creatures have prevented proper understanding of the involved mechanics and controls. Also, the present state of the art in digital computers has allowed the implementation of only simplified dynamic models for walking machines. Specifically, in these systems, pertubations in body attitude and altitude and the corresponding rates are corrected by small changes in the stride to produce a limit cycle stability. Coordination is much more complex as compared to statically stable machines. However, grouping of legs permits extension of control strategies developed for one or two legs to four legged dynamic locomotion (Raibert *et al* 1986). As mentioned earlier, a detailed critique of such machines is outside the scope of this paper. Further work in this direction is detailed in (Glower and Özgüner 1986; Goldberg and Raibert 1987; Agrawal and Waldron 1988).

The present generation of walking machines almost exclusively emulates the mechanism of walking in insects. Control in such machines is obviously simpler, and is well within the reach of modern technology. Also, the statically stable crawl typified by insects is better suited to heavy machines with rigid structures like the present day machines. A brief account of the evolution of a comprehensive theory for gait analysis for statically stable systems follows.

Biological Systems
A vast variety of examples of legged locomotion can be found in nature and many quantitative studies of animal gaits have been undertaken with a view to discovering strategies for control and coordination in locomotion. See for example, (Alexander 1984; Delcomyn 1981; Gambaryan 1974; Gray 1968; Hildebrand 1960; Pandy *et al* 1988; Pearson and Franklin 1984). The earliest studies of gaits date back to 1872 when Muybridge used successive photographs to study the locomotion of animals (Muybridge 1957) and later human locomotion (Muybridge 1955). However, Hildebrand (1965) was probably the first researcher to analyze gaits quantitatively. Since then, many reports have been published on this subject (Alexander 1984; Delcomyn 1981; Gambaryan 1974; Gray 1968) motivated in some cases by an interest in legged robots (Pandy *et al* 1988; Pearson and Franklin 1984).

Wilson's report (1966) on insect gaits is particularly informative. He developed a model for insect gaits by proposing a few simple rules. A wave of protraction runs from posterior to anterior, and *contralateral* (on opposite sides of the body) legs of the same segment alternate in phase. The protraction time is constant and the retraction time is varied to control the frequency. Further, the intervals between the steps of the hind and middle *ipsilateral* (on the same side of the body) legs, and those between the middle and the fore ipsilateral legs are constant. Though these rules are not observed in all species (Wilson 1966, Wilson 1976), most of these rules have been validated qualitatively with a variety of insects when walking on smooth horizontal surfaces. The classical alternating *tripod* gait exhibited by fast moving locusts and cockroaches is a good example (Delcomyn 1981). More recently, a cinematic analysis of locusts undertaken to study locomotion characteristics on uneven terrain has been described (Pearson and Franklin 1984). Studies by Cruse (1976) on stick insects have shown that the posterior legs are placed close to the support sites of the immediately anterior ipsilateral legs. This *follow-the-leader* behavior was observed in horses by Hildebrand and in domestic (Nubian) goats (Pandy *et al* 1988) and may be even a general strategy for walking on uneven terrain.

Walking Vehicles
Tomovic (1961) and Hildebrand (1965) were the first to study gaits quantitatively. Hildebrand developed the concepts of a *gait formula* and a *gait diagram* to describe symmetric gaits of horses (see (Song and Waldron 1988) for definitions of new terms). McGhee (1968) started the development of a general mathematical theory of locomotion based on a finite state concept. A leg was defined as a sequential machine with an output state 1 representing the support phase, where the leg is in contact with the supporting surface, and an output state 0 representing the transfer phase. In physiology, the support phase is the period of retraction and the transfer phase is the period of protraction. McGhee and his coworkers developed several mathematical tools such as *event sequences, gait formulae and gait matrices* to study stepping patterns (see McGhee and Frank (1968), McGhee and Jain (1972), and McGhee(1984)).

The main emphasis of investigations on statically stable gaits was the relationship between the stepping patterns and the static stability engendered by the gait. To this end, McGhee and Frank (1968) defined the *static stability margin* for gaits as a measure of the static stability of gaits (see Figure 1). A computationally more tractable measure called the *longitudinal stability margin*, as shown in Figure 1 , was also proposed by them. They identified the statically stable gaits for a quadruped as creeping gaits, of which the regular crawl gait (crawl gaits were defined by Hildebrand (1965)) was shown to be the most optimal in terms of the longitudinal stability margin. McGhee and Jain (1972) attempted to explain the bias shown by animals towards certain gaits through a characteristic called *regular realizability*.

The study of gaits has led to the definition and classification of different gaits. In particular, *periodic* and *regular* gaits were found to possess interesting properties. In a periodic gait, it is sufficient to describe one locomotion cycle in order to specify the gait. In a regular gait, the fraction of time in a locomotion cycle that a leg spends in support phase, or the *duty factor*, is the same for all legs. Other important quantities are the *time period* of a cycle, and the *stroke*, or the distance through which a foot translates relative to the vehicle body through the locomotion cycle. Bessenov and Umnov (1973) used numerical experimentation for hexapods to demonstrate that the optimal gaits are regular and *symmetric*. A gait is symmetric if the motion of contralateral adjacent legs are exactly half a cycle out of phase. These optimal gaits were described as "wavy" gaits. This agreed with Wilson's observation of a metachronal pattern of steps in insect walking. The *wave gait* is a symmetric, periodic and regular gait in which any adjacent ipsilateral pair of legs differ in phase by the same amount. The regular crawl gait, which was shown to be optimal for quadrupeds (McGhee and Frank 1968) could be classified as a wave gait. It was discovered through numerical experimentation that for a given duty factor, wave gaits were optimally stable gaits for hexapods as well as quadrupeds. More recently, Song and Waldron (1987) carried out a detailed analysis and classification of gaits, in which the wave gait was analytically proven to be the optimal gait for hexapods. The variation in load carying capacity with load and speed for walking vehicles was investigated by Huang and Waldron (1987) for wave gaits.A more extensive treatment of statically stable gaits can be found in the work reported in references (Kumar and Waldron 1988; Song and Waldron 1988).

Statically stable gaits and stepping patterns on even terrain are now extensively researched and are well-understood. The focus of research on gaits has shifted from optimization based on static stability to more practical considerations such as foothold selection and motion planning (Lee and Orin 1988). In particular, terrain - adaptive walking has become an extremely important area of research. If an accurate description of the terrain geometry is available, a pre-defined sequence of steps can be identified for traversing a path (Qui and Song 1988). One such example is the *large obstacle gait* (Song and Waldron 1988). However this is clearly an ideal situation. In a practical case, an intelligent system is required for terrain - adaptive locomotion (Kumar and Waldron 1989).

The follow-the-leader gait was studied by Özgüner *et al* (1984) as a method of decreasing the burden of foothold selection on the operator. It was demonstrated on the OSU Hexapod. The need for a gait which could be automatically adapted to varying terrain conditions lead to the development of aperiodic gaits, for which the large obstacle gait is a good example. An aperiodic gait called the *free gait*, was proposed by Kugushev and Jaroshevskij (1975) to address the problem of automatic foothold selection in a real world situation. It was later modified by McGhee and Iswandhi (1979) for the OSU Hexapod. The free gait algorithm sought to maximize the stability as well as the "availability" of legs by appropriately specifying the footholds in space as well as time for each leg. The utility of the free gait was subsequently demonstrated through simulations of the ASV (Kwak 1986) and on the actual machine (Patterson *et al* 1983; Klein *et al* 1987). However, the free gait required the terrain to be discretized into cells each of which had to be classified as GO or NO-GO. Further, it did not address the problem of locomotion on rough terrain. Another problem was that the algorithm involved a heavy penalty in terms of computational load. An adaptive gait controller that was designed to overcome these shortcomings was proposed by Kumar and Waldron (1989) for motion planning on uneven terrain. A periodic, optimally stable gait (that is, the wave gait) was selected for locomotion, but transitory aperiodicity and asymmetry in the support patterns were allowed to enable the system to circumvent variants in the terrain relief. Once more static stability was an important criterion, but the measures of static stability allowed for deviations from ideal conditions of motion along the axis of symmetry and a two-dimensional terrain (Kumar and Waldron 1988a).

Though work on terrain-adaptive walking has been reported, all of the past work has involved numerical descriptions of the terrain-vehicle systems accompanied by research on quantitative analysis of walking and numerical measures of kinematic optimality of gaits. The possibility of symbolic descriptions of the task, vehicle-terrain interaction and control of the task must also be investigated. Such techniques are perhaps better suited to terrain description and adaptive gait control and might prove to more powerful.

4. DYNAMICS AND CONTROL

Introduction
In this section, we describe the dynamics of the vehicle-terrain system and the problems associated with the control of walking robots. With few

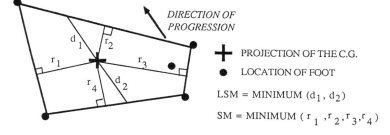

Figure 1   Measures of Stability for Statically Stable Gaits

exceptions (Glower and Özguner 1986; Shih and Frank 1987), quasi-static models of the system have been used for analysis (McGhee and Orin 1976, Kumar and Waldron 1988b). In other words, the legs have been assumed to massless compared to the vehicle body. This was a reasonable assumption for statically stable vehicles of the type being currently built. Further, the legs as well as the body were assumed to be rigid.

We first discuss the kinematics of walking vehicles. Next, a discussion of the equations of motion and the nature of static indeterminacy in the problem is presented. A critique of past and current research in the areas of trajectory planning and coordination follows. A parallel is drawn between walking vehicles and other robotic systems with parallelism in actuation, and the common features in the problem of coordination of such systems are described.

Kinematics of Legged Systems

Consider an earth-fixed reference frame $\mathbf{E}$, and a body-fixed reference frame $\mathbf{B}$ with the origin at the center of gravity, O. Let the transformation from $\mathbf{E}$ to $\mathbf{B}$ be denoted by $^B\Gamma_E$ and its inverse by $^E\Gamma_B$. $\mathbf{v}_O$ is the velocity of the center of gravity of the vehicle and ω_B is the angular velocity of the vehicle. The position of a point P on a foot is denoted by $\mathbf{r}_P$ and the velocity $\mathbf{v}_P$. Quantities measured relative to $\mathbf{B}$ possess a leading superscipt, B, and position or velocity vectors relative to O are denoted by a trailing subscript P/O.

The position of a point P on a foot, at the center of contact between the foot and the ground is given by

$$\mathbf{r}_P = \mathbf{r}_O + \mathbf{r}_{P/O}$$
$$= \mathbf{r}_O + {}^B\mathbf{r}_{P/O}$$

The velocity of P, $\mathbf{v}_P$ can be found by differentiating this equation:

$$\mathbf{v}_P = \mathbf{v}_O + \mathbf{v}_{P/O}$$
$$= \mathbf{v}_O + {}^B\mathbf{v}_{P/O} + \omega_B \times {}^B\mathbf{r}_{P/O}$$

Therefore,

$$^B\mathbf{v}_{P/O} = \mathbf{v}_P - \mathbf{v}_O - \omega_B \times {}^B\mathbf{r}_{P/O}$$

If the leg were in transfer phase, its desired absolute velocity, $\mathbf{v}_P$, would be computed independent of the kinematic constraints induced by the vehicle-terrain interaction, by a trajectory planning algorithm as detailed by Orin (1982). The above equation would then yield $^B\mathbf{v}_{P/O}$. On the other hand, if P belongs to a leg in support phase, $\mathbf{v}_P$ must equal zero, since P must be at rest with respect to the ground. In such a situation:

$$^B\mathbf{v}_{P/O} = -\mathbf{v}_O - \omega_B \times {}^B\mathbf{r}_{P/O}$$

Thus the desired leg velocity with respect to the vehicle body can be computed in either case. Once the desired foot velocity is known, the desired joint velocities can be computed quite simply by inverting the "leg Jacobian". This is analogous to the resolved motion rate control algorithms which involves the inversion of the "arm Jacobian" to determine the desired joint rates from the desired velocity of the hand. Decoupling the three translational degrees of freedom for each leg by using a pantograph mechanism results in efficient rate decomposition algorithms (Waldron et al 1984; Hirose 1984).

If the angular velocity of the vehicle is zero, the foot contacts of all the legs in support phase have the same velocity with respect to the body. This velocity is equal and opposite to the vehicle velocity. In this simple case, all the foot velocities are related to the gait parameters, β (duty factor), R (stroke), and T (cycle time period):

$$v_{P/O} = \frac{R}{\beta T}$$

The legs stroke in a direction parallel to the relative velocity vector which is, in general, arbitrarily directed. If ω_B is not zero, $v_{P/O}$ is constantly changing in direction and magnitude and is different for different legs. Notice also that even if the vehicle body linear and angular velocity are constant, the relative leg velocity varies with position. In fact, no leg velocity may equal $v_{P/O}$ which makes the above equation invalid. However, an instantaneous stroke may be defined based on the instantaneous relative velocity ($v_{Pi/O}$, for a point P_i on the ith foot) for each leg to assist gait planning (Kumar and Waldron 1989).

Dynamics of Legged Vehicles
Based on the assumptions outlined earlier, the equations of motion assume a very simple form. The desired accelerations and current velocities are known in a typical case. Let the resultant forces and moments of the foot forces (reaction forces exerted by the ground on the feet) be denoted by **R** and **M** respectively in the reference frame **B**. If ω and **v** are the angular velocity and the velocity of the center of gravity of the vehicle and **H** is the angular momentum of the body in the body fixed frame, and **g** is the acceleration due to gravity,

$$\mathbf{R} = \frac{d}{dt}(m\mathbf{v}) + \omega \times m\mathbf{v} - {}^B\Gamma_E \, m\mathbf{g}$$

$$\mathbf{M} = \frac{d}{dt}(\mathbf{H}) + \omega \times \mathbf{H}$$

It is convenient to model the interaction of the feet with the ground by frictional point contact which was used by Salisbury and Roth (1983) for their analysis of multifingered grippers. According to the point contact model, the contact interaction can be represented by a pure force through an appropriate point, which may be called the center of contact. Note this center of contact may not be the geometric center of contact of the feet. However, this model does not take into account the frictional characteristics of the foot-terrain interface. A foot force is valid only if the force does not tend to pull the terrain (the component of force normal to the terrain can only be along one of two directions). In addition, the foot force must satisfy the appropriate laws of friction. If Coulomb's model of friction is accepted, the angle formed by the components of the foot force normal and tangential to the ground must be within a certain limit so that the foot does not slip (McGhee and Orin 1976; Orin and Oh 1981).

If $\mathbf{F}_i$ denotes the contact forces at the ith contact, whose center is at $\mathbf{r}_i$ (x_i, y_i, z_i) in the reference frame **B**, the following equations must be satisfied:

$$\sum_{i=1}^{i} \mathbf{F}_i = \mathbf{R} \qquad \sum_{i=1}^{n} (\mathbf{r}_i \times \mathbf{F}_i) = \mathbf{M}$$

where n is the number of feet on the ground. The above equations represent a system of linear equations which must be solved for the F_i. This is because, to produce a desired acceleration at the current velocity the foot forces given by the expressions above must be exerted on the body. The actual implementation can use a variety of schemes to control the support legs such as force control or active compliance control schemes (Klein *et al* 1983). Nevertheless, force set-points do need to be supplied to such control systems. Since the foot forces are not completely determined by the above equation, the problem is statically indeterminate. The determination of these foot forces subject to the frictional constraints is an important problem in coordination and is discussed next.

Coordination

Algorithms which decompose the external (including inertial) force system into foot forces are called coordination algorithms. The word coordination is used to mean the level of control at which the positions, rates, forces, or torques to be commanded from the actuator servos are computed, based on a relatively small number of command inputs from a supervisory control system. The corresponding analog in serial chain manipulators can be found in algorithms used to decompose desired end-effector rates into joint rates (Paul 1981).

The kinematics of a walking vehicle involves simple closed chains and multiple frictional contacts between the actively coordinated articulated legs and the passive terrain. Multifingered grippers and multiple cooperating arms belong to the same class of systems. Walking machines, along with multifingered grippers and multiarm systems can also be treated as parallel manipulation systems. Redundancy in manipulation systems with parallelism has been studied with reference to multiple arms (Zheng and Luh 1988), multifingered grippers (Abel *et al* 1985; Holzmann and McCarthy 1985; Hollerbach and Narasimhan 1986; Salisbury and Roth 1983; Kerr and Roth 1986; Kumar and Waldron 1988c) and walking vehicles (Orin and Oh 1981; Klein and Chung 1987; Gardner, Srinivasan and Waldron 1988; Kumar and Waldron 1988b). *Kinematic redundancy* in serial manipulators, in which the number of actuators is greater than the dimension of the task space, has been shown to be mathematically isomorphic to *static indeterminacy* (redundancy) in parallel systems such as walking robots (Waldron and Hunt 1987).

Optimization is a logical choice for analysis of under-determined systems and there are several mathematical techniques that can be employed for this purpose. McGhee and Orin (1976) combated the problem of static indeterminacy by optimizing the foot forces to obtain a minimum energy consumption. Linear programming was also used by Orin and Oh (1981). However, simulations showed that the associated computational load was unacceptable for real-time performance. In addition, the solution may exhibit unacceptable chatter behavior (Klein and Chung 1987). Methods based on the pseudo inverse have been described by Klein, Olson and Pugh (1983). Experiments on the OSU Hexapod demonstrated improved stability and better performance on even terrain. However, these methods fail to yield satisfactory solutions which satisfy the frictional constraints unless an iterative procedure is employed (Waldron 1986; Song and Waldron 1988). This problem becomes worse in uneven terrain. A comparison of different methods for walking vehicles is presented by Kumar and Waldron (1988b). Further, they also suggest using a compliant model for the legs instead of assuming that they are rigid. The displacement equations necessitated by geometric compatibility automatically resolve the indeterminacy in the problem. The basic idea here is that the leg compliances can be selected so that legs with poorer footholds are made softer, while legs with good footholds are rigid. Clearly the desired compliance can be electronically simulated using feedback. Though the resulting force distribution does not completely satisfy the frictional constraints, the legs in which these constraints are violated support a very small fraction of the load and therefore do not jeopardize the stability of the vehicle.

Another problem arises at the planning level of control. In order to predict whether or not a given (planned) configuration is stable, it is essential to determine if a valid set of foot (finger) forces (that does not violate the frictional constraints) can be commanded to maintain equilibrium. However, in this case, it is sufficient to be able to determine if *a* valid solution exists. In other words, it is not necessary to find an optimal solution to the problem. Kumar and Waldron (1988b) used the first phase of the Simplex Method for linear programming to this end. Recall that Phase I of the method merely seeks a feasible solution while Phase II optimizes the feasible solution (Gass 1985). The number of computations is considerably reduced. Further, the planning algorithms are not required to be implemented in real-time.Thus, this approach would be adequate in terms of computing time as well as efficacy.

For real-time control, an efficient technique for obtaining a solution to the equations of motion while satisfying the friction angle constraints still remains to be found. Perhaps, the answer lies in a parallel computing scheme for linear programming. Further, it is necessary to perform a dynamic analysis for faster vehicles. This is especially true for heavy vehicles - even if the accelerations are small, the inertial forces would be large. The idea of varying leg compliances could still be implemented in the dynamic case. However, now the leg impedance or admittance must be specified and controlled.

5. MOTION PLANNING AND CONTROL

Machine Intelligence in Mobile Robots

In most of the mobile robotic systems developed so far, a common feature has been the recognition for the need of a hierarchical structure of intelligence and hierarchical decomposition of the task (of locomotion) and the model of the perceived world. The higher levels define tasks or subgoals for the lower levels and monitor their status. As we go up the hierarchy, on a time scale there is a decrease in the frequency of updating sensory information, and on the length scale, the world taken into consideration is larger but with fewer details. At each successive level down the hierarchy, there is a decrease in the generality and scope of the search and greater resolution. Four levels may be identified in the hierarchy - the *planner* or the route layout module, the *navigator* or the path selection module, the *pilot* or the guidance module and the *controller* (Isik and Meystel 1988). The *cartographer* is concerned with maintaining maps for the path selection and guidance modules at the required resolution and range and with appropriate detail.

Planner

The route layout is done at the highest level. It involves planning a route for the vehicle taking into account the general characteristics of the robot's ability to adapt to different terrains and surmount various obstacles. At this level, the basic element of locomotion (leg or wheel) is not of concern. In a completely autonomous system, it may be the only level which interacts with a human operator. This interaction may be limited to off-line specification of the task. The planner prescribes a set of subgoals for the next lower level. It works on a model of the world which extends typically to a hundred body lengths. Presently, there is little or no research in evidence in this regime of machine intelligence.

Navigator

The navigator is primarily concerned with path selection. It uses the terrain preview data and information from other sensors to chart a "best" course for several "vehicle body lengths" to realize the subgoal command from the upper level. Obstacle avoidance is an integral part of such a process. It in turn prescribes subgoals for the pilot level to meet the requirements of the selected path. It maintains a terrain map and a model of the world confined to a few (typically ten) body lengths which possesses a higher level of resolution than the planner's model.

Most of the work on autonomous navigation has been at the path selection level. It is specific to wheeled or walking vehicles only to the extent that the characteristics of the vehicle have to be known (dimensions of the body, size of the tire or foot, maximum stride length and so on). However most of the work reported in this field has been with reference to wheeled robots. This problem is similar to the problem of planning manipulator transfer movements without explicit programming of the motion (Lozano-Peréz 1983). Given the initial and final location (starting point and subgoal) of the mobile robot, the optimal path that circumvents obstacles has to be found. The concept of shrinking the robot to a single reference point while expanding the obstacle regions (Udupa 1977) has proved to be useful. Lozano-Peréz (1981) has formalized this approach with the concept of a configuration space. This involves approximating the obstacles by polyhedra. In two dimensions the shortest collision free path is composed of straight lines joining the origin to the destination through a set of vertices of obstacle polygons. A *V-graph* or visibility graph in which each link represents a straight line between two points which can "see" each other, can be built and a search routine is used to obtain the optimal path (Davis and Camacho 1984; Nillson 1980). A similar method was employed to navigate SHAKEY, an integrated mobile wheeled system built at the Stanford Research Institute. In 3-dimensions however, this method gets a little more complicated. Lozano-Peréz (1983) uses *slices* (a projection of any space into a lower dimension space) to overcome this problem.

Thompson (1977) describes a path planning module for the JPL rover. This approach avoids the construction of a V-graph for the whole space but builds (expands) the graph as and when needed. Similarly, Koch *et al* (1985) propose the concept of *sectors* (sectors of a circle containing the current position of the vehicle and the goal with the vehicle at the center) which yields a subset of the visibility graph. Thus they overcome the problem of combinatorial explosion.

Methods of representing the robot world have received considerable attention, since the actual search for a suitable path can be easily performed by standard algorithms like the A^* algorithm (Nilsson 1980). Brooks (1983a; 1983b) modeled the free space (space outside the obstacles) as a union of generalized cones. This eventually leads to a more efficient utilization of space. He used generalized cones to build a connectivity graph which was then the input to the path search routine. HILARE was a three wheeled robot built in France (Giralt *et al* 1979) whose model of the world was also represented by a connectivity graph. Typically, the nodes of a connectivity graph would be places or rooms and links, traversible boundaries. The rooms could be further decomposed into polygonal cells representing free space.

In legged locomotion, modeling of the environment typically involves the interpretation of terrain elevation data (Klein *et al* 1987). Two general approaches can be identified. The first method involves object identification or feature extraction (Poulos 1986). The environment is partitioned into regions corresponding to physical features. This is followed by a symbolic description of the regions and matching the descriptions with a preprogrammed set of object descriptions. This is a complex problem which has been studied with reference to structured environments with some success. However, the complexity greatly increases in an unstructured terrain and the required computations are prohibitively time consuming.

The lesser used region classification technique is based on inferring local topographic features for each *voxel* (VOLume piXEL) from the discrete elevation data, which are then used to build a symbolic model of the terrain. This alleviates problems of symbolic description and eliminates problems of matching descriptions to known patterns. Olivier and Özguner (1986) propose a similar method in which they characterize voxels by fitting local quadric surfaces through each voxel employing the least squares technique with the elevation data. The Hessian matrix and its eigenvalues are then used to classify each voxel as a peak, pit, ridge, ravine, flat area, safe hill, or an unsafe hill. Poulos (1986) pursued this method and demonstrated its feasiblity for real-time operations on a Symbolics® machine.

Richbourg *et al* (1987) and Ross and McGhee (1987) used a cost function for each voxel in conjunction with a *wavefront* method. The basic idea here is if the space is divided into regions each of which can be associated with a cost of traversal, the locally optimal path can be found by a combination of refraction and defraction optics (Snell's law). This analogy to optics is immediately seen if the cost rate is equated to the refractive index in optics.

Vegetative cover and varying soil properties still pose problems as elevation data only yields geometric information. Further, though a good optical terrain scanner can provide a geometric description of the immediate terrain, better sensors are required for global path selection. Another problem arises due to self-occlusion, in which the terrain is obscured from the sensors by local peaks and high-rises leading to uncertainty. Clearly, more work on reasoning and planning in such environments is needed.

The path selection algorithms must also be used at a lower level of guidance to enable fine tuning of the planned path. This is because the lower level would typically have a more accurate and detailed description of the terrain, and obstacles which were not evident at higher levels would become apparent at the lower levels. However, the same basic algorithms that are used for path selection can be used for such a local path modification strategy as long as higher resolution maps and models are available.

Pilot

The pilot plans a sequence of elementary acts of motion in space and time to generate a path between the goals prescribed by the path selection level. Minor deviations from the prescribed path may be tolerated to circumvent small obstacles. The pilot has a short term memory and its perception of the world is confined to one or two body lengths. This level may be virtually absent in wheeled locomotion where the motion is almost fully constrained or specified by the path selection level. However, a legged locomotion system is free to select footholds in space and time. That is, when and where it "samples" the terrain. Thus optimization of gait parameters, optimal foothold selection, dynamic balancing and attitude control of the vehicle body are tasks which must be performed at this level. The choice of legs as elements of locomotion introduces a new level of intelligence and a new degree of complexity into motion planning for the body. This level has also been described as the guidance module by some researchers. The pilot treats the leg as a finite state machine (McGhee 1968) and delegates the lower level task of planning trajectories for the legs and the associated problem of leg collision avoidance to the controller.

Thomson (1977) has identified a guidance module in wheeled vehicles which translates the planned path into a set of commands for the actuators and uses feedback to control the vehicle limiting the deviation from the prescribed path to a few evasive movements. However, this description would seem to encompass the function of the next lowest level (controller) too. In general, the pilot's function could be stated as being limited to walking vehicles.

The operation of the pilot has been more actively pursued for legged locomotion. In the four-level hierarchy of the University of Paris hexapod (Kessis *et al* 1985), the gait level and the plan interpreter would appear to constitute the guidance module. Hirose (1984) describes a gait control level partitioned into a local motion-trace generator (which responds to the commands of the navigation planning module) and an adaptive gait controller. For the ASV, the free gait has been used as an automatic heuristic technique to optimize the vehicle walk over unstructured terrain (McGhee and Iswandhi 1979) and the guidance algorithm is based on this free gait (Patterson, Reidy and Brownstein 1983). This enables automatic selection of footholds and gait parameters but the process is quite inefficient and slow. Current efforts are directed towards *supervisory* control of the ASV which is aimed at implementing the "horse-intelligence" in the "rider-horse-system" analogy. This will at least enable supervisory control in rough terrain with man providing the "rider-intelligence".

Kumar and Waldron (1989) classify the functions at this level into 6 categories. These are local path modification or local navigation, foothold selection, gait optimization, adaptive gait control, automatic body posture regulation and checking vehicle stability. They use a finite state approach to planning and use a numerical representation of the terrain vehicle system. In this study, algorithms based on heuristics are developed for foothold selection, posture regulation and gait control. The efficacy of these algorithms has been demonstrated through simulations (Kumar 1987).

Controller

The lowest level, the controller, is the only level which interacts directly with the actuators. It represents the "spinal" level associated with control of individual joints in natural systems and involves real-time servo control loops and sensory feedback at the actuator level. This is the only area in this hierarchy which is reasonably well researched and documented and is the foundation for all robotic technology. In legged locomotion systems, this also entails leg trajectory planning using proximity sensors, and actuation, and also incorporates "cerebellar" intelligence to a certain extent. The corresponding level in wheeled systems controls the actuation of the wheels. This level involves very little intelligence and it is also possible to identify the controller as a "plan execution" module and exclude it from this model of the intelligence structure. Since the technology involved is not unique to walking vehicles it has not been discussed in detail here. However, information about specific systems can be found elsewhere (Waldron *et al* 1984; Sutherland and Ullner 1984; Orin 1982; Hirose and Umetani 1980).

6. CONCLUDING REMARKS

In the foregoing, we have reviewed the principal problem areas associated with legged locomotion systems with particular reference to statically stable walking machines, and have cited many of the principal sources of information in the literature. The emphasis has, of course, been on scientific information. For this reason, references to some machines which have not been scientifically described have not been included.

Some of the problem areas cited above are unique to legged locomotion systems. More usually, problem areas have features in common with other types of robotic systems, although there may be differences in the emphasis. For this reason, research in legged locomotion has drawn heavily on robotics research in general. There is, correspondingly, great potential for transfer of results and techniques generated for legged systems, to other types of robotic systems.

REFERENCES

Abel, J.M., Holzmann, W., and McCarthy, J.M. 1985. On grasping objects with two articulated fingers. *IEEE J. Robotics and Automation*. Vol. RA-1. No. 4 : 211-214.

Agarwal, S.K. and Waldron, K.J. 1988 (Dec.). Impulse model of an actively controlled running machine. *ASME Winter Annual Meeting.* Chicago (to be presented).

Alexander, R.M. 1984. The gaits of bipedal and quadrupedal animals. *Int. J. Robotics Research*, Vol. 3, No. 2, 1984, pp. 49-59.

Bessonov, A.P. and Umnov, N.V. 1973. The analysis of gaits in six-legged vehicles according to their static stability. *First Symp. Theory and Practice of Robots and Manipulators.* Amsterdam: Elsevier Scientific Publishing Company.

Brooks, R.A. 1983a. Solving the find-path problem by good representation of free space. *IEEE Trans. Systems, Man, Cybernetics.* SMC-13 (3) pp. 191-197.

Brooks, R.A. 1983b. Planning collision-free motions for pick-and-place operations. *Int. J. Robotics Research.* Vol. 2. No. 4 : 19-44.

Cruse, H. 1976. The control of body position in the stick insect (*carausius morosus*), when walking over uneven terrain. *Biological Cybernetics.* No. 24: 25-33.

Davis, R.H. and Camacho, M. 1984. The application of logic programming to the generation of paths for robots. *Robotica.* Vol. 2: 93-103.

Delcomyn, F. 1981. Insect locomotion on land. *Locomotion and Energetics in Arthropods.* Eds. C.F. Herried and C.R. Fourtner. Plenum Press. New York.

Gambaryan, P.P. 1974. *How Mammals Run.* John Wiley and Sons. New York.

Gardner, J.F., Srinivasan, K., and Waldron, K.J. 1988. A new method for controlling force distribution in redundantly actuated closed kinematic chains. *1988 ASME Winter Annual Meeting.* Chicago.

Gass, S.I. 1985. *Linear Programming.* McGraw-Hill. New York.

Giralt, G., Sobek, R., and Chatila, R. 1979. A multi-level planning and navigation system for a mobile robot. *Proc. Sixth IJCAI.* Vol. 1. Tokyo. Japan.

Glower, J.S., and Özguner, G. 1986. Control of a quadruped trot. *IEEE Conference on Robotics and Automation.* San Fransisco: 1496-1501.

Goldenberg, K.Y. and Raibert, M.H. 1987. Conditions for symmetric running in single and double support. *IEEE Conference on Robotics and Automation.* Raleigh. North Carolina: 1890-1895.

Gray, J.1968. *Animal Locomotion.* London:WiedenField andNicolson.

Hildebrand, M. 1960. How animals run. *Scientific American.*. 148-157.

Hildebrand, M. 1965. Symmetrical gaits of horses. *Science.* Vol. 150.

Hildebrand, M. Analysis of tetrapod gaits. 1976. *Neural Control of Locomotion.* Eds. R.M. Herman, *et al.* Plenum Press. New York.

Hirose, S. and Umetani, Y. Sept 1980. The basic motion regulation system for a quadruped walking machine. *ASME Paper 80-DET-34.*

Hirose, S. 1984. A study of design and control of a quadrupedal walking vehicle. *Int. J. Robotics Research.* Vol. 3. No. 2: 113-133.

Hirose, S., Masui, T., Hidekazu, K., Fukuda, Y. and Umetani, Y. 1985. TITAN III: a quadruped walking vehicle. *Robotics Research* 2. ed. H. Hanafusa and H. Inoue. MIT Press: 325-332.

Hollerbach, J. and Suh, K.C. 1985. Redundancy resolution of manipulators through torque optimization. *IEEE Conf. on Robotics and Automation.* St. Louis, Missouri: 1016-1021.

Hollerbach, J. and Narasimhan, S. 1986. Finger force computation without the grip Jacobian. *IEEE Conf. on Robotics and Automation.* San Fransisco: 871-875.

Holzmann, W., and McCarthy, J.M. 1985. Computing the friction forces associated with a three fingered grasp. *IEEE J. Robotics and Automation.* Vol. RA-1. No. 4.

Huang, M. and Waldron, K.J. 1987. Relationship between payload and speed in legged locomotion. *IEEE Conf. on Robotics and Automation.* Raleigh. North Carolina: 533-536.

Isik, C. and Meystel, A.M. 1988. Pilot level of a hierarchical controller for an unmanned Robot. *IEEE J. Robotics and Automation.* 4(3): 241-255.

Kato, T., Takanishi, A., Jishikawa, H., and Kato, I. 1983. The realization of quasi-dynamic walking by the biped walking machine. *Fourth Symp. on Theory and Practice of Robots and Manipulators.* A. Morecki, G. Bianchi, and K. Kedzior. eds. Warsaw: Polish Scientific Publishers: 341-351.

Kerr, J. and Roth, B. 1986. Analysis of multifingered hands. *Int. J. Robotics Research.* Vol. 4. No. 4: 3-17.

Kessis, J.J., Rambaut, J.P., Penné, J., Wood, R. and Mattar, N. 1985. Hexapod walking robots with artificial intelligence capabilities. *Theory and Practice of Robots and Manipulators.* A. Morecki, G. Bianchi, and K. Kedzior. eds. Kogan Page, London: 395-402.

Klein, C.A., Olson, K.W., and Pugh, D.R. 1983. Use of force and attitude sensors for locomotion of a legged vehicle over irregular terrain. *Int. J. Robotics Research.* Vol. 2. No. 2: 3-17.

Klein, C.A. and Chung, T.S. 1987. Force interaction and allocation for the legs of a walking vehicle. *IEEE J. Robotics and Automation.* RA-3. No. 6.

Klein, C.A., Kau, C.C., Ribble, E.A., and Patterson, M.R. 1987. Vision processing and foothold selection for the ASV walking machine. *SPIE Conf. - Advances in Intelligent Robotics Systems.* Cambrige. MA.

Koch, E., Yeh, C., Hillel, G., Meystel, A. and Isik, C. 1985. Simulation of path planning for a system with vision and map updating. *IEEE Conf. on Robotics and Automation.* St. Louis, Missouri: 146-160.

Kugushev, E.I., and Jaroshevskij, V.S. 1975. Problems of selecting a gait for an integrated locomotion robot. *Proc. Fourth IJCAI*, Tilisi. Georgian SSR. USSR: 789-793.

Kumar, V. 1987. Motion planning for legged locomotion systems on uneven terrain. *Ph.d. Dissertation.* The Ohio State University, Columbus, Ohio.

Kumar, V., and Waldron, K.J. 1988a. Analysis of omnidirectional gaits for walking vehicles on uneven terrain. *Seventh Symp. on Theory and Practice of Robots and Manipulators.* A. Morecki, G. Bianchi, and K. Kedzior. eds.: 37-62.

Kumar, V., and Waldron, K.J. 1988b. Force distribution in walking vehicles.*Trends and Developments in Mechanisms, Machines and Robotics.* ASME. DE-15. Vol. 3. A. Midha. ed.: 473-480.

Kumar, V. and Waldron, K.J. 1988c. Force distribution in closed kinematic chains. *IEEE J. Robotics and Automation.* in press.

Kumar, V., and Waldron, K.J., 1989. Adaptive Gait Control for a Walking Robot. *J. Robotic Systems* (in press).

Kwak, S. A Computer Simulation Study of a free gait motion coordination algorithm for rough-terrain locomotion by a hexapod walking machine. *Ph.d. Dissertation.* The Ohio State University, Columbus.

Lee, W. J. and Orin, D.E. 1988. The kinematics of motion planning for multilegged vehicles over uneven terrain. *IEEE J. Robotics and Automation.* 4(2):204-212.

Lozano-Peréz, T. 1983. Spatial Planning: A configuration space approach. *IEEE Trans. on Computers.* Vol. C-32. No. 2.

Lozano-Peréz, T. 1981. Automatic planning of manipulator transfer movements. *IEEE Trans. on Systems, Man, Cybernetics.* SMC-11 (10): 681-689.

McGhee, R.B. 1968. Finite state control of quadruped locomotion. *Mathematical Biosciences.* No. 2: 57-66.

McGhee, R.B. and Frank, A.A. 1968. On the stability properties of quadruped creeping gaits. *Mathematical Biosciences.* No. 3: 331-351.

McGhee, R.B., and Jain, A.K. 1972. Some properties of regularly realizable gait matrices. *Mathematical BioSciences.* No. 13: 179-193.

McGhee, R.B. 1977. Control of legged locomotion systems. *Proc.Joint Automatic Control Conference.* San Fransisco: 205-215.

McGhee, R.B., and Iswandhi, G. 1979. Adaptive locomotion of a multilegged robot over rough terrain.*IEEE Transactions on Systems, Man and Cybernetics.* SMC-9(4): 176-182.

McGhee, R.B. 1984. Vehicular legged locomotion. *Advances in Automation and Robotics.* ed. G.N. Saridis. Greenwich. Connecticut: Jai Press.

Mosher, R.S. Exploring the potential of a quadruped. 1969. *Int. Automotive Engineering Congress.* Paper no. 690191. NewYork: SAE.

Muira, H., and Shimoyama, I. 1984. Dynamic walk of a biped. *Int. J. Robotics Research.* Vol. 3. No. 2: 60-74.

Muybridge, E.1955. *The Human Figure in Motion*, Dover, New York, 1955.

Muybridge, E. 1957. *Animals in Motion*, Dover, New York, 1957.

Nilsson, N. 1980. *Principles of Artificial Intelligence.* Tioga Publishing. California.

Okhotsimski, D.E., Gurfinkel, V.S., Devyanin, E.A., and Platonov, A.K. 1977. Integrated walking robot development. *Machine Intelligence.* Vol. 9. Eds. J.E. Hayes, D. Michie and L.J. Mikulich.

Olivier, J.F. and Özguner, F. 1986. A navigation algorithm for an intelligent vehicle with a Laser Range Finder. *IEEE Conf. on Robotics and Automation..* San-Fransisco.

Orin, D.E. 1982. Supervisory control of a multilegged robot. *Int. J. Robotics Research.* Vol. 1. No. 1: 79-81.

Orin, D.E. and Oh, S.Y. 1981 (June). Control of force distribution in robotic mechanisms containing closed kinematic chains," *J. Dynamic Systems, Measurements, and Control*, Vol. 102: 134-141.

Özguner, F., Tsai, S.J., and McGhee, R.B. 1984. An approach to the use of terrain-preview information in rough-Terrain locomotion by a hexapod walking machine. *Int. J. Robotics Research.* 3(2): 134-146.

Pandy, M.G., Kumar, V., Berme, N., and Waldron, K.J. 1988. The dynamics of quadrupedal locomotion. *J. Biomechanical Engineering.* 110(3): 230-237.

Patterson, M.R., Reidy, J.J., and Brownstein, B.B. 1983. Guidance and actuation techniques for an adaptively controlled vehicle. *Final Report, Contract MDA 903-82-c-0149*. Battelle Columbus Division, Ohio.

Paul, R.P. 1981. *Robot Manipulators, Mathematics, Programming and Control.* The MIT Press, Cambridge.

Pearson, K.G., and Franklin, R. 1984. Characteristics of leg movements and patterns of coordination in locusts walking on rough terrain. *Int. J. Robotics Research.* Vol. 3.No. 2: 101-112.

Poulos, D.D. 1986. Range image processing for local navigation of an autonomous land vehicle. *M.Sc. Thesis.* Naval Postgraduate School. Monterey, California.

Qui, X. and Song, S.M. 1988. A strategy of wave gait for a walking machine traversing a rough planar terrain. *Trends and Developments in Mechanisms, Machines and Robotics.* ASME. DE-15. Vol. 3. A. Midha. ed.: 487-496.

Raibert, M.H., and Sutherland, I.E. 1983. Machines that walk. *Scientific American.* 248 (2): 44-53.

Raibert, M.H.1985. *Legged Robots that Balance.* MIT Press. Cambridge. Massachusetts.

Raibert, M.H., Chepponis, M., Brown, H.B. Jr. 1986. Running on four legs as though they were one. *IEEE J. Robotics and Automation.* RA-2 (2): 70-82.

Raibert, M.H. 1986. Running with symmetry. *Int. J. Robotics Research.* 5(4): 3-19.

Richbourg, R.F., Rowe, N.C., Zyda, M.J. and McGhee, R.B. 1987. Solving global two-dimensional routing problems using Snell's law and A* search. *IEEE Conf. on Robotics and Automation.* Raleigh. North Carolina: 1631-1636.

Ross, R.S., Rowe, N.C., and McGhee, R.B. 1987. Dynamic multivariate terrain cost maps for automatic route planning. *AAAI Workshop on Planning Systems for Autonomous Mobile Robots.* Seattle. Washington.

Russell, M. 1983. Odex 1: The first Functionoid. *Robotics Age.* 5(5): 12-18.

Salisbury, J.K., and Roth, B. 1983. Kinematic and force analysis of articulated mechanical hands. *J. Mechanisms, Transmissions, and Automation in Design.* Vol. 105: 35-41.

Shih, L. and Frank, A.A. 1987. A study of gait and flywheel torque Effect on legged machines using a dynamic compliant joint model. *IEEE Conference on Robotics and Automation*, Raleigh, N.Carolina: 527-532.

Sitek. G. 1976. Big Muskie. *Heavy Duty Equipment Maintenance.* Vol. 4:16-23.

Song, S.M, and Waldron, K.J. 1987. Geometric design of a walking machine for optimal mobility. *J. Mechanisms, Transmissions, Automation in Design.* Vol. 109. No. 1: 21-28.

Song, S.M, and Waldron, K.J. 1987. An analytical approach for gait study and its application on wave gaits. *Int. J. Robotics Research.* Vol. 6. No. 2.

Song, S.M, and Waldron, K.J. 1988. *Machines That Walk: The Adaptive Suspension Vehicle.* MIT Press. Cambridge. Massachusetts.

Sutherland, I. and Ullner, M.K. 1984. Footprints in the asphalt. *Int. J. Robotics Research.* Vol. 3. No. 2: 29-36.

Tomovic, R. 1961. A general theoretical model of creeping displacement. *Cybernetica IV*: 98-107 (English Translation).

Thompson, A.M. 1977. The navigation system of the JPL robot. *Proc. Fifth IJCAI.* Cambridge. MA: 749-757.

Udupa, S.M. 1977. Collision detection and avoidance in computer Controlled Manipulators. *Proc. Fifth IJCAI.* M.I.T.: 737-748.

Vukobratovic, M., Frank, A.A., and Juricic, D. 1970. On the stability of biped locomotion. *IEEE Trans. on Biomedical Engineering.* Vol. 17. No. 1: 25-36.

Waldron, K.J., Song, S., Wang, S., and Vohnout, V.J. 1984. Mechanical and geometric design of the Adaptive Suspension Vehicle. *Theory and Practice of Robots and Manipulators.* Proc. Romansy '84: Fifth CISM-IFToMM Symp., Ed. A. Morecki, G. Bianchi, and K.Kedzior: 295-306.

Waldron, K.J., Vohnout, V.J., Pery, A., and McGhee, R.B. 1984. Configuration design of the Adaptive Suspension Vehicle. *Int. J. Robotics Research.* Vol. 3. No. 2: 37-48.

Waldron, K.J. and McGhee, R.B. 1985. The Adaptive Suspension Vehicle Project. *Unmanned Systems.* Summer. 1985.

Waldron, K.J. 1986. Force and motion management in legged locomotion. *IEEE J. on Robotics and Automation.* Vol. RA-2. No. 4.

Waldron, K.J. and McGhee, R.B. 1986. The Adaptive Suspension Vehicle. *IEEE Control Systems Magazine.* 6(6): 7-12.

Waldron, K.J. and Hunt, K.H. 1988. Series-parallel dualities in actively coordinated mechanisms. *Robotics Research 4.* R.Bolles and B. Roth. eds.: 175-182.

Wilson, D.M. 1966. Insect walking. *Annual Review Entomology.* Vol. 11.

Wilson, D.M. 1976. Stepping patterns in tarantula spiders.*J. Experimental Biology,* No. 47: 133-151.

Zheng, Y.F. and Luh, J.Y.S. 1988. Optimal load distribution for two industrial robots handling a single object. *IEEE Conf. on Robotics and Automation.* Philadelphia. PA: 344-349.

Zuk, D.M., and Dell'Eva, M.L. 1983. 3-D vision system for the Adaptive Suspension Vehicle. *Final Report.* Ann Arbor, Michigan: Environment Research Institute of Michigan.

A Bibliography on Robot Kinematics, Workspace Analysis, and Path Planning

J. M. McCarthy and R. M. C. Bodduluri
Dept. of Mechanical Engineering
University of California, Irvine
Irvine, CA 92715

1. Introduction

This bibliography lists articles dating from 1964 through 1987 that study the geometric relation between the joint angles of a linked system of rigid bodies, termed a robot, and the position in space of a distinguished body of the system, called the end effector. Three different lines of research that have common interest in this relationship are brought together in this list. The first, kinematics, derives and solves the mathematical equations that relate joint variables of the robot to the position of its end effector; the second, workspace analysis, studies the range of positions reachable by the robot; and the last, path planning, determines the joint movements needed for the robot to perform a task.

 The relation between the robot's joint angles and its end effector position is defined mathematically by equating the sequence of relative positions of its links to the desired end effector position. If we view these equations as a map from joint angle space to end effector position space, then robot kinematics, workspace analysis, and path planning are united by interest in and use of the geometric properties of this map. Kinematics studies the formulation of the map itself, while workspace analysis characterizes its range which is the set of positions attainable by the end effector, and path planning focuses on its domain and determines the inverse image of end effector movements in order to compute useful joint trajectories.

 The geometric emphasis of this bibliography eliminates from the list those articles concerned with the dynamic properties of a robot in its workspace, as well as those studying potential field techniques for path planning. It includes, however, all serial and parallel robots, redundant robots and cooperating robots, and embeds mechanism kinematics in the study of robots as general multi-degree of freedom jointed systems.

2. The Group of Positions of a Rigid Body

The set of positions of a rigid body in space is a mathematical object known as a Lie Group, Belinfante and Kolman (1972), Brockett (1983), Karger and Novak (1985). We may distinguish between this set and the various ways of parameterizing it, which are termed representations of the group, Rooney (1977, 1978). In kinematics, four different representations have been used to specify the position of a rigid body in space: 1. point coordinate transformations in three dimensions, consisting of a rotation followed by a translation, Hunt (1978), Bottema and Roth (1979), Chen (1987), the matrix form of this representation is often denoted $SO(3) \times R^3$; 2. homogeneous point coordinate transformations in four dimensions, known as 4×4 Homogeneous transforms, Uicker et al (1964), Hartenberg and Denavit (1964), Paul (1981), Herve (1982), Craig (1986); 3. line coordinate transformations which may be written as 6×6 matrices operating on six dimensional vectors, called screws, or equivalently as the dual orthogonal matrix transformation of three dimensional vectors of dual numbers, Piper and Roth (1969), Yang (1969), Woo and Freudenstein (1970), Yuan and Freudenstein (1971), Yuan et al (1971), Veldkamp (1976), Hiller and Woernle (1984), Pennock and Yang (1985), McCarthy (1986a), Gu and Luh (1987), Sugimoto (1987); and 4. a line transformation algebra called Dual Quaternions, Yang and Freudenstein (1964), Sandor (1968), Bottema and Roth (1979).

No matter which representation is used, the resulting kinematic equations define a complicated mapping. They fold the joint space and cover the set of attainable end effector positions with several sheets. The result is that different sets of joint angles define the same position of the end effector, and there are several different joint trajectories that provide a specific end effector movement, Pieper and Roth (1969), Duffy (1980). The first problem in robot path planning is to invert the kinematic equations in order to determine these joint trajectories, Paul (1981), Craig (1986). Points on the fold are called singularities and must be considered while planning trajectories, Featherstone (1983), Litvin et al (1986), Aboaf and Paul (1987).

3. Screw Theory

The manifold of attainable end effector positions is six dimensional and, generally, the kinematic equations have six independent partial derivative vectors that define the manifold's tangent hyperplane, Waldron et al (1985). Singularities of the manifold are the points at which these partial derivatives no longer span the tangent space, Hunt (1986), Hsu and Kohli (1987). For example, the positions at the extreme reach of the robot are singularities.

Each six dimensional partial derivative vector of the kinematic map can be manipulated slightly to obtain a pair of three dimensional vectors that are the partial angular and linear velocities of the end effector. These vectors are called screws, Dimentberg (1965), Roth (1967), Yang (1974), and a linear combination of screws is called a screw system, Davies and Primrose (1971), Hunt (1978), Sugimoto and Duffy (1982). In the language of screw theory the robot is in a singular configuration when its partial derivative screws form a five system or less, Sugimoto et al (1982). This algebraic requirement yields, in the case of extreme reach of a robot, the geometric fact that the axes of each joint have a common normal, Shimano and Roth (1976), Baker (1978), Derby (1981), Sugimoto and Duffy (1981a,b).

The higher derivatives of the kinematic equations define the local properties of this manifold, McCarthy and Ravani (1986), Ghosal and Roth (1987a,b).

4. Workspace Analysis

In general the workspace of a robot is the image of the attainable positions of the end effector, independent of its representation. Because the kinematic equations are continuous, the topology of the joint space is preserved in each representation. However, the representation $SO(3) \times R^3$ conveniently separates the set of points the robot can reach, known as its reachable workspace, from the orientations the end effector can achieve at each of these points, termed the dextrous workspace, Vertut (1974), Roth (1975), Kumar and Waldron (1981), Gupta (1986), McCarthy (1987b). The workspace depends directly on the dimensions of the robot and therefore can be used as the basis for robot design procedures, Fichter and Hunt (1975), Shimano and Roth (1978), Tsai and Soni (1981), Gupta and Roth (1982), Freudenstein and Primrose (1984), Tsai and Soni (1984), Yoshikawa (1985), Lin and Freudenstein (1986), Vijaykumar et al (1986), Davidson and

Hunt (1987b). Simply mapping the workspace of an existing robot poses a challenge, Hansen et al (1983), Lee and Yang (1983), Selfridge (1983), Tsai and Soni (1983), Yang and Lee (1983), Cwiakala and Lee (1985), Kumar and Patel (1986), Yang and Chiueh (1986), especially if the various sheets are to be separated, Kohli and Spanos (1985), Spanos and Kohli (1985), Hunt (1987), Kohli and Hsu (1987), Rastegar and Deravi (1987), or the dextrous workspace is considered, Paul and Stevenson (1983), Yang and Lai (1985), Davidson and Hunt (1987a), Davidson and Pingali (1987). The workspace of closed chains such as formed by cooperating robots poses its own challenges Bajpai and Roth (1986), Kerr and Roth (1986), Litvin et al (1986), Luh and Zheng (1987).

Another representation for the set of reachable positions of an end effector has been explored for planar systems, Sarkisyan et al (1973), Gupta and Roth (1975), Smeenk (1975), Bottema and Roth (1979), Desa and Roth (1979, 1981a,b), Chen et al (1982), Ravani and Roth (1983), McCarthy (1986b), Young and Duffy (1987a,b). A similar representation for the orientation of a body is obtained using quaternions, Ravani and Roth (1984), Canny (1985), McCarthy (1987a). Both of these representations generalize to define spatial positions of a rigid body by means of the algebra of Dual Quaternions, Ravani and Roth (1984), McCarthy and Ravani (1986).

5. Path Planning

If an external constraint is placed on the workspace of a robot so that a set of otherwise attainable positions is forbidden, then the inverse image of the constraint in the robot's joint space must be computed in order to plan movements. This image of the constraint is an obstacle in joint space and planning movements that avoid these obstacles is a challenge attracting an increasing amount of research.

Suppose the constraint is a solid obstacle located within the reachable workspace of a robot, then the boundary of the associated joint space obstacle is the inverse image of positions of the end effector that touch the obstacle. If other links of the robot can also hit the obstacle then the positions of these links that contact the obstacle add to its joint space image. Thus the problem of mapping joint space obstacles reduces to computing the positions attainable by a moving body while it maintains contact with a fixed obstacle, Whitney (1969), Udupa (1977), Brooks (1983a), Grechanovsky and Pinsker (1983), Lozano-Perez (1983a,b), O'Dunlaing et al (1983), Faverjon

(1984), Hopcroft et al (1985), Red and Truong-Cao (1985), Cai and Roth (1987), Lozano-Perez (1987), Lumelsky (1987), Singh and Wagh (1987).

6. The Piano Movers Problem

Planning the movement of a single rigid body to avoid an obstacle is called the "Piano Mover's Problem" and has been formulated in many ways, Howden (1968), Brooks (1983b), Schwartz and Sharir (1983a,b,c, 1984), Sharir and Ariel-Sheffi (1984), Brooks and Lozano-Perez (1985), Canny (1985), Donald (1985), Oommen and Reichstein (1987), Schwartz et al (1987), Yap (1987). If we consider the location of a reference point on the body and its orientation about this point as generalized joint variables, termed configuration variables, then the computation of the joint space obstacle, or more appropriately the configuration space obstacle, can be done by "growing" the actual obstacle by an amount dictated by the size of the robot and its orientation, Lozano-Perez and Wesley (1979), Lozano-Perez (1981). This approach can be applied directly to the problem of guiding a mobile robot through a known field of obstacles, Nilsson (1969), Thompson (1977), Moravec (1979), Gouzenes (1984), Chang et al (1987), Laumond (1987).

Notice that this choice of configuration variables and the use of $SO(3) \times R^3$ as the representation of positions of rigid body result in the identity as the direct kinematic map. If the moving body is constrained to be the end effector of a manipulator then the inverse kinematic formulas define the joint trajectory that avoids the obstacle. Of course, now the problem must be solved for each link of the robot.

The planning problem can be further complicated if the body is held by a redundant robot, Maciejewski and Klein (1985), cooperating robots, Luh and Zheng (1987), or by a robot with many degrees of freedom, Craig (1986), Erdmann and Lozano-Perez (1987), Faverjon and Tournassoud (1987).

7. Complexity

The difficulty of robot path planning is described in computational terms by an estimate of the relation between the size of the planning problem and the amount of memory space required by a computer to execute the algorithm. This relation, known as computational complexity of the algorithm, has been shown to be a polynomial in the number of obstacles for the general path planning problem, and it is as hard as any known problem that is polynomial in memory space, Reif (1979), Hopcroft et al (1984a,b, 1985), Schwartz et al (1987). It is, therefore, classified as PSPACE-hard. The degree of the polynomial relationship depends on the formulation of the problem and the algorithm used to solve it.

8. Conclusion

The ability of a robot to determine its own configuration and movement to perform a task, to avoid collisions, or to cooperate with another robot or person is fundamental to all our imagined uses of a robot, Roth (1978), Lozano-Perez (1983b). This ability requires that the robot understand the relation between its position in space and the values of its joint variables, not only when standing freely but also when constrained in some way. The papers listed here detail the evolution our understanding of this same relationship.

References

Classical

Ball, R. S., 1900, *A Treatise on the Theory of Screws*, Cambridge University Press.

Hamilton, W. R., 1969, *Elements of Quaternions*, Chelsea Press.

Study, E., 1903, *Die Geometrie der Dynamen*, Leipzig.

1964

Uicker, J. J., Denavit, J., and Hartenberg, R. S., 1964(June),"An Iterative Method for the Displacement Analysis of Spatial Mechanisms," *ASME J. Appl. Mech.*, pp.309-314.

Hartenberg, R. S., and Denavit, J.,1964, *Kinematic Synthesis of Linkages*, McGraw-Hill, NY.

Yang, A. T., and Freudenstein, F., 1964, "Application of Dual-Number Quaternion Algebra to the Analysis of Spatial Mechanism," *ASME J. Appl. Mech.*, 86:300-308.

1965

Dimentberg, F. M., 1965, *The Screw Calculus and its Applications in Mechanics*, Moscow, USSR: Izd. Nauka. (English Trans. 1968. Foreign Tech. Div. WP-AFB, Ohio).

1967

Roth, B., 1967, "On the Screw Axes and Other Special Lines Associated with Spatial Displacements of a Rigid Body," *ASME J. Eng. for Ind.*, 89B(1):102-110.

1968

Howden, W. E., 1968, "The Sofa Problem," *Comput. J.*, 11:299-301.

Sandor, G. N., 1968, "Principles of a General Quaternion-Operator Method of Spatial Kinematic Synthesis," *J. Appl. Mech.*, 35(1):40-46.

1969

Nilsson, N., 1969, "A Mobile Automaton: An Application of Artificial Intelligence Techniques," *Proc. 1st Int. Joint Conf. Artificial Intell.*, Washington, D.C., pp. 509-520.

Pieper, D., and Roth, B., 1969, "The Kinematics of Manipulators under Computer Control," *Proc. 2nd World Congr. on the Theory of Machines and Mechanisms*, Zakopane, Poland, 2:159- 169.

Whitney, D. E., 1969, "State Space Modes of Remote Manipulation Tasks," *IEEE Trans. Automatic Control*, AC-14(6):617-623.

Yang, A. T., 1969, "Analysis of an Offset Unsymmetric Gyroscope with Oblique Rotor Using 3x3 Matrices with Dual Number Elements," *ASME J. Eng. for Ind.*, 91(3):535-542.

1970

Woo, L., and Freudenstein, F., 1970, "Application of Line Geometry to Theoretical Kinematics and the Kinematic Analysis of Mechanical Systems," *J. Mech.*, 5:417-460.

1971

Yuan, M. S. C., and Freudenstein, F., 1971, "Kinematic Analysis of Spatial Mechanisms by Means of Screw Coordinates, Part 1–Screw Coordinates," *ASME J. Eng. for Ind.*, 93:61-66.

Yuan, M. S. C., Freudenstein, F., and Woo, L. S., 1971, "Kinematic Analysis of Spatial Mechanisms by Means of Screw Coordinates, Part 2–Analysis of Spatial Mechanisms," *ASME J. Eng. for Ind.*, 93:67-73.

Davies, H. T., and Primrose, E. J. F., 1971, "An Algebra for the Screw Systems of Pairs of Bodies in a Kinematic Chain," *Proc. 3rd World Congr. Theory of Machines and Mechanisms*, pp. 199-212.

1972

Belinfante, J. G. F., and Kolman, B, 1972, A Survey of Lie Groups and Lie Algebras: with Applications and Computational Methods, *SIAM*, Phila., PA.

1973

Sarkisyan, Y. L., Gupta, K. C., and Roth, B., 1973, "Kinematic Geometry Associated with the Least-Square Approximation of a Given Motion," *ASME Trans. J. Eng. for Ind.*, pp. 503-510.

1974

Yang, A. T., 1974, "Calculus of Screws," in *Basic Questions of Design Theory*, (ed. W. R. Spillers), North-Holland/American Elsevier, New York.

Vertut, J., 1974, "Contribution to Analyze Manipulator Morphology Coverage and Dexterity," *On the Theory and Practice of Manipulators*, Vol. 1, Springer-Verlag, pp. 277-289.

1975

Fichter, E. F., and Hunt, K. H., 1975, "The Fecund Torus, Its Bitangent Circles and Derived Linkages," *Mechanism and Machine Theory*, 10:167-176.

Gupta, K. C., and Roth, B., 1975, "A General Approximation Theory for Mechanism Synthesis," *ASME Trans., J. Appl. Mech.*, pp. 451-457.

Roth, B., 1975(Oct.), "Performance Evaluation of Manipulators from a Kinematic Viewpoint," *National Bureau of Standards*, NBS Spec. Publ. 459, pp. 39-61.

Smeenk, D. J., 1975, "Rational Motions of Special Spatial Four-Bars," *Mechanism and Machine Theory*, 10:177-188.

1976

Shimano, B., and Roth, B., 1976, "Ranges of Motion of Manipulators," *Second CISM-IFToMM Int. Symp. Theory and Practise of Robots and Manipulators*, pp. 17-26.

Veldkamp, G. R., "On the Use of Dual Numbers, Vectors, and Matrices in Instantaneous Spatial Kinematics," *Mech. Mach. Theory*, 11:141-156.

1977

Rooney, J., 1977, "A Survey of Representations of Spatial Rotation about a Fixed Point," *Environment & Planning*, B4:185-210.

Thompson, A. M., 1977, "The Navigation System of the JPL Robot," *Proc. 5th Int. Joint Conf. Artificial Intell.*, MIT, Cambridge, MA, pp749-757.

Udupa, S., 1977, "Collision Detection and Avoidance in Computer Manipulators," *5th Int. Joint Conf. Artificial Intelligence*, MIT, Cambridge, MA, pp. 737-748.

1978

Baker, J. E., 1978, "On the Investigation of Extrema in Linkage Analysis, using Screw System Algebra," *Mechanism and Machine Theory*, 13:333-343.

Hunt, K. H., 1978, *Kinematic Geometry of Mechanisms*, Clarendon Press, Oxford.

Rooney, J., 1978, "A Comparison of Representations of General Spatial Screw Displacement," *Environment & Planning*, B5:45-88.

Roth, B., 1978, "Robots," *Applied Mechanics Reviews*, 31(11):1511-1519.

Shimano, B., and Roth, B., 1978, "Dimensional Synthesis of Manipulators," *Third CISM- IFToMM Int. Symp. Theory and Practise of Robots and Manipulators*, Udine, Italy, pp. 166- 187.

1979

Bottema, O., and Roth, B., 1979, *Theoretical Kinematics*, North Holland Publ. Amsterdam.

DeSa, S., and Roth, B., 1979, "Symmetrical Algebraic Motions in the Plane," *ASME J. Mech. Des.*, 101:15-19.

Lozano-Perez, T., and Wesley, M., 1979, "An Algorithm for Planning Collision-Free Paths Among Polyhedral Obstacles," *Comm. ACM*, 22(10):560-570.

Moravec, H. P., 1979, "Visual Mapping by a Robot Rover," *Proc. 6th Int. Joint Conf. Artificial Intell.*, Tokyo, pp. 598-600.

Reif, J., 1979, "Complexity of the Mover's Problem and Generalizations," *Proc. 20th Symp. Found. of Computer Science*, pp. 421-427.

1980

Duffy, J., 1980, *Analysis of Mechanisms and Robotic Manipulators*, Wiley, New York.

1981

Derby, S., 1981, "The Maximum Reach of Revolute Jointed Manipulators," *Mech. Mach. Theory*, 16:255-261.

DeSa, S., and Roth, B., 1981a, "Kinematic Mappings. Part 1: Classification of Algebraic Motions in the Plane," *ASME J. Mech. Des.*, 103:585-591.

DeSa, S., and Roth, B., 1981b, "Kinematic Mappings. Part 2: Rational Algebraic Motions in the Plane," *ASME J. Mech. Des.*, 103:712-717.

Kumar, A., and Waldron, K. J., 1981, "The Workspaces of a Mechanical Manipulator," *ASME J. Mech. Des.*, 103:665-672.

Lozano -Perez, T., 1981, "Automatic Planning of Manipulator Transfer Movements," *IEEE Trans. Sys. Man and Cyber*, SMC-11(10):681-698.

Paul, R. P., 1981, *Robot Manipulators: Mathematics, Programming, and Control*, MIT Press, Cambridge, MA.

Sugimoto, K., and Duffy, J., 1981a, "Determination of Extreme Distances of a Robot Hand– Part 1. A General Theory," *ASME J. Mech. Des.*, 103(3):631-636.

Sugimoto, K., and Duffy, J., 1981b, "Determination of Extreme Distances of a Robot Hand– Part 2. Robot Arms with Special Geometry," *ASME J. Mech. Des.*, 103(4):776-783.

Tsai, Y. C., and Soni, A. H., 1981, "Accessible Region and Synthesis of Robot Arms," *ASME J. Mech. Des.*, 103(4):803-811.

1982

Chen, Y., Bottema, O., and Roth, B., 1982, "Rational Rotation Functions and the Special Points of Rational Algebraic Motions in the Plane," *Mechanism and Machine Theory*, 17(5):335- 348.

Gupta, K. C., and Roth, B., 1982, "Design Considerations for Manipulator Workspace," *ASME J. Mech. Des.*, 104(4):704-711.

Herve, J. M., 1982, "Intrinsic Formulation of Problems of Geometry and Kinematics of Mechanisms," *Mechanism and Machine Theory*, 17(3):179-184.

Sugimoto, K., and Duffy, J., 1982, "Application of Linear Algebra to Screw Systems," *Mechanism and Machine Theory*, 17(1):73-83.

Sugimoto, K., Duffy, J., and Hunt, K. H., 1982, "Special Configurations of Spatial Mechanisms and Robot Arms," *Mechanism and Machine Theory*, 17(2):119-132.

1983

Brockett, R. W., 1983, "Robotic Manipulators and the Product of Exponentials Formula," *Proc. of the MTNS-83 Int. Sym.*, Beer-Sheva, Israel, pp. 120-129.

Brooks, R. A., 1983a, "Planning Collision-Free Motions for Pick and Place Operations," *Int. J. Rob. Res.*, 2(4):19-44.

Brooks, R. A., 1983b, "Solving the Find-Path Problem by Good Representation of Free Space," *IEEE Trans. on Sys. Man. and Cyb.*, SMC-13.

Featherstone, R. 1983, "Position and Velocity Transformations between end-effector coordinates and joint angles," *Int. J. Rob. Res.*, 2(2):35-45.

Grechanovsky, E., and Pinsker, I.S., 1983, "An Algorithm for Moving a Computer-Controlled Manipulator while Avoiding Obstacles," *8th Int. Joint Conf. Artificial Intelligence*, Karlsruhe, W.G., pp. 807-813.

Hansen, J. A., Gupta, K. C., and Kazerounian, S. M. K., 1983, "Generation and Evaluation of the Workspace of a Manipulator," *Int. J. Rob. Res.*, 2(3):22-31.

Lee, T. W., and Yang, D. C. H., 1983, "On the Evaluation of Manipulator Workspaces," *ASME J. Mech., Trans., and Auto. in Des.*, 105(1):70-77.

Lozano-Perez, T., 1983a, "Spatial Planning: A Configuration Space Approach," *IEEE Trans. on Computers*, C-32(2):108-120.

Lozano-Perez, T., 1983b, "Task Planning," in *Robot Motion: Planning and Control*, ed. Brady, et al., MIT Press, Cambridge, MA.

O'Dunlaing, Sharir, M., Yap, C. K., 1983, "Retraction: A New Approach to Motion Planning," *Proc. 15th ACM STOC*, pp. 207-220.

Paul, R. P., and Stevenson, C. N., 1983, "Kinematics of Robot Wrists," *Int. J. Rob. Res.*, 2(1):31-38.

Ravani, B., and Roth, B., 1983, "Motion Synthesis using Kinematic Mappings," *ASME Trans., J. Mech., Trans., Auto., in Des.*, 105:460-467.

Schwartz, J. T., and Sharir, M., 1983a, "On the Piano Movers Problem I. The Case of a Two- Dimensional Rigid Polygonal Body Moving Amidst Polygonal Barriers," *Comm. Pure and Appl. Math.*, 36(3):345-398.

Schwartz, J. T., and Sharir, M., 1983b, "On the Piano Movers Problem II. General Techniques for Computing Topological Properties of Real Algebraic Manifolds," *Advances in Applied Mathematics*, 4:298-351.

Schwartz, J. T., and Sharir, M., 1983c, "On the Piano Movers Problem III. Coordinating the Motion of Several Independent Bodies: The Special Case of Circular Bodies Moving Amidst Polygonal Barriers," *Int. J. Rob. Res.*, 2(3):46-75.

Selfridge, R. G., 1983, "The Reachable Workarea of a Manipulator," *Mech. Mach. Theory*, 18(2):131-138.

Tsai, Y. C., and Soni, A. H., 1983, "An Algorithm for the Workspace of a General n-R Robot," *ASME J. Mech., Trans., Auto. in Des.*, 105(1):52-57.

Yang, D. C. H., and Lee, T. W., 1983, "On the Workspace of Mechanical Manipulators," *ASME Trans., J. Mech., Trans., and Auto. in Des.*, 105(1):62-69.

1984

Faverjon, B., 1984, "Obstacle Avoidance Using an Octree in the Configuration Space of a Manipulator," *Proc. Int. Conf. on Robotics*, Atlanta, GA.

Freudenstein, F., and Primrose, E. J. F., 1984, "On the Analysis and Synthesis of the Workspace of a Three-Link Turning-Pair Connected Robot Arm," *ASME Trans., J. Mech., Trans., Auto. in Des.*, 106:365-370.

Gouzenes, L., 1984, "Strategies for Solving Collision-Free Trajectories Problems for Mobile and Manipulator Robots," *Int. J. Rob. Res.*, 3(4):65.

Hiller, M., and Woernle, C., "A Unified Representation of Spatial Displacements," *Mech. Mach. Theory*, 19(6):477-486.

Hopcroft, J., Joseph, D., and Whitesides, S., 1984a, "Movement Problems for 2-Dimensional Linkages," *SIAM J. Comput.*, 13(3):610-629.

Hopcroft, J.E., Schwartz, J.T., and Sharir, M., 1984b, "On the Complexity of Motion Planning for Multiple Independent Objects; PSPACE-Hardness of the Warehouseman's Problem," *Int. J. Rob. Res.*, 3(4):76-88.

Ravani, B., and Roth, B., 1984, "Mappings of Spatial Kinematics," *ASME J., Mech., Trans., and Auto. in Des.*, 106(3):341-347.

Schwartz, J. T., and Sharir, M., 1984, "On the Piano Movers Problem V. The Case of a Rod Moving in Three-dimensional Space Amidst Polyhedral Obstacles," *Comm. Pure and Appl. Math.*, 37:815-848.

Sharir, M. and Ariel-Sheffi, E., 1984, "On the Piano Mover's Problem IV. Various Decomposable Two-Dimensional Motion-Planning Problems," *Comm. Pure and Appl. Math.*, 37:479-493.

Tsai, Y. C., and Soni, A. H., 1984, "The Effect of Link Parameter on the Working Space of General 3R Robot Arms," *Mech. Mach. Theory*, 19(1):9-16.

1985

Brooks, R. A., and Lozano-Perez, T., 1985, "A Subdivision Algorithm in Configuration Space for Findpath with Rotation," *IEEE Trans. Sys. Man and Cyber*, SMC-15(2):224-233.

Canny, J., 1985, "A Voronoi Method for the Piano-Movers Problem," *Proc. IEEE Rob. and Auto. Conf.*, St. Louis, MO., pp. 530-535.

Cwiakala, M., and Lee, T. W., 1985, "Generation and Evaluation of Manipulator Workspace Based on Optimum Path Search," *ASME J. Mech., Trans., Auto., in Des.*, 107(2):245-255.

Donald, B. R., 1985, "On Motion Planning with Six Degrees of Freedom: Solving the Intersection Problems in Configuration Space," *Proc. IEEE Rob. and Auto. Conf.*, St. Louis, MO., pp. 536-541.

Hopcroft, J., Joseph, D., and Whitesides, S., 1985, "On the Movement of Robot Arms in 2- Dimensional Bounded Regions," *SIAM J. Comput.*, 14(2):315-333.

Karger, A., and Novak, J., 1985, *Space Kinematics and Lie Groups*, Gordan and Breach, New York, NY.

Kohli, D., and Spanos, J., 1985, "Workspace Analysis of Mechanical Manipulators Using Polynomial Discriminants," *ASME J. Mech., Trans., Auto. in Des.*, 107(2):209-215.

Maciejewski, A. A., and Klein, C. A., 1985, "Obstacle Avoidance for Kinematically Redundant Manipulators in Dynamically Varying Environments," *Int. J. Rob. Res.*, 4(3):109-117.

Pennock, G. R., and Yang, A. T., 1985, "Application of Dual-Number Matrices to the Inverse Kinematics Problem of Robot Manipulators," *ASME J. Mech., Trans., Auto. in Des.*, 107(2):201-208.

Red, W.E., and Truong-Cao, H-V., 1985, "Configuration Maps for Robot Path Planning in Two Dimensions," *ASME J. Dyn. Sys., Meas., and Cont.*, 107:292-298.

Spanos, J., and Kohli, D., 1985, "Workspace of Regional Structures of Manipulators," *ASME J. Mech., Trans., Auto., in Des.*, 107(2):216-222.

Waldron, K. J., Wang, S. L., and Bolin, S. J., 1985, "A Study of the Jacobian Matrix of Serial Manipulators," *ASME J. Mech., Trans., Auto., in Des.*, 107(2):230-238.

Yang, D. C. H., and Lai, Z. C., 1985, "On the Dexterity of Robotic Manipulators–Service Angle," *ASME J. Mech., Trans., Auto., in Des.*, 107(2):262-270.

Yoshikawa, T., 1985, "Manipulability of Robotic Mechanisms," *Int. J. Rob. Res.*, 4(2):3-9.

1986

Bajpai, A., and Roth, B., 1986, "Workspace and Mobility of a Closed-Loop Manipulator," *Int. J. Rob. Res.*, 5(2):131-142.

Cai, C, and Roth, B., 1986, "On the Planar Motion of Rigid Bodies with Point Contact," *Mech. Mach. Theory*, 21(6):453-466.

Craig, J., 1986, *Introduction to Robotics: Mechanics and Control*, Addison-Wesley, Reading, MA.

Dupont, P. E., and Derby, S., 1986, "Planning Collision Free Paths for Redundant Robots Using a Selective Search of Configuration Space," *ASME paper no. 86-DET-145*.

Faverjon, B., 1986, "Object Level Programming of Industrial Robots," *Proc. IEEE Rob. and Auto. Conf.*, St. Louis, MO., pp. 1406-1412.

Gupta, K. C., 1986, "On the Nature of Robot Workspace," *Int. J. Rob. Res.*, 5(2):112-121.

Hunt, K. H., 1986, "Special Configurations of Robot-Arms via Screw Theory. Part 1. The Jacobian and its Matrix of Cofactors," *Robotica*, 4:171-179.

Kerr, J., and Roth, B., 1986, "Analysis of Multifingered Hands," *Int. J. Rob. Res.*, 4(4):3-7.

Kumar, A., and Patel, M. S., 1986, "Mapping the Manipulator Workspace Using Interactive Computer Graphics," *Int. J. Rob. Res.*, 5(2):122-130.

Lin, C. C. D., and Freudenstein, F., 1986, "Optimization of the Workspace of a Three-Link Turning Pair Connected Robot Arm," *Int. J. Rob. Res.*, 5(2):104-111.

Litvin, F.L., Yi, Z., Parenti-Castelli, V., and Innocenti, C., 1986, "Singularities, Configurations, and Displacement Functions for Manipulators," *Int. J. Rob. Res.*, 5(2):52-65.

McCarthy, J. M., 1986a, "Dual Orthogonal Matrices in Manipulator Kinematics," *Int. J. Rob. Res.*, 5(2):45-51.

McCarthy, J. M., 1986b, "On the Relation Between Kinematic Mapping of Planar and Spherical Kinematic," *ASME J. Appl. Mech.*, 53(2):457-459.

McCarthy, J. M., and Ravani, B., 1986, "Differential Kinematics of Spherical and Spatial Motions using Kinematic Mapping," *ASME J. Appl. Mech.*, 53(1):15-22.

Vijaykumar, R., Waldron, K. J., and Tsai, M. J., 1986, "Geometric Optimization of Serial Chain Manipulator Structures for Working Volume and Dexterity," *Int. J. Rob. Res.*, 5(2):91-103.

Yang, D. C. H., and Chiueh, T. S., 1986, "Work-area of Six-joint Robots with Fixed Hand Orientation," *Int. J. Rob. and Auto.*, 1(1):23-32

1987

Aboaf, E. W., and Paul, R. P., 1987, "Living with the Singularity of Robot Wrists," *Proc. IEEE Rob. and Auto. Conf.*, Raleigh, NC, pp. 1713-1717.

Cai, C., and Roth, B., 1987, "On the Spatial Motion of Rigid Bodies with Point Contact," *Proc. IEEE Rob. and Auto. Conf.*, Raleigh, NC, pp. 686-695.

Chang, T. S., Qiu, K., and Nitao, J. J., 1987, "An Obstacle Avoidance Algorithm for an Autonomous Land Vehicle," *Int. J. Rob. and Auto.*, 2(1):21-25.

Chen, C. H., 1987, "Applications of Algebra of Rotations in Robot Kinematics," *Mech. Mach. Theory*, 22(1):77-83.

Davidson, J. K., and Hunt, K. H., 1987a, "Robot Workspace of a Tool Plane: Part 1. A Ruled Surface and Other Geometry," *ASME J. Mech., Trans., Auto. in Des.*, 109(1):50-60.

Davidson, J. K., and Hunt, K. H., 1987b, "Rigid Body Location and Robot Workspaces: Some Alternative Manipulator Forms," *ASME J. Mech., Trans., Auto., in Des.*, 109(2):224-232.

Davidson, J. K., and Pingali, P., 1987, "Robot Workspace of a Tool Plane: Part 2. Computer Generation and Selected Design Conditions for Dexterity," *ASME J. Mech., Trans., Auto. in Des.*, 109(1):61-71.

Erdmann, M, and Lozano-Perez, T, 1987, "On Multiple Moving Objects," *Algorithmica*, 2:477-521.

Faverjon, B., and Tournassoud, P., 1987, "A Local Based Approach for Path Planning of Manipulators with a High Number of Degrees of Freedom," *Proc. IEEE Rob. and Auto. Conf.*, Raleigh, NC, pp. 1152-1159.

Ghosal, A., and Roth, B., 1987a, "Instantaneous Properties of Multi-Degree of Freedom Motions–Point Trajectories," *ASME J. Mech. Trans. and Auto. in Des.*, 109(1):107-115.

Ghosal, A., and Roth, B., 1987b, "Instantaneous Properties of Multi Degree of Freedom Motions–Line Trajectories," *ASME J. Mech. Trans. and Auto. in Des.*, 109(1):116-124.

Gu, Y. L., and Luh, J. Y. S., 1987, "Dual-Number Transformation and Its Applications to Robotics," *IEEE J. Rob. and Auto.*, RA-3(6):615-623.

Hsu, M. S., and Kohli, D., 1987, "Boundary Surfaces and Accessibility Regions for Regional Structures of Manipulators," *Mech. Mach. Theory*, 22(3):277-289.

Hunt, K. H., 1987, "Special Configurations of Robot-Arms via Screw Theory. Part 2. Available End-Effector Displacements," *Robotica*, 5:17-22.

Kohli, D., and Hsu, M. S., 1987, "The Jacobian Analysis of Workspaces of Mechanical Manipulators," *Mech. Mach. Theory*, 22(3):265-275.

Laumond, J.P., 1987, "Finding Collision-Free Smooth Trajectories for a Non-Holonomic Mobile Robot," *10th Int. Joint Conf. Artificial Intelligence*, Milan, Italy, pp. 1120-1123.

Lozano-Perez, T., 1987, "A Simple Motion-Planning Algorithm for General Robot Manipulators," *IEEE J. Rob. and Auto.*, RA-3(3):224-238.

Luh, J. Y. S., and Zheng, Y. F., 1987, "Constrained Relations between Two Coordinated Industrial Robots for Motion Control," *Int. J. Rob. Res.*, 6(3):60-70.

Lumelsky, V. J., 1987, "Effect of Kinematics on Motion Planning for Planar Robot Arms Moving Amidst Unknown Obstacles," *IEEE J. Rob. and Auto.*, RA-3(3):207-223.

McCarthy, J. M., 1987a, "The Differential Geometry of Curves in an Image Space of Spherical Kinematics," *Mech. Mach. Theory*, 22(3):205-211.

McCarthy, J. M. (ed.), 1987b, *Kinematics of Robot Manipulators*, MIT Press, 211pp.

Oommen, B. J., and Reichstein, I., 1987, "On the Problem of Translating an Elliptic Object Through a Workspace of Elliptic Obstacles," *Robotica*, 5:187-196.

Payandeh, S., and Goldenberg, A. A., 1987, "Formulation of the Kinematic Model of a General (6 DOF) Robot Manipulator Using a Screw Operator," *J. of Robotic Systems*, 4(6):771-797.

Rastegar, J., and Deravi, P., 1987, "Methods to Determine Workspace, Its Subspaces with Different Numbers of Configurations and all the Possible Configurations of a Manipulator," *Mech. Mach. Theory*, 22(4):343-350.

Singh, J. S., and Wagh, M. D., 1987, "Robot Path Planning Using Intersecting Convex Shapes: Analysis and Simulation," *IEEE J. Rob. and Auto.*, RA-3(2):101-108.

Schwartz, J. T., Sharir, M., and Hopcroft, J. (Eds.), 1987, *Planning, Geometry, and Complexity of Robot Motion*, Ablex Publ., Norwood, NJ.

Sugimoto, K., 1987, "Kinematic and Dynamic Analysis of Parallel Manipulators by Means of Motor Algebra," *ASME J. Mech., Trans., Auto. in Des.*, 109:3-7.

Yap, C. K., 1987, "How to Move a Chair Through a Door," *IEEE J. Rob. and Auto.*, RA-3(3), 172-181.

Young, L., and Duffy, J., 1987a, "A Theory for the Articulation of Planar Robots: Part I– Kinematic Analysis for the Flexure and the Parallel Operation of Robots," *ASME J. Mech., Trans., Auto. in Des.*, 109(1):29-36.

Young, L., and Duffy, J., 1987b, "A Theory for the Articulation of Planar Robots: Part II– Motion Planning Procedure for Interference Avoidance," *ASME J. Mech., Trans., Auto. in Des.*, 109(1):37-41.

Design and Control of Direct-Drive Robots - a Survey

K. Youcef-Toumi
Department of Mechanical Engineering
Massachusetts Institute of Technology
Cambridge, MA 02139

This synopsis introduces the direct-drive concept which has proven to have good promise for high speed, high precision manipulation. Several direct-drive robots that have been developed and built at research centers and industry are presented. The related applications and the performance achieved by such manipulators is also discussed. An important point is made in regard to the design and control issues involved in developing such direct-drive manipulators. Issues pertaining to mechanical design, namely actuation, sensing and arm design are discussed from static and dynamic point of view. Also the control system design, particularly for high speed trajectory control and force control is summarized. The intent of this synopsis is then to review the theoretical and experimental work that has been conducted in the direct-drive technology.

1 The Direct-Drive Concept

Conventional electromechanical robot arm actuator usually consists of a motor that drives a load through a gearing mechanism. This gearing mechanism which can be a harmonic drive, a gear train, a chain or a lead screw is used to match the impedances of drives and their corresponding loads. These mechanisms, while used extensively, introduce significant problems in the drive system and therefore degrade the control performance of the robots. Some of these problems are backlash, friction, compliance and wear. These in turn result in poor accuracy, poor dynamic response, regular maintenance, and poor torque control capability.

Direct Drive is a drive method in which an actuator is coupled directly to its load without gearing. A basic construction is shown in Figure 1.1. The direct coupling of the motor shaft to the link completely eliminates the mechanical backlash, reduces friction and increases mechanical stiffness at the drive mechanism. The concept has been used in several high performance drives including radar tracking systems and video tape recorders.

A direct-drive robot arm is a manipulator in which high torque electric motors drive the arm linkage directly without the use of reducers. Manipulator arms driven by pneumatic and hydraulic actuators may be excluded from this definition. Pneumatic and hydraulic actuators have significant friction and compressibility in the actuators and the transmissions between the actuators and the power sources. Thus, their behavior is quite different from the direct-drive electric motors that convert electrical power into torque. Thus the complete elimination of reducers provides significant improvement in the robot design and control performance. High accuracy can be achieved because of the zero mechanical backlash and low friction in the drive system. Speed and dynamic response is also improved because of the low friction and high mechanical stiffness of the drives.

These features of the direct-drive arms meet critical requirements for many advanced robot applications. High speed is a crucial requirement in high production rate manufacturing. Traditional robots are not applicable to such production lines because of severe tact time requirements. Fast but simple positioning devices such as pneumatic cylinders have been used in high production rate applications but the limited flexibility is a major problem for these simple devices. Thus, direct drive arms with higher speed and greater flexibility are appropriate for high production rate manufacturing.

There is also an increasing need for high accuracy robots, particularly in the micro-electronics industry. Assembly of optical fiber, for example, requires accuracy on the order of 3 to 10 um. Such high accuracy operations are out of the range of today's industrial robots with reducers. As the size of electronic devices and the density of chips increases, still higher accuracy is required for positioning devices. Direct-drive methods will meet such increasing requirements. Other applications in the computer industry may require between 0.1 and 0.5 um in precision. In addition, in the wafer technology resolutions on the order of .01 um may be required.

Both high speed and precision are required for some advanced robot applications. Sheet metal cutting using lasers requires speeds of over 1 m/s with 3 to 5 G in acceleration, as well as tolerancing errors of less than 0.1

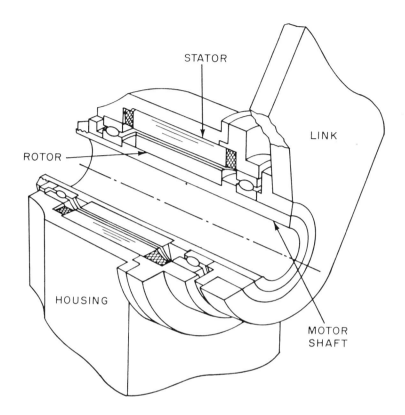

Figure 1-1: Basic construction of a direct-drive joint

mm or .05 mm. In order to meet such severe requirements, the mechanical design of manipulators with appropriate control system designs have to be considered.

2 Direct-Drive Robots

The basic design concept of a direct-drive arm was first established by Asada and Kanade (Asada and Kanade 1981, 1983) and then reduced to practice through the development of the first prototype direct drive arm at Carnegie Melon University in 1981 (Asada, Kanade and Takeyama 1983). The CMU-DD arm, has six revolute joints all of them directly driven by DC torque motors attached at individual joint shafts. The prototype robot has found significant advantages over its traditional counterpart. Positioning repeatability was improved due to the low friction and low backlash at the direct-drive mechanism. The simple mechanisms also reduced uncertainties in dynamic modelling and lead to an accurate model of the arm for control purposes. A feedforward compensation based on the actual dynamic model significantly increased the trajectory control accuracy. A more recent direct-drive robot has been developed at CMU and is described in reference (Kanade and Schmitz 1985). At the Electrical Engineering Laboratory in Japan, a similar direct drive robot arm was developed using DC torque motors in which the whole dynamic torque compensation was implemented (Takase, Hasegawa and Suehiro 1984). At the Massachusetts Institute of Technology improved direct-drive robots were developed (Asada and Youcef-Toumi 1987). The M.I.T. DD Arm Model I is a three degree of freedom serial link manipulator shown in Figure 2.1. This arm, currently installed at M.I.T.'s Artificial Intelligence Laboratory, was built in the period of 1982 to 1983. This robot has a unique kinematic structure that allows the elimination of gravity loads at individual motors. The M.I.T. model II developed in 1983 is shown in Figure 2.2. A pair of drive mechanisms which forms a closed link kinematic chain was employed in this case. The upper two motors located at the base-frame drive the two input links of the parallelogram mechanism which cause a two-degree of freedom motion of the upper arm. In this arm construction the motor at the elbow joint of the Model I design is removed and mounted on the base frame. The driving torque developed by this actuator is then transmitted through the parallelogram mechanism. This remote drive mechanism results in a reduction of arm weight significantly and shows an improved dynamic performance. Even though high speed and high accelerations were achieved with Model II, the mechanical stiffness of the whole arm was not high enough and thereby has effected the overall control performance. Figure 2.3 shows an improved design of the M.I.T. DD Arm Model III which uses the same ideas of parallel drive mechanisms however the arm is made out of graphite composite material. In general, graphite composite material has a density about half that of aluminum and stiffnesses and rigidities can be four to five times higher than that of aluminum. Thus the arm can be expected to be lightweight and stiffer. In this arm design the lowest natural frequency of the arm was increased to about 70 Hz, which is an order of magnitude higher than that of the M.I.T. DD Arm Model II which was about 14 Hz (Asada, Youcef-Toumi and Ramirez 1984). The required stiffness in each link were obtained by considering graphite composite fiber orientations and stacking sequences. The resulting arm is thus capable of moving an 8 kg. payload at 3 G with a mechanical endpoint deflection of less than 0.1 mm. Both the M.I.T. Models II and III have been designed for the specific application of high-speed laser cutting application, and both have achieved speeds and accelerations on the order of 10 m/s and 5G respectively.

Another prototype, the M.I.T. Direct-Arm Model IV, was developed for high speed, high accuracy assembly applications, and is shown in Figure 2.4. In this design, two motors are aligned on the vertical axis and the parallelogram mechanism is in the horizontal plane. This type of kinematic construction is often referred to as SCARA type. The SCARA type of design is an attractive design for direct-drive robots. The AdeptOne direct-drive robot is the first commercialized direct-drive robot which used this particular structure (Currim and Mayer 1985). This robot was developed in 1984. This arm has four degrees of freedom, two motors located at the base of the manipulator which produce the horizontal motion of the upper arm. The other two motors are on the forearm, one to produce a translational motion in the vertical axis, and the other to rotate the end-effector about the vertical axis. The maximum speed achieved by this particular manipulator is on the order of 9 m/s. Also, the repeatability of this robot is determined to be $\pm.0254$ mm. These achievements are an order of magnitude better than what conventional robots with gearing mechanisms

Figure 2.1: 3 d.o.f. MIT Direct-Drive Arm Model I

can achieve. The Matsushita Electric Corporation, Ltd. in Osaka, Japan also developed a direct-drive manipulator. This robot has a construction similar to the M.I.T. Arm Model IV and has a reach of about 60 cm, Photo 2.1. One feature of the Matsushita robot is that the motors when combined with high performance drive amplifiers show an excellent linearity in producing large torques. A special sensor developed specifically for this direct-drive robot is a laser interferometer type of encoder which also has high resolution. The Matsushita robot shows a repeatability less than $\pm.01$ mm. A larger sized direct-drive robot developed by Shin Meiwa Industries, Ltd. is shown in Photo 2.2. This robot is a five degree of freedom robot specifically designed for high speed, high accuracy laser cutting applications. The laser beam is guided inside the arm links and reaches the laser gun which is attached to the end effector. This robot is capable of tracing complex spacial curves at high speeds and able to maintain high

Figure 2.2: 3 d.o.f. MIT Direct-Drive Manipulator Model II

accuracy. In order to trace curves with small radii, extremely high accelerations are required. The Shin Meiwa direct-drive laser cutting robot has achieved accelerations over 5G at the arm tip. Yokogawa Hokushin Electric Corporation, Japan, has also developed an articulated direct-drive robot arm with six degrees of freedom. The joints of this particular robot are driven by high torque variable reluctance stepping motors with integral extra high resolution shaft encoders. Using the direct-drive motors and microprocessors based servo-controller, repeatabilities of less than .01 mm were obtained (Kuwahara, Ono, Nikaido and Matsumoto 1985).

Several other research centers have recently begun research and development of direct-drive robots. These centers include AT&T Bell laboratories, University of California at Berkley, University of California at

Figure 2.3: 3 d.o.f. M.I.T. Direct-Drive Manipulator Model III

Santa Barbara, University of Minnesota, and University of Southern California. Efforts have been concentrated in the development of motors and sensors, arm design techniques and control theory.

3 Design And Control Issues

3.1 Mechanical Design

3.11 Actuation and Sensing

One key component in the design and control of direct-drive robots is the motor and the drive amplifier. The motors must generate an order of magnitude of higher torque than that of conventional drive systems. Also,

Figure 2.4: 2 d.o.f. M.I.T. Direct-Drive Manipulator Model IV

the fluctuations in torque or speed must be minimized in order to achieve accurate control. The motors used in direct-drive robots are of three types: direct-current torque motors, brushless DC torque motors, and variable reluctance motors (Electro-Craft). These three motors have been investigated and described in reference (Asada and Youcef-Toumi 1987).

The development of high torque compact motors is indispensable in arm design. The largest DC motor used in the CMU-DD arm has a peak torque

Photo 2.1: The Matsushita Direct-Drive Robot
for high speed assembly
Pana Robo HDD-1
(Courtesy of Matsushita Industrial Co., Ltd.)

of 150 Nm. The motor diameters were about 56 centimeters (Asada, Kanade and Takeyama 1983). Direct-drive arms that have been developed lately employ much stronger and compact motors. The motor used in the MIT direct drive arm (Asada and Youcef-Toumi 1983a, 1984a) was a brushless motor in which the peak torque was increased to 660 Nm and the diameter reduced to 35 cm. The brushless motor (Asada and Youcef-Toumi 1983a, 1987; Davis and Chen 1984) has strong permanent magnets of 26 MGOe in BHmax and has shown excellent dynamic responses (Asada and Youcef-Toumi 1984a). Motornetics Corporation has developed high torque brushless variable reluctance motors for direct drive arms (Welburn 1984). The motor has a dual stator annular rotor configuration and can not only exert a large torque but also reduce energy loss. Thus the motor has a larger motor constant (Asada and Youcef-Toumi 1987), the ratio of output torque to the square root of power. The motor constant is one of the critical performance indices for direct-drive motors. Commercialized brushless DC torque motors that are developed by Motion Control Systems, Inc. and Shin Meiwa Industries in Japan, are the same type of motors that were used in the M.I.T. Direct-Drive robots. One main difference between the variable reluctance motors and the brushless motors is that the variable reluctance motor construction reduces the internal loss of magnetic energy and therefore results in a larger motor constant. There is also a trade-off in speed, weight and output torque fluctuation for these two types of motors.

In addition to motors, all the components such as sensors and controllers appropriate for direct drive applications must be developed. The positioning accuracy of a direct drive robot depends highly upon the accuracy of the position transducers used. The CMU arm has optical encoders with 13 - 15 bit resolution (Asada, Kanade and Takeyama 1983). One of the MIT arms employed resolvers of 16 bit resolution (Asada and Youcef-Toumi 1983a) and the other has double-speed resolvers of 19 bit resolution (Asada and Ro 1984; Asada and Youcef-Toumi 1987).

Another critical component that determines the control performance of direct drive arms is velocity measurement. Since the arm mechanism has very low mechanical damping the control system must provide sufficient damping through velocity feedback or equivalent control action. The velocity measurement of a direct drive motor is, however, more difficult than that of geared motors since the speed range is much lower. Two types of velocity measurements have been used in direct drive arms. One is to differentiate position transducer signals (Asada, Kanade and Takeyama 1983; Takase, Hasegawa and Suehiro 1984; and Welburn 1984) and the other is to use a

Direct-Drive Robots 293

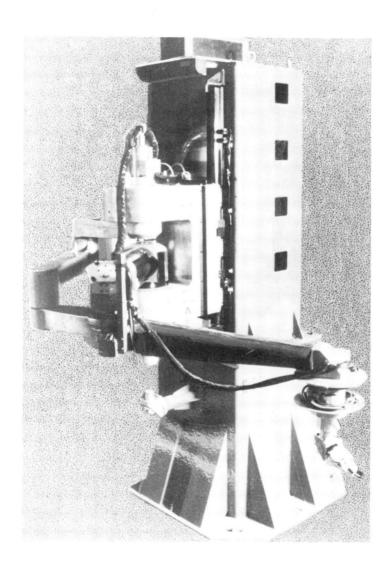

Photo 2.2: The Shin Meiwa Direct-Drive Robot
for high speed laser cutting applications
(Courtesy of Shin Meiwa Industry Co., Ltd.)

tachometer generator (Asada and Ro 1984; Asada and Youcef-Toumi 1983a, 1984a). To measure slow speeds with a small time delay, high resolution position transducers are necessary for the former case. The latter method requires a high sensitivity tachometer with low ripple.

The drive amplifiers used to drive these direct-drive are usually PWM type of amplifiers. Although these amplifiers are designed to provide high performance, nonlinearities do exist. The motor current is distributed to form the phase current signals which are multiplied by weighting functions that are rotor position dependent. Each phase current is amplified by a gain and fed to the PWM block of the amplifier. The PWM part of the amplifier will turn the switches of the power stage on and off alternatively. These switches are turned on and off by gate signals such as a base current of bipolar transistors. In order to prevent short circuit, the base signals of the switches are turned off simultaneously during a very short time. This time is referred to as the anti-coincident time, and will manifest itself as a deadband in the power amplifier. The size of the deadband will depend on the ratio of the anti-coincident time to the period of the PWM carrier (Kondoh, Yamamoto, Okuda and Youcef-Toumi 1986). This electrical deadband is inherent in this amplifier design but can be reduced.

3.12 Arm Design

The first issue in arm design of direct-drive robots is power dissipation. Consider a manipulator where n actuators are used to drive the arm linkage. Each actuator produces a torque T which is combined with other motor torques through the link mechanism to generate a force F at the endpoint to do work. The ability to produce a useful force F without overheating the actuators is of great importance in the design of direct-drive arms, since all gearing has been eliminated. In order to study these characteristics, the system consisting of the drive mechanisms and the arm linkage must be considered. Analytical tools were developed for evaluating mechanisms in terms of the power dissipation in the actuators and the end point force generated (Asada and Youcef-Toumi 1984a, 1987). This is particularly important in evaluating the overheating of direct-drive actuators under static load conditions. In addition, in applications with high duty cycle, the actuators tend to overheat.

These tools are shown to be useful in selecting an appropriate arm linkage design for a specific application. As an example, by analyzing the internal power dissipation in the motors when the manipulator exerts a force at the endpoint, it was shown that the parallel drive mechanism has a lower power dissipation than a serial drive mechanism that uses the same motors and covers the same workspace, (Asada and Youcef-Toumi 1984a, 1987).

Another issue is dynamic complexity which is a major concern in the control of manipulator arms. This problem is critical in manipulator designs for high-speed applications which have particularly prominent dynamic complexity. A design method for reducing dynamic complexity and improving dynamic performance involves the coordination of the controller and manipulator design efforts. In this method, the mechanical construction of a manipulator arm is modified so that the resulting dynamic behaviour can be improved and becomes desirable for controls. This is an important aspect particularly in the design of high-speed direct-drive arms and other manipulators with low gear reductions. In general, the dynamics of a manipulator arm depends upon the mass properties of individual arm links and the kinematic structure of the arm linkage. These would then be modified so that the resulting dynamic behavior can be improved or becomes desirable for control in joint space or task space (Asada 1983; Asada and Youcef-Toumi 1983b, 1984a, 1984b, 1987; Chung, Cho, Chung and Kang 1986; Khatib and Burdick 1985; Yang and Tzeng 1985, 1986; Youcef-Toumi 1985; Youcef-Toumi and Asada 1985, 1987).

Design techniques for arm linkage design in order to have reduced dynamic complexities and consequently reducing the control difficulties have been explored. Some particular forms of the dynamic equations in which the inertial matrix is reduced to a diagonal and/or configuration invariant form have been examined. Design theory was developed for achieving such desired manipulator dynamics.

The decoupled and configuration invariant matrix were first accomplished in (Asada and Youcef-Toumi 1983a, 1984a) in which a special five bar link parallel drive mechanism was devised for a direct-drive arm. The feasibility and practical usefulness were also demonstrated through the development of special direct-drive arms (Asada and Youcef-Toumi 1984a; Youcef-Toumi 1985). One approach was adopted in references (Yang and Tzeng 1985, 1986) in which the appropriate kinetic energy and potential energy are calculated and the designer examines all the terms involved in the energy expression to decide what mass distributions should be selected to achieve a simple dynamic behavior for the given manipulator. In references (Youcef-Toumi and Asada 1985, 1987) the conditions for the kinematic structure and mass properties of the manipulator arm which provides the decoupled and/or configuration invariant inertia were determined.

3.2 Control System Design

3.21 Trajectory Control

Motion planning of industrial robots has evolved from simple point-to-point playback of the end-effector to complex trajectory following. The difference spot welding and arc welding by industrial robots lies in the fact that the later application requires the end effector to follow a desired trajectory in space. Thus arc welding requires a more complex trajectory planning. A greater challenge has been raised recently in the application of laser cutting of sheet metal which requires both high-speed maneuvering of the end-effector and accurate cutting. Specifically the speed, acceleration and precision required are on the order of 1-3 m/s, 3.5 G and 0.1 mm to 0.05 mm respectively. At these high speeds the controller must compensate for the highly non-linear and coupled dynamic equations in order for the manipulator to track the desired trajectories. In addition, sampling time required can be as small as 1 msec.

A limited number of papers have been published in the area of trajectory control of direct-drive robots (Asada, Kanade and Takeyama 1983; An, Atkeson and Hollerbach 1986; Kanade, Khosla and Tanaka 1984; Khosla and Kanade 1986). All of these results were obtained on direct-drive robots with open kinematic chain type of structures. These structures exhibit significant coupling and interactions between the different joints. Nevertheless, the main control algorithm used by these researchers is based on the feedforward action that can be effective. The first experimental results (Asada, Kanade and Takeyama 1983) showed a promise to the direct-drive concept. Maximum joint speed ranged from 180 rad/s to 360 rad/s. Positioning accuracy measured by such responses was -.287°. The experimental results published recently (An, Atkeson and Hollerback 1986; Kanade, Khosla and Tanaka 1984) were also obtained using model-based feedforward control. The performance of model-based feedforward controllers depends greatly on model accuracy. This model usually consists of the robot dynamic equations, used to calculate the torques/forces necessary to drive the robot along the desired trajectory. These models are highly non-linear and are a function of robot parameters. The parameters of the model include the dimensions, inertial parameters, actuators, and other relevant system parameters. In reference (Khosla and Kanade 1986), the link inertial parameters were estimated from detailed drawings of the geometry solid model of the robot. The approach adapted in reference (An, Atkeson and Hollerbach 1986) is to estimate the model inertial parameters through arm excitation and CAD data. These approaches provide satisfactory results when appropriate algorithms and adequate computing hardware are used.

An alternative approach to achieving satisfactory tracking performance was to consider both the robot arm mechanism design and the controllers design. The M.I.T. Direct-Drive Arms for laser cutting applications were designed with these issues in mind through appropriate design and mass redistribution techniques, the arm dynamics were made decoupled and inertia invariant (Asada and Youcef-Toumi 1984a, 1984b, 1987; Youcef-Toumi 1985). For these specially designed manipulators, each joint operates independently because of the manipulator inertial matrix is decoupled with constant elements. A satisfactory performance characterized by 3 m/sec, 4 G and less than 0.4 mm in trajectory tracking is achieved with a PD type of controller with constant coefficients (Youcef-Toumi and Kuo 1987).

3.22 Force Control

Several methods were suggested for the control of manipulators under constrained motion. These approaches include hybrid position/force control, stiffness control and impedance control. A few other techniques have been reported in the literature and (Whitney 1985) gives an excellent overview of force control.

Papers on force control with direct-drive manipulators are a few. The first papers which investigated the torque control capability of direct-drive manipulators are presented in references (Asada and Yamamoto 1984; Asada, Youcef-Toumi and Lim 1984; Takase, Hasegawa and Suehiro 1984). The basic issue is to design a control system so that the direct-drive actuator which consists of an amplifier and a motor behaves as a torque source. Two basic methods were used. One is to employ a direct measurement technique whereby a torque sensor is incorporated inside the motor and installed between the rotor and the motor shaft. In this case, any torque that is developed between the stator and the rotor will be transmitted through the torque sensor and then applied to the motor shaft. This is a direct torque measurement because the torque transmitted to the motor shaft is measured directly. This is the advantage of this method. However, special care has to be taken in order to design the torque sensor appropriately so that resonances that are introduced in the motor are outside the bandwidth of the actuator. The other approach was to measure the motor currents in addition to the instantaneous motor position. These currents will be commutated in order to estimate the instantaneous torque output. This method requires no additional hardware and can provide excellent results, particularly when amplifiers possess current feedback. A calibration of the torque constant, however, has to be done a priori in order for this technique to be used. These two methods are described in more details in reference (Asada and Youcef-Toumi 1987).

The control of force of direct-drive manipulators of the endpoint has also been done and a limited number of papers have appeared in the literature. Reference (Maples and Becker 1986) presents some force control strategies and classification of algorithms. These algorithms were categorized into joint and cartesian based algorithms. Also, this classification distinguishes whether the control algorithm involves an inner torque, velocity, or position loop. Such techniques have been used in performing successful assembly of parts with 0.0254 mm in clearances using an AdeptOne direct-drive robot. In reference (Youcef-Toumi and Li 1987) experiments using direct-drive manipulators for end-point force control have also shown to achieve greater performance, particularly when using an inner velocity loop for disturbance rejection and surface tracking properties. In addition, experiments in force control, speed of response, impact control, and surface following have been conducted. Reference (An and Hollerbach 1987a) investigated different types of force control algorithms, specifically the hybrid force/position control, stiffness control, and the operational space methods. Analyses on whether stability can occur due to kinematics were conducted, and experimental results with the MIT Direct-Drive Arm Model I were performed. In reference (An and Hollerbach 1987b), force control was achieved by a combination of joint torque control in conjunction with end-point force control. The dynamic stability of such a control has been analyzed and also experiments have been conducted on the MIT Direct-Drive Arm Model I.

The force control experiments conducted on direct-drive robots testify to the special features that these type of manipulators have. Specifically, the force control performance achieved is superior to that of conventional manipulators.

4 Conclusion

This synopsis has presented the most current research in direct-drive robot design and control. It emphasized the performance of such robots and described the key issues in sensors, actuators, amplifiers, arm linkage design and control system design.

References

An, C. H., Atkeson, C. and Hollerbach, J. 1986 "Experimental Determination of the Effect of Feed-Forward Control of the Trajectory of Tracking Errors". In *Proc. of IEEE Conference on Robotics and Automation.*, April.

An, C. H., Hollerbach, J. M. 1987a "Dynamic Stability Issues in Force Control of Manipulators" In *Proc. of IEEE Conference and Automation..*

An, C. H. and Hollerbach, J. M. 1987b "Kinematic Stability Issues in Force Control of Manipulators" In *Proc. of IEEE Conference and Automation..*

Asada, H. 1983 "A Geometrical Representation of Manipulator Dynamics and Its Applications to Arm Design". *Trans.* ASME *Journal of Dynamic systems, Measurement, and Control*, Vol. 105.

Asada, H. and Kanade, T. 1981 "Design Concept of Direct-Drive Manipulators Using Rare-Earth DC Torque Motors" In *11th Int. Symp. on Industrial Robots..*

Asada, H. and Kanade, T. 1983 "Design of Direct-Drive Mechanical Arms". ASME *Journal of Vibration, Acoustics, Stress and Reliability in Design*, Vol. 105, No. 3, pp. 312-316.

Asada, H., Kanade, T., and Takeyama, I. 1983 "Control of a Direct-Drive Arm". *ASME Journal of Dynamic systems, Measurement and Control*, Vol. 105, No. 3, pp. 136-142.

Asada, H. and Ro, I. H. 1984 "A Linkage Design for Direct-Drive Robot Arms". In *ASME Mechanisms Conference, No. 84-det-143.*

Asada, H. and Yamamoto, H. 1984 "Torque Feedback Control of M.I.T. Direct-Drive Robot". In Proc.of the 12th ISIR. Sweden, October.

Asada, H. and Youcef-Toumi, K. 1983a "Analysis and Design of Semi-Direct-Drive Robot Arms". In *Proc. of the 1983 American Control Conference*, pages 757-764, San Francisco.

Asada, H. and Youcef-Toumi, K. 1983a "Analysis of Multi-degree of Freedom Actuator Systems for Robot Arm Design". In *Proc. of the ASME Winter Annual Meeting*.

Asada, H. and Youcef-Toumi, K. 1984a "Analysis and Design of a Direct-Drive Arm With a Five-Bar-Link Parallel Drive Mechanism". *ASME Journal of Dynamic systems, Measurement and Control*, Vol. 106, No. 3 pp. 225-230.

Asada, H. and Youcef-Toumi, K. 1984b "Decoupling of Manipulator Inertia Tensor by Mass Distribution". In *ASME Mechanisms Conference, No. 84-det-40*.

Asada, H., Youcef-Toumi, K. and Lim, S. K. 1984 "Joint Torque Measurement of a Direct-Drive Arm". In *Proc. of the Conference on Decision and Control.*, December.

Asada, H., Youcef-Toumi, K. and Ramirez, R. 1984 "Design of M.I.T.Direct-Drive Arm". In *International Symposium on Design and Synthesis, Japan Society of Precision Engineering*, Tokyo, Japan, July.

Asada, H. and Youcef-Toumi, K. 1987 "Direct -Drive Robots -- Theory and Practice" M.I.T. Press, Cambridge, Massachusetts,June.

Chung, W. K., Cho, H. S., Chung, M. J., and Kang, Y. K. 1986 "On the Dynamic Characteristics of Balanced Robotic Manipulators". In *Proc. of Japan-USA Symposium of Flexible Automation*, Osaka, Japan.

Currim, R. and Mayer, G. 1985 "The Architecture of the ADEPTONE Direct-Drive Robot". In *Proceedings of the 1985 American Control Conference*, Boston, Massachusetts, June.

Davis, S. and Chen, D. 1984 "High Performance Brushless DC Motors for Direct-Drive Robot Arms". In *Proc. of IEEE International Conference,* Tokyo, Japan, October.

Electro-Craft "DC Motors, Speed Controls, Servo-Systems" Engineering Handbook by Electro-Craft Corporation, Minnesota, Fifth Edition.

Kanade, T., Khosla, P. and Tanaka, N. 1984 "Real Time Control of CMU Direct-Drive Arm II Using Customized Inverse Dynamics". In *23rd IEEE Conference on Decision and Control,* December.

Kanade, T. and Schmitz, D. 1985 "Development of CMU Direct-Drive Arm II". In *Proc. of the 1985 American Control Conference,* Boston, Massachusetts, June.

Khatib, O. and Burdick, 1985 "Dynamic Optimization in Manipulator Design: The Operational Space Formulation", ASME Winter Annual Meeting.

Khosla, P. and Kanade, T. 1986 "Real Time Implementation and Evaluation of Model-Based Controls on CMU DD Arm II". In *Proc. IEEE Conference on Robotics and Automation.,* April.

Kondoh, T., Yamamoto, H., Okuda, H. and Youcef-Toumi, K. 1986 "Practical Issues in the Design and Control of Direct-Drive Robots" ASME Winter Annual Meeting.

Kuwahara, H., Ono, Y., Nikaido, M. and Matsumoto, T. 1985 "A Precision Direct Drive Robot Arm" In *Proceedings of the 1985 American Control Conference,* Boston, Massachusetts, June.

Maples, J. A. and Becker, J. J. 1986 "Experiments in Force Control of Robotic Manipulators". In *Proc. of IEEE Conference on Robotics and Automation,* San Francisco, California.

Takase, K., Hasegawa, T., and Suehiro, T. 1984 "Design and Control of a Direct-Drive Manipulator". In *International Symposium on Design and Synthesis, Japan Society of Precision Engineering,* Tokyo, Japan, July.

Welburn, R. 1984 "Ultra High Torque Motor System". In *Proc. of ROBOT 8 Conference, Society of Mechanical Engineers..*

Whitney, D. E. 1985 "Historical Perspective in State-of-the-Art Robot Force Control". In *Proc. of the IEEE Conference on Robotics and Automation*, St. Louis.

Yang, D. C. H., Tzeng, S. W. 1985 "Simplification and Linearization of Manipulator Dynamics by Design". In *Proc. of the 9th Applied Mechanics Conference*, October.

Yang D.C.H., and Tzeng, S.W. 1986 "Simplification and Linearization of Manipulator dynamics by the Design of Inertia Distributions". In *International Journal of Robotics Research*, 5(3).

Youcef-Toumi, K. 1985 "Analysis, Design and Control of Direct-Drive Manipulators". In Doctor of Science Mechanical Engineering Dept., Massachusetts Institute of Technology, May.

Youcef-Toumi, K. and Asada, H. 1985 "The Design of Arm Linkages with Decoupled and Configuration-Invariant Inertia Tensors: Part II: Actuator Relocation and Mass Distribution". *ASME Winter Annual Meeting*.

Youcef-Toumi, K. and Asada, H. 1987 "The Design of Open Loop Manipulator Arms with Decoupled and Configuration-Invariant Inertial Tensors". *ASME Journal of Dynamic systems, Measurement and Control.*, September.

Youcef-Toumi, K. and Kuo, A. T. Y., 1987 "High Speed Trajectory Control of a Direct-Drive Manipulator". In *Proc. of the IEEE Conference on Decision and Control.*

Youcef-Toumi, K. and Li, D. 1987 "Force Control of Direct-Drive Manipulators for Surface Following". In *Proc. of the IEEE Conference on Robotics and Automation* .

Computationally Efficient Kinematics for Manipulators with Spherical Wrists Based on the Homogeneous Transformation Representation[1]

Richard P. Paul and Hong Zhang

Reviewed by

R. Featherstone
Philips Laboratories
North American Philips Corporation
Briarcliff Manor, NY 10510

This paper describes general methods for calculating in symbolic form the direct kinematics, inverse kinematics, Jacobians and inverse Jacobians of 6-DoF manipulators with spherical wrists. The techniques are based on those described in [Paul 1981]; the main improvement being the introduction of explicit intermediate results (the $\mathbf{U}$ matrices) which allow the authors to be more precise in describing the techniques, and allow them to show how the intermediate results from the direct kinematics can be used to simplify the calculation of the Jacobian. The techniques are described by way of a worked example using the joint arrangement of the PUMA 560 robot.

Symbolic techniques offer speed and other advantages over numerical techniques, and this paper is one of the most comprehensive treatments of general methods for symbolic kinematics equations currently available.

The method for direct kinematics produces a sequence of assignment statements which are, in effect, a symbolically simplified version of the general matrix equations of direct kinematics ($\mathbf{T}_6 = \mathbf{A}_1 \mathbf{A}_2 \ldots \mathbf{A}_6$), which can be evaluated much more quickly than the general matrix equations.

The direct kinematics is calculated as follows. First, the general matrix equation is broken down into a sequence of simpler equations: $\mathbf{U}_6 = \mathbf{A}_6$, $\mathbf{U}_5 = \mathbf{A}_5 \mathbf{U}_6$, ..., $\mathbf{T}_6 = \mathbf{U}_1 = \mathbf{A}_1 \mathbf{U}_2$. These equations are then expanded to scalar equations for the individual components of the $\mathbf{U}$ matrices, and symbolic expressions are substituted for the elements

---

[1] Int. Jnl. Robotics Research, vol. 5, no. 2, pp. 32–44, 1986.

of the **A** matrices. Since most of the elements of a typical **A** matrix are zero, a considerable amount of simplification is possible. Finally, all unproductive equations are removed from the list, so that the remaining equations represent the minimum number of calculations necessary to compute the overall solution ($\mathbf{T}_6$).

There are a couple of tricks which the authors use to improve efficiency. The first is to introduce compound **A** matrices for groups of consecutive parallel joint axes (like joints 2 and 3 of the PUMA robot), and the second is to calculate only the right-hand three columns of each of the **U** matrices and use the property that the first column of a homogeneous transformation matrix is the vector cross product of the second and third columns.

The solution to the inverse kinematics problem involves solving a sequence of equations of the form $\mathbf{V}_i = \mathbf{U}_{i+1}$ for the value of θ_{i+1} for $i = 0 \ldots 5$. $\mathbf{V}_i$ is defined by $\mathbf{V}_i = \mathbf{A}_i^{-1} \mathbf{V}_{i-1}$, $\mathbf{V}_0 = \mathbf{T}_6$, so the numeric value of $\mathbf{V}_i$ is known given $\mathbf{T}_6$ (the input) and the numeric values of $\theta_1 \ldots \theta_i$. For each equation above, one or two scalar equations are extracted which will allow the value of θ_{i+1} to be determined from the numeric values of the elements of $\mathbf{V}_i$ and the symbolic expressions for the elements of $\mathbf{U}_{i+1}$. The value of θ_{i+1} can then be used to compute $\mathbf{V}_{i+1}$, and so on.

This method depends for its success on the existence of expressions in each $\mathbf{U}_i$ which will allow a solution for θ_i. The existence of a closed-form solution is clearly a necessary condition for this to be so, but it is not clear whether it is a sufficient condition, or whether the condition that the robot should have a spherical wrist is either necessary or sufficient. In any case, it is left to the user to find these expressions if they exist.

It seems to me that only certain patterns can occur in the expressions in the **U** matrices, and it would be interesting to pursue this to see if there might be a fixed number of soluble equations to look for, thereby allowing the complete automation of the procedure. The beginnings of such an analysis appear in [Paul 1981].

One shortcoming of this method is that no attempt has been made to minimise the number of transcendental function calls, particularly sines and cosines.[2] Each time an angle is calculated and its sine and cosine are needed in subsequent equations, this method calculates them from the angle. It is often possible to avoid these expensive calculations,

---

[2] Apparently, the authors used a table-lookup scheme to calculate sines and cosines, which may explain why they did not attempt to minimise calls to these functions.

possibly at the cost of a square-root calculation, as demonstrated in [Hollerbach and Sahar 1983].

The Jacobian is calculated by formulating for each joint the velocity vector induced by that joint in the following link. The vectors are initially expressed (symbolically) in the coordinate frame of the following link, and have to be transformed to the coordinate system of link 6. This coordinate system is chosen because its origin coincides with the wrist point, and because the transformation from link i coordinates to link 6 is simply the matrix $\mathbf{U}_{i+1}$, whose value is known from the calculation of the direct kinematics. The consequence of having the origin at the wrist point is that position and orientation are decoupled, so the Jacobian takes on the following form:

$$\mathbf{J} = \begin{bmatrix} \mathbf{J}_{11} & 0 \\ \mathbf{J}_{12} & \mathbf{J}_{22} \end{bmatrix}.$$

This simplifies the process of inversion, which can be accomplished using the formula

$$\mathbf{J}^{-1} = \begin{bmatrix} \mathbf{J}_{11}^{-1} & 0 \\ -\mathbf{J}_{22}^{-1}\mathbf{J}_{12}\mathbf{J}_{11}^{-1} & \mathbf{J}_{22}^{-1} \end{bmatrix}.$$

$\mathbf{J}_{11}$ is simple enough to be inverted symbolically, but $\mathbf{J}_{22}$ is better inverted numerically.

References

[Hollerbach and Sahar 1983]
    Hollerbach J. M. and Sahar, G., Wrist-Partitioned Inverse Kinematic Accelerations and Manipulator Dynamics, *Robotics Research*, vol. 2, no. 4, pp. 61–76, 1983.

[Paul 1981]
    Paul, R. P., *Robot Manipulators: Mathematics Programming and Control*, MIT Press, Cambridge, Mass., London, 1981.

Development of Holonic Manipulator[1]
Michitaka Hirose, Yasushi Ikei, and Takemochi Ishii

Reviewed by

Hirochika Inoue
Department of Mechanical Engineering
University of Tokyo
7-3-1 Hongo, Bunkyo-ku
Tokyo, Japan

This paper presents a very interesting concept in robot manipulator implementation. In usual manipulator systems, the control computer and electronic circuits are separated from the mechanical manipulator itself. However in the approach of this paper, control computers and various electronic circuits are distributively embedded inside the manipulator structure, and are connected by means of a computer network inside the manipulator.

A central feature of this implementation is a hybrid integrated circuit (IC) which forms a single joint control unit. It consists of a couple of 16 bit microcomputers (Intel 8097), memory, communication hardware, I/O ports, and a power amplifier for pulse width modulation. In order to decrease the size and volume of the electronic circuits, specific hybrid IC modules have been developed, and thus the entire system can be enclosed within four small hybrid IC modules.

The paper describes the design of a four-degree-of-freedom arm with a multi-fingered hand. The joint control unit is built into structural members which serve as each manipulator link. The hand has three fingers, each of which has two joints. The finger tips and the palm contain 82 optical sensors. A hand control unit not only controls finger joints but also processes this sensor information. All the control units are connected by a loop type computer network inside the manipulator, resulting in a significant reduction of wires. In their im-

---

[1]Japan-USA Symposium on Flexible Automation, Osaka, pp. 269-273, 1986.

plementation, only six wires (two wires for power supply, four wires for computer communication) are enough to connect all the electronics.

All the control units communicate with each other by means of synchronous serial packet communication over the network inside the manipulator. The communication speed of the network is 4 Mbps, and the minimum packet size is 16 bits. The joint computers not only control the joint servo but also can compute the inverse dynamics of the manipulator. The communication speed and the computational performance of the joint control units are such that the calculation of inverse dynamics can be performed in real time. In their experiments, the recursive Newton-Euler equations are divided into forward kinematics and backward torque calculation. By assigning each CPU to one stage of those calculations, the inverse dynamics can be calculated once every 10 msec.

According to Hirose and Ishii, the approach described in this paper is motivated by the philosophy of an autonomous distributed control system or *holonic system*. The central idea of this concept is to have harmonized components which not only act independently or autonomously, but also act cooperatively as one component within the total system. The term "holon" is composed of two words, *holos* meaning *whole*, and the suffix *on* meaning *particle*. The holonic approach attempts to fuse two contradicting characteristics, autonomy and dependency.

A robot is a mechatronic device which consists of mechanical and electronic components. In the design of an integrated mechanism with sensors and electronics such as a robot manipulator, the design of the wiring system becomes increasingly important as the machine is equipped with more and more sensors, electronics, and computers. An implementation technique that places electronic components inside the mechanical structural members seems to be required to make robots simpler, smarter, and more reliable.

This paper suggests a new and promising approach for such implementation, viewing the distributed electronic circuits of manipulator as a computer network. Not only a fusion of mechanical and electronic components, but also a design of distributed computer architecture that connects mechatronic parts will be a key for the design of a sophisticated smart manipulator in the future.

The reviewed paper describes the design and implementation of an early prototype of the holonic manipulator. A companion paper (Ishii, Hirose, and Ikei, 1988) introduces the advanced version which is equipped with specially developed hybrid IC modules. Unfortunately, this paper is written Japanese, but the figures and pictures present the design philosophy very clearly.

References

Ishii, T. Hirose, M. and Ikei, Y., 1988. Implementation of Embedded Computer in Holonic Manipulator, *Circuit Technology*, Vol.3, No.2.

Numerical Simulation of Time-Dependent Contact and Friction Problems in Rigid Body Mechanics[1]
Per Lötstedt

Reviewed by
Matthew T. Mason
Computer Science Department and Robotics Institute
Carnegie-Mellon University
Pittsburgh, PA 15213

To simulate a Newtonian rigid-body system is easy if the system is very simple, but difficult if the system is more complex. Lötstedt describes a new approach to simulation of complex planar rigid-body systems, including unilateral contacts, sliding friction, and impact. Three example simulations, including a falling tower of five blocks, demonstrate Lötstedt's approach.

The primary contributions of the paper fall into two areas:

1. The *contact problem*: to find a set of contact forces that maintain contact constraints and satisfy relevant friction and impact laws.

2. *Dynamic simulation*: to design a numerical integration technique with good convergence properties.

This review focuses on the contact problem. With the "usual" assumptions—rigid bodies subject to Newton's laws with Coulomb friction—the contact problem is fundamentally ill-posed. The difficulty goes beyond the familiar static redundancy problem. Newtonian mechanics of rigid bodies with Coulomb friction admits *ambiguities,* where the body accelerations are not uniquely determined; and *inconsistencies,* where no set of body accelerations satisfies the theory. Additional difficulties arise when frictional impact occurs—there are two competing definitions of the coefficient of restitution, and many different ideas about laws of impulsive friction, none of which are completely satisfactory.

In the present paper, Lötstedt adopts an expedient to resolve the difficulties: Coulomb's law is modified. The original statement of the law

---

[1] SIAM J. Sci. Stat. Comput., v5 n2, June 1984, pp 370–393.

says that the tangential force is proportional to the normal force. Lötstedt's modification states that the tangential force is proportional to an *estimate* of the normal force, extrapolated from previous values. This trick resolves the ambiguity and inconsistency problems, though, of necessity, it does not reproduce the behavior expected of Coulomb's law. For impact problems, Lötstedt assumes a fixed impulsive coefficient of friction μ_I, then solves for the impulsive force that minimizes the system kinetic energy, yielding a perfectly plastic collision.

We will briefly retrace Lötstedt's derivations, with a simplified notation. The goal, both for finite forces and for impulsive forces, is to transform the contact problem into a linear least-squares problem. To begin, we assume frictionless contacts. Let q be a vector including position and orientation coordinates for each body. Each contact constraint is described by a vector function $\phi_i(q) = 0$. Let $G(q)$ be the matrix of partial derivatives $\partial \phi_i/\partial q_j$, for those ϕ_i corresponding to active contacts. Hence the columns of G describe the contact normals of the currently active contacts. Let λ be a vector of normal force magnitudes at the active contacts, so that $G\lambda$ gives the contact forces and moments applied on all the bodies. The equations of motion can be written

$$M\,d\dot{q} = f(q, \dot{q}, t)\,dt + G(q)\lambda(t)\,dt$$

where M is a matrix of body inertias, and f is a given applied force. The problem is to find values of λ that enforce the contact constraints and satisfy the friction laws. The relations governing the contact require that the contact normal forces are non-negative, the normal acceleration at a contact is non-negative, and, at each contact, either the force or the normal acceleration is zero. These can be written

$$\lambda \geq 0$$
$$G^T d\dot{q} \geq 0$$
$$\lambda^T G^T d\dot{q} = 0$$

We use the equations of motion to eliminate $d\dot{q}$, obtaining

$$\lambda \geq 0$$
$$G^T M^{-1}(f + G\lambda) \geq 0 \qquad \text{NLCP}$$
$$\lambda^T G^T M^{-1}(f + G\lambda) = 0$$

NLCP is a non-linear complementarity problem, which, as Löstedt shows, is equivalent to two different least-squares problems:

$$\text{minimize } \|f + G\lambda\|^2_{M^{-1}}, \; \lambda \geq 0 \qquad \text{KT}$$

and
$$\text{minimize } ||d\dot{q} - M^{-1}f\,dt||^2_M, \; G^T d\dot{q} \geq 0 \qquad \text{PLC}$$

where $||x||^2_{M^{-1}} = x^T M^{-1} x$. KT, the Kuhn-Tucker condition for NLCP, determines the total body forces $G\lambda$ uniquely, although the contact forces λ are not uniquely determined in cases of static redundancy. Given the body forces, the accelerations are determined, and the velocities and positions can be obtained by numerical integration. PLC is dual to KT, and is a restatement of Gauss' *principle of least constraint* (Lanczos 1949). PLC gives the body accelerations directly.

The extension to frictional contact is straightforward. For contacts at which no sliding takes place, Lötstedt introduces an additional tangential constraint. For contacts at which sliding is occurring, Lötstedt applies the modified Coulomb law, which introduces a tangential force as a function of previous normal forces. Since the tangential force is independent of the current normal force, the tangential force can be included as a given force in the term f above. For impact, Lötstedt assumes a fixed coefficient of friction μ_I relating tangential and normal impulses, and finds the impulsive forces minimizing the kinetic energy after impact. This again yields a least-squares formulation, but is restricted to perfectly plastic collisions.

Lötstedt's paper touches on a number of interesting issues. The reformulation of friction and impact models sheds light on the ambiguities and inconsistencies in rigid body mechanics. The application of non-linear programming techniques to the simulation of rigid body systems is novel. By altering the models of friction and impact, Lötstedt's system covers most of the bases, including multiple frictional contact, and simultaneous frictional collisions. Lötstedt also discusses generalization to three dimensions: many aspects of the system generalize readily, but sliding friction presents some problems.

For related work on simulation of rigid-body systems, see Featherstone (1987) and Hoffman and Hopcroft (1987). For recent work on ambiguity and inconsistency in rigid body mechanics, see Lötstedt (1981), Rajan et al. (1987) and Mason and Wang (1988). Also, see Featherstone (1986) for an inconsistency in frictionless mechanics.

Acknowledgments

This review grew out of discussions with Yu Wang. This work was supported by the National Science Foundation under grant number DMC-8520475.

References

R. Featherstone. 1986. "The dynamics of rigid body systems with multiple concurrent contacts," *Robotics Research: the Third International Symposium,* MIT Press, Cambridge, Massachusetts.

R. Featherstone. 1987. *Robot Dynamics Algorithms,* MIT Press, Cambridge, Massachusetts.

C. Hoffman and J. Hopcroft. 1987. "Simulation of Physical Systems from Geometrical Models," *IEEE J. Robotics and Automation,* v3 n3, pp. 194–206.

C. Lanczos. 1949. *The Variational Principles of Mechanics,* University of Toronto Press.

P. Lötstedt. 1981. "Coulomb friction in two-dimensional rigid body systems," *Zeitschrift für Angewandte Mathematik und Mechanik 61,* pp. 605–615.

M. T. Mason, and Y. Wang. 1988. "On the inconsistency of rigid-body frictional planar mechanics," *IEEE International Conference on Robotics and Automation,* Philadelpha, Pennsylvania.

V. T. Rajan, R. Burridge, and J. T. Schwarz. 1987. "Dynamics of a rigid body in frictional contact with rigid walls: motion in two dimensions," *IEEE International Conference on Robotics and Automation,* Raleigh, North Carolina.

Analysis of Multifingered Hands[1]
Jeffrey Kerr and Bernard Roth

Reviewed by

Yoshihiko Nakamura
Center for Robotics Systems in Microelectronics
University of California
Santa Barbara, CA 93106

This paper is concerned with three major problems of multifingered robot hands, that is, optimization of internal forces, inverse kinematics taking account of non-holonomic constraints, and work space computation. While reading this fairly long paper, one must be struck with admiration by the authors' perspective view which enabled to pick up these big problems as the research topics. A researcher often recognizes that choosing good research topics is more difficult and requires more knowledge and experiences than solving them. These problems are peculiar problems in multifingered robot hands and substantial for controlling dexterous robot hands.

The first problem is on optimization of internal forces. This can be viewed as a further extension of the work done by Salisbury [1982]. He showed internal forces are represented as a homogeneous solution of a linear equation and proposed to use them for keeping finger forces positive in the pushing directions. The authors are proposing to use them for satisfying various practical constraints such as static friction and joint torque limits. Kerr and Roth's most significant contribution in this first problem is to have formulated this problem as an optimization problem for *determining how hard to squeeze an object with the fingers in order to ensure that the object is grasped stably*. This motivation made it possible to bridge internal forces and grasping stability theoretically. Approximating friction cones by pyramids which are tangent to the cones, they represented all the constraints as linear constraints. The optimal internal forces were defined as the internal forces that makes the finger forces farthest from all the constraint planes, and solved using *the Simplex method*.

The second problem is on inverse kinematics taking account of non-holonomic constraints at contact points. To *determine the motion of each finger joint required to achieve a desired motion of the object* is much more complicated than to determine the motion of arm joints to do so. It is because, even in the ideal case, the motion of the object usually causes the rolling of fingertip relative to the object surface. This is a problem of non-holonomic constraint, that is, the equation relating two bodies (the object and

---

[1] The International Journal of Robotics Research, Vol.4, No.4, pp.3-17, 1986.

the fingertip) are expressed in terms of the their velocities rather than in terms of the their positions. Moreover, the equation becomes a function of the object shape and the fingertip shape. The authors explicitly derived a first-order nonlinear time-varying differential equation which is used to determine the joint motion and should be numerically integrated to compute the object position.

The third problem is to compute the workspace of multifingered robot hands. To discuss this highly complicated problem, the authors assumed that (1) the type of contact is point contact with friction and the contact points do not change during the motion, (2) the workspace is defined for a particular configuration of contact points on the object and for particular locations of the contact points on the fingertips, (3) the workspace of each fingertip is known and fixed relative to the palm, and (4) collisions or intersections between the object and the finger links or between the finger links themselves are ignored. The total workspace was defined as *the volume swept out by one particular point fixed in the object as the object is moved through all possible configurations.* Although the assumptions are strong except for the third one, Kerr and Roth succeeded in deriving a sufficiently analytical conclusion. They itemized the possible surfaces which might restrict the workspace of fingers. Considering the combination of them, the shape of workspace was topologically classified for planar-hands and spatial-hands. For planar case with two fingers, one surface to restrict the workspace is determined according to *the Grashof criteria* which is used for the analysis of four-bar mechanisms, and the workspace was classified into eight topologically different shapes as shown in Fig.1. For spatial case with three fingers, seven possible surfaces were specified and it was shown that the surface of the total workspace is made up of pieces of these seven surfaces. These results can be used to numerically determine the workspace.

As stated by the authors in the footnote, stable grasping has been studied by many researches and several definitions have been used. Unfortunately, however, it seems that the relationship between different stabilities is not very clear. Grasping stability has been discussed as the problem of generating restoring forces [Hanafusa and Asada, 1977; Nguyen, 1986; Jameson and Leifer, 1986; Fearing, 1986]. Hanafusa et al. [1985] discussed the determination of internal forces based on the motivation similar to that of authors', where a criterion to obtain enough finger forces was suggested in order to maintain the robustness against the external disturbing forces. These two papers are emphasizing the relationship between the internal forces and the stability or robustness of grasping and trying to find larger finger forces to ensure the stability or robustness. On the other hand, Cutkosky and Wright [1986] pointed out that the large internal forces sometimes mean the low-stable prehension because even small position errors may cause large disturbing forces. Nakamura et al. [1987] studied the grasping stability taking account of the dynamics of object and suggested to find small finger forces as long as they satisfy frictional constraints. It will be one of the important future problems in this field to clarify the relationship between these different stabilities and systematically formulate the total stability of grasping.

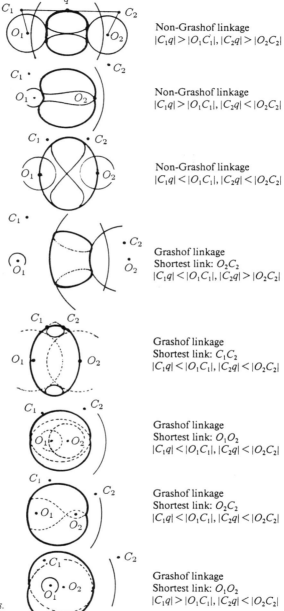

Fig. 1 Workspaces for planar hands with circular finger-workspace boundaries.

The connection between the second problem and tactile sensors is interesting. As mentioned by the authors, the use of tactile sensors simplifies the computation of the object motion. Contrarily, the requirements for tactile sensors will be able to be identified based on the similar discussion. This will provide another aspect to the research of tactile sensors. The computation of workspace needs a lot of further research to deal with weaker and practical conditions. The trajectory planning of the object will be done considering the computed workspace. Another future planning problem will be to determine finger trajectories allowing the discontinuous motion of contact points. Many important problems in this field could be suggested related to Kerr and Roth's paper. This is clearly showing the weight of this paper. The largest contribution would be to have illuminated the starting line and the track of the future research of robot hands.

References

Cutkosky, M.R., and Wright, P.K., 1986. Friction, Stability and the Design of Robotic Fingers. *International Journal of Robotics Research,* Vol.5, No.4, pp.20-37.

Fearing,R.S., 1986. Simplified Grasping and Manipulation with Dextrous Robot Hands. *IEEE Journal of Robotics and Automation,* Vol2, N0.4, pp.188-195.

Hanafusa, H., and Asada, H., 1977. Stable Prehension by a Robot Hand with Elastic Fingers. *Proc. 7th International Symposium on Industrial Robots,* pp.311-368, Tokyo.

Hanafusa, H., Yoshikawa, T., Nakamura, Y., and Nagai, K., 1985. Structural Analysis and Robust Prehension of Robotic Hand-Arm System. *Proc. '85 International Conference on Advanced Robotics,* pp.311-318, Tokyo.

Jameson, J.W., and Leifer, L.J., 1986. Quasi-Static Analysis: A Method for Predicting Grasp Stability. *Proc. 1986 IEEE International Conference on Robotics and Automation,* pp.876-883.

Nakamura, Y., Nagai, K., and Yoshikawa, T., 1987. Mechanics of Coordinative Manipulation by Multiple Robotic Mechanisms. *Proc. 1987 IEEE International Conference on Robotics and Automation,* pp.991-998.

Nguyen, V., 1986. The Synthesis of Stable Grasps in the Plane. *Proc. 1986 IEEE International Conference on Robotics and Automation,* pp.884-889.

Salisbury, J.K., 1982. *Kinematic and Force Analysis of Articulated Hands.* Ph.D. Thesis, Stanford University, Dep. Mechanical Engineering.

Efficient Parallel Algorithm for Robot Inverse Dynamics Computation[1]

C.S. George Lee and Po Rong Chang

Reviewed by

David E. Orin
Department of Electrical Engineering
The Ohio State University
Columbus, OH 43210

Over the past decade, a number of researchers have been addressing the problem of developing efficient algorithms for Inverse Dynamics. The motivation for this work stems from the desire to implement advanced robot control schemes such as the "computed torque" technique, which are based on computation of Inverse Dynamics, in real time. The main thrust of this paper is to develop a parallel algorithm for Inverse Dynamics which reduces the complexity of the computation from $O(n)$ to $O(\log_2 n)$, where n is the number of degrees of freedom of the robot manipulator. The most efficient serial algorithm for Inverse Dynamics, which is based on a recursive Newton-Euler formulation (Luh, Walker, and Paul 1980) and is of $O(n)$, is used as the starting point. These equations are then reformulated into a homogeneous linear recurrence form, and a classical method (the "recursive doubling" algorithm (Kogge and Stone 1973)) for parallel solution of linear recurrence problems is utilized.

While a number of others have considered efficient serial algorithms for Inverse Dynamics and their mapping on a parallel set of processors, few have considered parallel algorithms which reduce the order of the complexity. Lathrop (1985) was one of the first to do so, and many of the fundamental concepts of this paper are the same. This paper extends the work in a number of important ways. The authors consider the mapping of the parallel algorithm onto a set of p processors where $p \leq n$. The lower time bound for the computation is shown to be $O(k_1 \lceil n/p \rceil + k_2 \lceil \log_2 p \rceil)$ where k_1 and k_2 are constants, and a particular mapping and interconnection scheme among p processors

---

[1] *IEEE Trans. on Systems, Man, and Cybernetics*, Vol. SMC-16, No. 4, pp. 532-542, July/August 1986.

is shown which achieves the lower time bound. The interconnection scheme used is the inverse perfect shuffle which belongs to a class of network structures in which the number of links between processors is proportional to p.

One of the first results of the paper is the reformulation of the most efficient serial algorithm for Inverse Dynamics into homogeneous linear recurrence form. To do so, it is necessary to express the dynamics of each link in base coordinates rather than in link coordinates. Interestingly enough, this form reverts back to an earlier serial algorithm for Inverse Dynamics which was considered to be less efficient (Orin et al. 1979). The well-taken point is that parallel algorithms are not always derived directly from the most efficient serial forms.

A homogeneous linear recurrence problem is described in the paper as the problem of computing $x(1), x(2), \ldots, x(n)$ from $a(i)$, $1 \leq i \leq n$ where $x(0) = a(0)$ and

$$x(i) = x(i-1) * a(i). \tag{1}$$

The asterisk denotes an associative operator which can be a matrix product or addition, a vector dot product or addition, etc. Recursive doubling may be applied to such a problem by successive splittings of the computation. For instance, $x(3) = a(3) * a(2) * a(1) * a(0)$ can be split into two suboperations as $x(3) = [a(3) * a(2)] * [a(1) * a(0)]$, whose evaluations can be performed simultaneously on two processors. Generalizing the concept indicates that a total of $\lceil \log_2(n+1) \rceil$ steps are needed to obtain the result.

When recursive doubling is applied to Inverse Dynamics, the computational delay of the resulting algorithm is given in the paper as $27 \lceil log_2 n \rceil + 116$ scalar multiplications and $24 \lceil \log_2 n \rceil + 9 \lceil \log_2(n+1) \rceil + 84$ scalar additions or 197 and 183 for the case of $n = 6$. This compares favorably with the serial algorithm of Luh, Walker, and Paul (1980), (852 mult. and 738 adds).

A major concern of the paper is that of mapping the algorithm onto an architecture of less than n processors. This is a classical problem of computer science, and results of previous theoretical work on computer algorithms are applied to derive the previously stated lower time bound of the problem. An efficient parallel algorithm with p-fold parallelism is also derived in the paper and is shown to achieve the lower bound on the computation. The solution is based on p-parallel blocks with pipelined elements (serial execution) within each parallel block. A significant result of the work should also be noted. In this case, it was necessary to relax the restriction of assigning one processor strictly to each joint, and such a case should in general be considered when developing parallel computation methods.

The final sections of the paper address issues in implementation. The inverse perfect shuffle is used as the interconnection scheme among the processors and the data flow for the recursive doubling algorithm on the architecture is shown. The authors point out, as did Lathrop (1985), that the parallel algorithm may be appropriate for VLSI implementation with modular processors (MP), which have a microprocessor-like architecture. With large numbers of these processors, the computation delay can be reduced to 15 multiplications and 38 additions while successive computations have only 1 multiplication and 2 additions of delay.

As dynamics control schemes for robot manipulators are used more in practice, then there will be an increasing need for faster computations of Inverse Dynamics. This need will be accentuated with high-speed robots where the dynamic effects are even more significant. Hopefully, the capability of VLSI-based parallel computer architectures will also increase. However, given the cost-performance tradeoffs which are always an issue in engineering development, there will be a continuing need for parallel developments in algorithms suitable for such architectures. The work of Lee and Chang is an important contribution to this on-going effort.

References

Kogge, P.M. and Stone, H.S. 1973 (August). A Parallel Algorithm for the Efficient Solution of a General Class of Recurrence Equations. *IEEE Transactions on Computers*, Vol. C-22, No. 8, pp. 786-793.

Lathrop, R.H. 1985 (Summer). Parallelism in Manipulator Dynamics. *International Journal of Robotics Research*, Vol. 4, No. 2, pp. 80-102.

Luh, J.Y.S., Walker, M.W., and Paul, R.P.C. 1980 (June). On-line Computational Scheme for Mechanical Manipulators. *ASME Journal of Dynamic Systems, Measurement, and Control*, Vol. 102, pp. 69-76.

Orin, D.E., McGhee, R.B., Vukobratović, M., and Hartoch, G. 1979. Kinematic and Kinetic Analysis of Open-Chain Linkages Utilizing Newton-Euler Methods. *Mathematical Biosciences*, Vol. 43, pp. 107-130.

A Study of Multiple Manipulator Inverse Kinematic Solutions with Application to Trajectory Planning and Workspace Determination[1]

Paul Borrel and Alain Liegois

Reviewed by

Marc Renaud
LAAS/CNRS
7, Av. Edouard Belin
31077 Toulouse, France

The concept of *aspect* proposed by Borrel in his Ph.D. thesis (Borrel 1986) and summarized in this paper appears to be a promising notion related to the global computation of the different configurations of a robot manipulator (redundant or not) corresponding to a partially or totally specified *pose* (position and orientation) of its end-effector. This research introduces the decomposition of the admissible domain in joint space into sub-domains, named *aspects*, which take into account the morphology of the robot manipulator and the definition of the task to be performed. A decomposition algorithm is then presented and two fields of application are investigated:

- The automatic generation of the workspace of the end effector;
- The predetermination of trajectories which avoid the mechanical limits of the manipulator.

Finally, the implementation of these methods and algorithms on a CAD system is given.

Perhaps the main point of the paper is the concept of *aspect* which is not yet well known in the international literature but can lead to interesting theoretical and practical developments.

The research presented in this paper deals with simple kinematic chain robot manipulators with n degrees of freedom and an end-effector which is regarded as having m degrees of freedom in operational space.

---

[1] IEEE Conf. Robotics and Automation, pp 1180-1185, 1986.

Let q be the column matrix, of dimension $n \times 1$, of the n joint variables, q_i, defining the configuration of the robot manipulator (q_i represents an angle for a revolute joint and a distance for a prismatic joint). Let x be the column matrix, of dimension $m \times 1$, of the m operational coordinates defining, partially or totally, the pose (position and orientation) of the end effector.

There always exists a function f, called the direct geometric model, such that

$$f : q \in D_q \subset R^n \longrightarrow x \in D_x = f(D_q) \subset R^m;$$

where D_q is the admissible domain in the joint space such that

$$D_q = \prod_{i=1}^{n} [q_{i_{\min}}, q_{i_{\max}}].$$

The problem to be solved consists of the computation of the configurations q corresponding to a given pose x. It is very difficult to find a solution in the general case. When a configuration q_0 corresponding to a pose x_0 is known, and when the Jacobian matrix J of the function f is non-singular at the point q_0, the theorem of implicit functions allows us to say that there locally exits only one solution q, in the neighborhood of q_0, corresponding to a given pose x in the neighborhood of x_0.

This theorem then defines implicitly in the general case or explicitly in the case of robot manipulators with simple structure, a reciprocal or inverse function of f, in a neighborhood of x_0, of which the result, or image, belongs to the neighborhood of q_0.

The results presented in this paper allow the neighborhood of q_0 to be extended to a region A, named the *aspect*, of the domain D_q. A is more precisely defined as a connected sub-set of the domain D_q in which none of the m-order minors of the Jacobian matrix J go to zero, unless they are identically zero everywhere (Renaud 1981). Thus, the number of configurations corresponding to a given pose is less or equal to the number of aspects of the manipulator.

This research work doesn't address several problems which are however interesting:

1. Is it possible to characterize analytically the separating surfaces of the different aspects in order to avoid representation problems

arising with the utilisation of CAD systems for which it's not always easy to draw conclusions regarding the superposition or the non-superposition of some of the separating surfaces?

2. Is it possible to find the number of aspects for any robot and thus determining an upper bound on the number of configurations of the robot manipulator corresponding to any pose of its end effector?

3. Is it possible to characterize - analytically or by an other means - the images created by the function f of the aspects and is it possible to find their common parts in order to determine the number of configurations corresponding to a particular pose in operational space?

4. Is it posssible to compute the volumes of the aspects and thus to obtain the volume of the operational domain D_x (Gorla and Renaud 1984)?

References

Borrel, P. 1986. Contribution à la Modélisation Géométrique des Robots Manipulateurs. Application à la Conception Assistée par Ordinateur. Thèse de Doctorat d'Etat. USTL, Montpellier, France.

Renaud, M. 1981 (October). Geometric and Kinematic Models of a Robot Manipulator : Calculation of the Jacobian Matrix and its Inverse. 11th. ISIR. Tokyo, Japon.

Gorla, B. and Renaud, M. 1984. Modèles des Robots Manipulateurs. Application à leur Commande. Cepadues ed. Toulouse, France.

Mechanical Design of Robots [1]
Eugene I. Rivin

Reviewed by

Victor Scheinman
Automatix Incorporated
499 Seaport Court
Redwood City, CA 94063

So you've decided to design and build your own robot! Or maybe you want to know more about why the one you have shakes so much, or isn't as accurate as the specifications say it is. Well, Rivin has a good reference book for you. "Mechanical Design of Robots" addresses the kinematics, dynamics and general performance of robots from a structure and component design and selection viewpoint..

I came across this book while browsing in the MIT bookstore. I bought it on the spot because it is full of interesting discussion, clear conclusions and good references. Although a book intended for serious robot designers and mechanical engineering graduate students, anyone (including homebrewers) thumbing through the pages (325 of them) will certainly benefit and come away with a better understanding of mechanical design applied to robots and lots of useful pieces of information relating to robots, structures, materials, and novel components.

Getting from a performance specification to a working robot which meets that specification has generally been a poorly defined process with few clear design guidelines and limited appropriate analytical tools. A satisfactory result is rarely achieved without a tedious and time consuming iterative learning and refinement process. As robot performance requirements have increased, a rigorous mechanical design procedure has become more necessary. For example, stiffness cannot be increased just by beefing up a link or cross-section, or payload

---
[1] McGraw-Hill, 1988

and speed improved by using a larger motor, without other problems being created.

Mechanical design of robots is no longer a mechanical drawing and component selection process based on catalog data. It must be a well planned approach with emphasis on analytical techniques and consideration of the recent advances in motors, drive units, and materials. Although I don't completely agree with Rivin's design philosophy, which places the importance of good mechanical design above good software and electronics, the book makes an excellent case for that design approach with lots of detailed discussion and examples.

The definition section, or as Rivin titles it, "Basic Parameters of Robots", presents a commentary and useful typical values relating to the importance or uselessness of traditional specifications such as payload, speed, accuracy, and workspace, to which he adds more useful and complex robot characterizations such as stiffness, damping, effective speed and service index. The reader is led to these more meaningful properties in a brief but orderly manner, through several application examples and conclusions drawn within the text rather than just at the end of the chapters. For example, stiffness and compliance are first presented through their influence on robot motion with discussion of the effect of low stiffness on accurate continuous path operations and the need for controlled stiffness in many assembly operations.

The section on kinematic analysis is relatively brief, properly leaving extensive development and discussion to others and instead focusing on drawing conclusions about robot geometry and configuration. Workspace (gross reach region) and service index (orientation range at reach locations) and how these parameters relate to several common robot link and joint configurations are the focus issues of this chapter. A reference to the human arm "the best manipulator design available" is used to verify some of the conclusions about jointed manipulators.

In dealing with forces in manipulators the focus is not on just producing a mathematical solution to a problem but on explaining the results and drawing conclusions about them as they apply to robot structures and joint drive design. Emphasis is properly placed on the importance of gravity, inertia, and external interaction forces. Several manipulator configurations are examined, from a statics and dynamics approach. With respect to statics, the development is rather straightforward, showing the generally significant contribution of gravity to joint force and torque variations in vertical motion manipulator configurations. The dynamic development with discussion of the equations on a term by term basis shows the importance of the non-linear gravity terms and on manipulator design which reduces or removes joint coupling and minimizes these non-linear terms in the motion equations. This results in what I consider to be too much emphasis on mass counterbalancing as an engineering problem solution particularly when one considers that the next generation of robots will be characterized by average accelerations much greater than $1g$ and significantly higher natural structural frequencies.

Real or non-rigid manipulator structures are introduced with a rule of thumb relating to picking control system cutoff frequency at less than one-half the lowest structural natural frequency. Since many robot controllers handle this problem with notch filters, or don't even worry about it at all, it's a point of limited value, but serves to focus the reader on mechanical solutions to robot problems. More important in this section is the development and demonstration of kinematically induced damping and the excitation of this potential link (vibration) instability by the selected trajectory (motion profile) and joint drive components, notably "Harmonic Drives". Rivin concludes that mechanical solutions of structural damping treatment and joint and link stiffness are often the only viable way to stabilize certain high speed manipulators.

Over half the book is devoted to detailed discussion of robotic structures and components. Emphasis is placed on understanding the sources of structural compliance and designing for stiffness. Robot links are studied in detail with respect to shape, material selection, and mode of loading with a good presentation on the use of preload to improve stiffness in links as well as in joints and drive train components.

A final chapter on wrists gives a rather compact presentation and summary of a wide variety of wrist designs. The discussions focus on singularities and ranges of motion for the various configurations presented. Several of the recent novel non-degenerative wrist designs are presented with reasonably clear descriptions of their operation, features and limitations.

Lots of very good engineering numbers, such as typical mass ratios, accuracies and natural frequencies are presented with easy to understand comparative graphs and tables. Many different types of mechanical components are studied with accompanying detailed figures. The discussions get a bit wordy, but the reasoning is important in mechanical design because tradeoffs are not usually obvious.

The Calculation of Robot Dynamics Using Articulated-Body Inertias[1]
Roy Featherstone

Reviewed by
Michael W. Walker
Robot Systems Division
1101 Beal Avenue
University of Michigan
Ann Arbor, Michigan 48109-2110

This paper presents an efficient algorithm for the simulation of a robot manipulator. Simulation has always played an important role in system design and controller validation. More recently, real-time manipulator simulation has been used within long distant teleoperator control systems. In this application significant delays exist between the time the operator generates a desired manipulator position with a joystick and the time those positions are received by the manipulator controller at a remote site. This time delay problem causes significant degradation of the operators ability to carry out telemanipulation task. One solution to this problem is for the operator to plan motions with a real-time simulation of the manipulator [1,2]. The planned motions are continuously sent to the remote site for duplication of motion by the real manipulator. Thus, real-time manipulator simulation is becoming increasingly important.

Dynamic simulation involves the integration of the derivatives of the state variables of a system. For a manipulator these are the position and velocity of each joint of the manipulator. The forward dynamics problem for manipulators is the determination of the derivatives of these variables. No computations are required to obtain the time derivative of the position since it is the velocity which is a state variable and already known to the simulation. However, the time derivative of the velocity is the acceleration which is time consuming to compute. The basic result of the paper was the presentation of an algorithm that efficiently calculates this acceleration for serial link manipulators.

At least as important as the simulation method is the introduction of the spatial algebra. In this algebra velocities of links and the forces exerted on links are represented by 6-element column vectors called spatial velocities and spatial forces, respectively. They can be thought

---
1. "The International Journal of Robotics Research," vol. 2, no. 1, Spring 1983.

of as consisting of two 3-element column vectors. For spatial velocities, the first 3-element vector is the angular velocity of the link and the last 3-element vector is the linear velocity of the link. Both of these vectors are referred to same coordinate system. A somewhat confusing factor is the notion of the linear velocity of the link as opposed to the linear velocity of a point fixed on the link. One could think of the linear velocity of the link as the instantaneous velocity of a point located at the origin of the reference coordinate system, but fixed with respect to the link. Likewise, a system of forces and moments exerted on a link is equivalent to a single spatial force acting on the link. The first 3-element vector component is the sum of the forces acting on the link and the second 3-element vector is the total moment about the origin of the reference coordinate. A significant feature of this representation of velocities and forces is that the same 6 × 6 transformation matrix used to transform spatial velocities to a different coordinate system is used to transform spatial forces. Thus, if v_2 and f_2 are the spatial velocity and force referred to coordinate system 2 then the velocity and force referred to coordinate system 1, v_1 and f_1, is given by

$$v_1 = {}_1X_2 v_2$$
$$f_1 = {}_1X_2 f_2$$

where ${}_1X_2$ is the 6 × 6 transformation matrix from coordinate system 2 to coordinate system 1.

Link moments of inertia are represented as 6 × 6 spatial inertia matrices whose components consists of the zeroth, first and second moment of mass of the link with respect to the reference coordinate system. The same transformation matrix used for transforming velocities and forces is used to transform spatial inertia matrices. Thus, if I_2 is the spatial inertia of a rigid body with respect to coordinate system 2, then the spatial inertia of the object with respect to coordinate system 1, I_1, is

$$I_1 = {}_1X_2 I_2 \, {}_2X_1$$

Thus, the spatial algebra provides a notation which is particularly suited to the dynamic analysis of manipulators. It's properties allow one to simply formulate the equations of motion and to explore topics which might otherwise be impossible. Such is the case in the current paper.

The paper uses the spatial algebra to formulate a solution to the forward dynamics problem using the concept of articulated body inertias. Like the spatial inertia matrix, the articulated body inertias are

6 × 6 matrices. They relate the spatial force applied to a particular member of an articulated body and the spatial acceleration of that member, taking into account the effect of the rest of the articulated body. In the paper the author showed this relationship to be of the form

$$f = I^A a + P$$

where f is the spatial force, I^A is the 6 × 6 articulated body inertia, a is the acceleration, and P is the bias force. The bias force is a function of the current applied actuator torques contained in the articulated body. It is this equation from which the acceleration of the joint connecting the articulated body is determined as a function of the previous link acceleration, the actuator torque and the bias vector. This naturally leads to a recursive procedure for computing the joint accelerations. First the bias forces and articulated body inertias for each link are computed, starting at the last link in the chain and working back towards the base. Then, sequentially working out from the base to the end-effector, the accelerations are computed for each joint and each link. Thus, the algorithm complexity is linear in the number of joints of the manipulator.

The fact that the computational complexity is linear in the number of links in a manipulator is important because it ensures that the computations will remain manageable regardless of the number of links in the manipulator. Also, linearity in computational complexity usually implies a structure to the equations which facilitates implementation on special purpose hardware. For example, the algorithm could be efficiently implemented with hardware designed to handle the spatial algebraic computations.

The results presented in this paper is an important contribution not only from the standpoint of efficient manipulator dynamic simulation, but also because of the introduction of spatial algebra. It is likely that the algebra will be widely used in future robotics dynamics and controls research. More extensive results in manipulator simulation algorithms can be obtained from the author's book[3].

References

[1] Noyes, M., Sheridan, T.B., "A Novel Predictor for Telemanipulation through a Time Delay", *Proc. the Annual Conference on Manual Control*, NASA Ames Research Center, Moffett Field, CA, 1984.

[2] Conway, L., Volz, R., Walker, M.W. "Tele-Autonomous Systems: Methods and Architectures for Intermingling Autonomous and Telerobotic Technology," *Proc. of the IEEE International Conference on Robotics and Automation*, 1987, pp. 1121-1130.

[3] Featherstone, R. *Robot Dynamics Algorithms*, Kluever Academic Publishers.

IV

Motion and Force Control

Robot control, the engine through which the mechanism carries out tasks, is a central issue in the development of robot systems. The ability of a robot to execute a task is largely dependent on the performance of its control system.

Robot control is concerned with the execution and monitoring of robots' physical actions. A critical characteristic of robot control systems is their "real-time" nature. Dictated by the dynamic nature of the mechanism to be controlled, real-time behavior is also an important requirement in the control of the physical interaction of the robot with the world. By its real-time nature, robot control is the shortest link in the connection between perception and physical action.

Three problem areas in robot control can be identified. The problems of *dynamic control* are raised by the very nature of robot mechanisms: highly nonlinear and coupled multi-body systems. The second problem area, *motion coordination and trajectory generation*, is also a consequence of the multi-body nature of robot mechanisms. The third area is *interaction with the world*. This involves issues such as maneuvering the robot around obstacles and controlling its contact with objects during motion. The difficulty of these problems increases for systems, such as multi-arm robots, multi-fingered hands, and space-robots with a moving base, that involve large numbers of degrees of freedom.

The goal in *dynamic control* is to develop control algorithms that can cope with the system's nonlinear properties, with inertial coupling, and with the effects of centrifugal, Coriolis, and gravity forces.

The basic motivation for the use of dynamic control is *dynamic performance*. Simple PID controllers provide sufficient performance in many elementary industrial tasks. However, the performance of such controllers decreases rapidly when dynamic effects become significant. This problem area has received continuous attention over the past 15 years. Research in dynamic control has produced model-based nonlinear dynamic decoupling techniques and adaptive control methodologies. It has also dealt with issues of robustness, stability, and real-time implementation of dynamic control algorithms.

In *motion coordination and trajectory generation*, the objective is to produce a set of coordinated trajectories for the robot's individual degrees of freedom in order to accomplish the manipulation task. Trajectories are generated using the mechanism's geometric and kinematic descriptions. In recent work, dynamic considerations and actuator limitations have also been included in a search for optimal trajectories. In another line of work, the motion-coordination problem has been addressed within the general framework of dynamic control, where the robot's dynamics is described in terms of task coordinates, or operational coordinates.

The problem of *interaction with the world* covers wide areas of planning and control. Robot control systems have traditionally been associated with the execution of elementary tasks specified by the programming or planning system. Recent developments in planning and control clearly show a trend toward a tighter connection between perception and action. The goal is to develop real-time sensor-based control methodologies for a robot's operations in an evolving world in which there are both tolerances and uncertainties to deal with. Two broad areas of research have been addressed: gross motion control and force or compliant motion control. Local methods for collision avoidance have been used to provide real-time capabilities in dealing with obstacles. Force control has emerged as a basic means of expanding robots' capabilities in performing assembly tasks. Research in this area deals with issues of task description, passive compliance, active force control, and contact stability.

In part IV we have three survey articles (one by Hogan and Colgate, one Koditschek, and one by Slotine and Craig), each of which discusses about forty published papers in a targeted subarea of motion and force control. In addition, we have seven reviews of selected articles in motion and force control. Below, we give a brief synopsis of all of these pieces, arranged in an order suggested by the introduction above.

Craig reviews a paper by Arimoto and Miyazaki on the stability and robustness of PID feedback control. This paper is concerned with the stability analysis of the most straightforward control technique, PID, for robot manipulators.

Uncertainties in the parameters describing a manipulator's dynamics and load have motivated a number of efforts to apply adaptive control techniques to the area of robotics. The survey on "Adaptive Trajectory Control of Manipulators," by Slotine and Craig, discusses the earlier attempts based on the techniques developed for linear time-invariant single-input single-output systems and presents the more recent methodologies based on the linear parametrization of the dynamics of robot systems. This line of research is also discussed in Yoshikawa's review of a paper by Craig, Hsu, and Sastry on adaptive control of mechanical manipulators.

Hirzinger reviews a paper on satellite-mounted robot manipulators by Longman, Lindberg, and Zedd. The paper is concerned with the kinematic and dynamic problems associated with the motion of the satellite base. This is a research area of growing interest.

Patarinski reviews a paper by Krustev and Lilov on the generation of continuous paths for robot manipulators under joint position and velocity constraints. Kircanski reviews a paper on optimal trajectory planning for industrial robots by Johanni and Pfeiffer. These papers are representative of the attention that has been given to the problem of trajectory generation.

In the quest for higher interaction between the robot and the world, research in control systems is expanding beyond the traditional boundaries of planning and execution. The artificial potential field method is the subject of an extensive survey by Koditschek on robot planning and control techniques.

Asada reviews Mason's paper on compliance and force control of computer-controlled manipulators. This paper has provided the basic foundations for compliant motion task description. Planning and construction of sensor-based strategies for the control of compliant motion tasks is an active research area in robotics.

Salisbury reviews a paper by Maples and Becker on experiments in force control of robotic manipulators. A joint-position-based scheme is proposed to provide force-control capabilities for manipulators with imperfect torque-transmission mechanisms.

An important issue raised by the physical interaction of robots with their environments is the stability at contact. The problem of stability in contact tasks is the subject of a survey by Hogan and Colgate. This article presents stability analyses of various methodologies proposed in force control and discusses their limitations and their performance.

Stability Problems in Contact Tasks

Neville Hogan and Ed Colgate
Department of Mechanical Engineering
Massachusetts Institute of Technology
Cambridge, MA 02139

Introduction

The last several years have witnessed explosive growth in the study of robot force control. The development of successful strategies and implementations for force control is seen as a crucial step in enabling robots to perform tasks, such as deburring, drilling, and parts assembly, which require significant interaction with the environment. The implementation of high-bandwidth, high-accuracy force control, however, has proven to be quite difficult, primarily due to stability problems that occur upon contact with a rigid surface.

Like other types of robot control, a full scale implementation of force control calls for a heirarchical architecture. The breakdown of this heirarchy is a choice of the designer, but it is clearly useful to separate design of the higher levels, which must address task planning and supervision, from design of the lower levels, which must address issues of stability and performance. Much of the force control literature has been divided roughly along these lines.

Paul and Shimano (1976), Mason (1981), and Raibert and Craig (1981) were responsible for much of the early work in planning. In particular, the hybrid position/force control proposed by Raibert and Craig and based upon the theoretical framework of Mason has evolved into a fairly active field of research (Anderson and Spong 1987; Paul 1987; Shin and Lee 1985; Tilley, Cannon, and Kraft 1986; Yabuta, Chona, and Beni 1988). The basic idea behind hybrid control is that the physical constraints of the task should dictate those axes along which force is controlled and those axes along which position is controlled. For instance, in a surface grinding task, force should be controlled normal to the surface, and position tangential to the surface.

Early hybrid schemes implemented force control based upon a kinematic rather than a dynamic description of the robot. More recently, Khatib (1986, 1987) proposed an approach to hybrid control based upon the operational space formulation which accounts for the rigid body dynamics of a robot. Although efforts such as this indicate that the gap between those concerned with higher level issues and those concerned with lower level issues is beginning to close, it remains the case that most of the papers in the area of hybrid control have not addressed the lower level issues of stability or bandwidth in any detail. Yet, stability and high bandwidth are important issues in force control, and are the focus of this survey.

Contact Stability versus Performance

Whitney (1976) was the first to provide a stability analysis of a force controlled manipulator. He modeled a manipulator as a velocity-input integrator, and assumed that proportional position, velocity, and force feedback were implemented in discrete time. He modeled the environment as a spring and derived the following stability result:

$$0 < TGK_e < 1$$

where T is the sampling rate, G is the force feedback gain, and K_e is the combined stiffness of the sensor and the environment. For fixed T, this indicates that a tradeoff exists between G and K_e. In other words, high bandwidth force control requires a compliant sensor or environment.

In the years since Whitney published this result, the essential stability tradeoff between force feedback gain and stiffness of the environment[1] has been substantiated many times (Eppinger and Seering 1986, 1987; Hirzinger 1983; Kazerooni 1987; Roberts, Paul, and Hillberry 1985). It is now evident, based on the wealth of literature, that the problem is not tied to a particular geometry, but applies to all robots; moreover, it is now known that the problem transcends the digital implementation, that even for analog control the tradeoff exists (Wlassich 1986). In a recent review, Paul (1987) indicated that the contact problem with a rigid manipulator, rigid sensor, and rigid environment is still unsolved.

---

[1] The stiffness of the sensor is generally included as part of that of the environment.

Various explanations for contact instability have been offered. Kazerooni (1987) and An (1986) have both shown that the tradeoff may be attributed to unmodeled dynamics. Eppinger and Seering (1986, 1987, 1988) have carried out extensive stability analyses based on a series of increasingly sophisticated single-axis models. They have shown that stability problems arise due to the *non-colocation*[2] of sensors and actuators. There are two important consequences of their observations: first, that it is specifically those dynamic elements (modeled or not) separating the actuator and sensor which are most seriously implicated in contact instability; second, that, even if the robot is exquisitely modeled, high gain force control is a challenging proposition.

Eppinger and Seering (1987) also performed simulations which indicated that the force discontinuity that occurs upon contact with a surface may decrease the range of acceptable gains. However, not all nonlinearities necessarily degrade force control performance; Townsend and Salisbury (1987) have shown that coulomb friction can actually have a stabilizing effect.

Using a Nyquist stability analysis, Colgate and Hogan (1988) have derived the necessary and sufficient conditions for stability in contact with arbitrary passive objects and have shown that contact instability is determined by the impedance of the robot at the point of contact. An important consequence is that contact instability is not an exclusive property of force-feedback controlled systems (1987), but may arise in other controller designs. For example, joint-level position controllers which include integral action will exhibit instability when objects exceeding a certain mass are grasped.

Dynamic Compensation

Many techniques have been tested in efforts to improve force control fidelity. A number of these techniques involve the use of dynamic compensators (e.g., PID force controllers). Interpreting the literature in this area is complicated by the variety of manipulator models that are used. Although a rigid body description generally serves as the base model, it is frequently assumed that the force control loop is closed *around* some higher-bandwidth inner loop, such as a position (Kazerooni 1987; Roberts, Paul, and Hillberry 1985) or velocity (Haefner,

---
[2] The importance of non-colocation in the stability of feedback systems was first noted by Gevarter (1970) in the context of flexible space vehicles.

et al. 1986; Stepien, et al. 1987; Whitney 1976) controller. Others, however, have analyzed force feedback in the absence of such inner loops (An 1986; Eppinger and Seering 1986, 1987; Youcef-Toumi and Li 1987). A review of the different control architectures in which force feedback appears is presented by Whitney (1985). De Schutter (1987) has contrasted the use of position, velocity, and acceleration[3] controlled plants. His analysis, however, focuses on performance criteria such as steady state error and accuracy rather than contact stability.

Kazerooni (1987) has shown, for (inner loop) position controlled manipulators, that the gain of any force feedback compensator is inversely proportional to the stiffness of the position servo and to the stiffness of the environment, if stability is to be guaranteed by the multivariable Nyquist criterion. If the position controlled manipulator is non-backdriveable, as is generally the case, and the environment is a rigid surface, the conclusion is that even very low force feedback gains should be sufficient to create contact instability. Roberts, Paul, and Hillberry (1985) experimentally confirmed the tradeoff between servo/environment stiffness and the gain of an integral force controller. They found that low stiffness was necessary to achieve reasonable bandwidth.

Various dynamic compensators—PD, PI, and PID—have been implemented on velocity controlled manipulators with similar results. For instance, Haefner, et al. (1986) implemented a PID force controller for a deburring task. Although they achieved stable control, they found that the bandwidth was too low to be of practical value.

A similar tradeoff exists in the absence of an inner loop control; however, in this case the relative merits of various sorts of dynamic compensation have been more thoroughly investigated. Eppinger and Seering (1987) suggest that compensators which add lead (e.g., PD) should allow larger gains, whereas those which add lag (e.g., PI or slow filters) should have the opposite effect. However, both An (1986) and Youcef-Toumi and Li (1987) have achieved promising results with a first-order lag compensator. This inconsistency can perhaps be resolved by noting that these implementations required that the compensator rolloff occur at very low frequency, which Eppinger and Seering did not necessarily assume.

---

[3] In the case of a rigid body manipulator, acceleration control corresponds to the absence of an inner loop control.

Khatib and Burdick (1986) implement force control in the abscence of an inner loop, but also introduce a compensator for the "impact transition" stage when the manipulator first makes contact (typically at non-zero velocity) and chatter is most likely. This compensator, which uses velocity feedback to dissipate energy during the transition, has been shown to create stable impact behavior. Once stable contact has been established, pure proportional force control is implemented; any tradeoff between feedback gain and stability is not discussed.

Alternative Approaches

A number of other approaches to improved force control exist. For instance, Hogan (1985a, 1985b) has shown that impedance control can be used to control the force exerted on an environment without the need for force feedback. This approach is to implement an endpoint impedance which is a dynamic relationship between motions input by the environment and forces output by the robot. Certain classes of impedances have been shown to guarantee contact stability (Hogan 1988). Hogan (1987) and Wlassich (1986) have also investigated impedance control implementations which make use of force feedback. They have shown that force feedback acts primarily to modulate the inertia of the robot as seen by the environment, and that high gain force feedback does the equivalent of attempting to mask this inertia. One interesting result is that *positive* force feedback (which acts to increase the inertia) was shown to have a strong stabilizing effect on contact. The utility of this result in the context of high bandwidth force control has not yet been demonstrated.

Other approaches to improving force control involve robot design modifications. A number of investigators have suggested the use of passive compliance between the robot and environment (An 1986; De Schutter 1987; Paul and Shimano 1976; Whitney 1976). One example of passive compliance is the use of a remote center compliance (RCC) (Drake 1981), which is well-known as a solution to the peg-in-hole insertion problem. De Schutter (1987) has made the case that passive compliance is essential to reduce impact loads and therefore to increase task execution speeds. Of course, passive compliance is not satisfactory in all applications due to the positioning inaccuracies which may be created. Roberts, Paul, and Hillberry (1985) show that it is possible to compensate for this effect, but their implementation exhibits fairly low bandwidth.

Part of the value of force feedback from a wrist sensor is that, in theory, undesirable effects such as motor cogging and joint friction may be masked. If these effects can be eliminated through design, however, open loop force control becomes a conceivable alternative. Youcef-Toumi and Li (1987) have made the case that, due to the inherently simple dynamics of a direct-drive arm, improved force control may be achieved, possibly in an open loop fashion. Townsend (1988) has designed a cable-driven manipulator which is also intended to exhibit particularly good open loop force control.

Another design alternative is the use of a macro/micro architecture. Sharon, Hogan, and Hardt (1988) have argued that the use of a high bandwidth micro-robot mounted on the end of a standard macro-robot should enable the implementation of higher bandwidth force controllers. Tilley, Cannon, and Kraft (1986) have presented encouraging data with a similar system. Their manipulator is composed of a very fast wrist subsystem mounted on a very flexible arm. An integral force controller at the wrist was shown to exhibit reasonable stability and bandwidth.

Conclusions

In summary, high bandwidth control of the force exerted on a rigid environment remains elusive. Although stable force controllers have been implemented on a number of different robots with a wealth of different control architectures, performance has been limited in every instance. None of the sundry compensation schemes that have been introduced have achieved notable success from an applications standpoint; however, these studies have continued to improve the basic understanding of those issues which affect the performance of force controlled robots, and have provided direction for ongoing research. Many current efforts are addressing better dynamic modeling of robots as well as better design of robots for force control.

References

An, C.H. 1986. *Trajectory and Force Control of a Direct Drive Arm*. PhD thesis, M.I.T. Department of Electrical Engineering and Computer Science.

Anderson, R.J. and Spong, M.W. 1987. Hybrid Impedance Control of Robotic Manipulators. *Proceedings of the IEEE International Conference on Robotics and Automation*, pp. 1073–1080.

Colgate, J.E. and Hogan, N. 1987. On the Stability of a Manipulator Interacting with its Environment. *Proceedings of the Twenty-Fifth Annual Allerton Conference on Communication, Control, and Computing*, Monticello, IL, pp. 821–828.

Colgate, J.E. and Hogan, N. 1988. Robust Control of Dynamically Interacting Systems. *International Journal of Control*, vol. 48, no. 1, pp. 65–88.

De Schutter, J. 1987. A Study of Active Compliant Motion Control Methods for Rigid Manipulators Based on a Generic Scheme. *Proceedings of the IEEE International Conference on Robotics and Automation*, pp. 1060–1065.

Drake, S. 1981. *Using Compliance Instead of Sensory Feedback for High Speed Robot Assembly.* Technical Report CSDL-81-833, C.S. Draper Laboratory.

Eppinger, S.D. and Seering, W.P. 1986. On Dynamic Models of Robot Force Control. *Proceedings of the IEEE International Conference on Robotics and Automation*, pp. 29–34.

Eppinger, S.D. and Seering, W.P. 1987. Understanding Bandwidth Limitations in Robot Force Control. *Proceedings of the IEEE International Conference on Robotics and Automation*, pp. 904–909.

Eppinger, S.D. and Seering, W.P. 1988. Modeling Robot Flexibility for Endpoint Force Control. *Proceedings of the IEEE International Conference on Robotics and Automation*, pp. 165–170.

Gevarter, W.B. 1970. Basic Relations for Control of Flexible Vehicles. *AIAA Journal*, vol. 8, no. 4, pp. 666–672.

Haefner, K.B., Houpt, P.K., Baker, T.E., and Dausch, M.E. 1986. Real Time Robotic Position/Force Control for Robotic Deburring. F.W. Paul and K. Youcef-Toumi, editors, *Robotics: Theory and Applications*, ASME, New York.

Hirzinger, G. 1983. Direct Digital Robot Control Using a Force-Torque Sensor. *Proceedings of the IFAC Symposium on Real Time Digital Control Applications*, pp. 243–255.

Hogan, N. 1985a. Impedance Control: An Approach to Manipulation: Part I—Theory. *Journal of Dynamic Systems, Measurement, and Control*, vol. 107, pp. 1–7.

Hogan, N. 1985b. Impedance Control: An Approach to Manipulation: Part II—Implementation. *Journal of Dynamic Systems, Measurement, and Control*, vol. 107, pp. 8–16.

Hogan, N. 1987. Stable Execution of Contact Tasks Using Impedance Control. *Proceedings of the IEEE International Conference on Robotics and Automation*, pp. 1047–1054.

Hogan, N. 1988. On the Stability of Manipulators Performing Contact Tasks. *IEEE Journal of Robotics and Automation*, in press.

Kazerooni, H. 1987. Robust, Non-Linear Impedance Control for Robot Manipulators. *Proceedings of the IEEE International Conference on Robotics and Automation*, pp. 741–750.

Khatib, O. 1987. A Unified Approach for Motion and Force Control of Robot Manipulators: The Operational Space Formulation. *IEEE Journal of Robotics and Automation*, vol. RA-3, no. 1, pp. 43–53.

Khatib, O. and Burdick, J. 1986. Motion and Force Control of Robot Manipulators. *Proceedings of the IEEE International Conference on Robotics and Automation*, pp. 1381–1386.

Mason, M.T. 1981. Compliance and Force Control for Computer Controlled Manipulators. *IEEE Transactions on Systems, Man, and Cybernetics*, vol. SMC-11, no. 6, pp.418–432.

Paul, R.P. 1987. Problems and Research Areas Associated with the Hybrid Control of Force and Displacement. *Proceedings of the IEEE International Conference on Robotics and Automation*, pp. 1966–1971.

Paul, R.P. and Shimano, B. 1976. Compliance and Control. *Proceedings of the Joint Automatic Control Conference*, pp. 694–699.

Raibert, M. and Craig, J. 1981. Hybrid Position/Force Control of Manipulators. *Journal of Dynamic Systems, Measurement, and Control*, vol. 102, pp. 126–133.

Roberts, R.K., Paul, R.P., and Hillberry, B.M. 1985. The Effect of Wrist Force Sensor Stiffness on the Control of Robot Manipulators. *Proceedings of the IEEE International Conference on Robotics and Au-*

tomation, pp. 269–274.

Sharon, A., Hogan, N., and Hardt, D. 1988. High Bandwidth Force Regulation and Inertia Reduction Using a Macro/Micro Manipulator System. *Proceedings of the IEEE International Conference on Robotics and Automation*.

Shin, K.G. and Lee, C. 1985. Compliant Control of Robotic Manipulators with Resolved Acceleration. *Proceedings of the Twenty-Fourth Conference on Decision and Control*, pp. 350–357.

Stepien, T.M., Sweet, L.M., Good, M.C., and Tomizuka, M. 1987. Control of Tool/Workpiece Contact Force with Application to Robotic Deburring. *IEEE Journal of Robotics and Automation*, vol. RA-3, no. 1, pp. 7–18.

Tilley, S.W., Cannon, Jr., R.H., and Kraft, R. 1986. Endpoint Force Control of a Very Flexible Manipulator with a Fast End Effector. F.W. Paul and K. Youcef-Toumi, editors, *Robotics: Theory and Applications*, ASME, New York.

Townsend, W.T. 1988. *The Effect of Transmission Design on the Performance of Force Controlled Manipulators*. PhD thesis, M.I.T. Department of Mechanical Engineering.

Townsend, W.T. and Salisbury, J.K. 1987. The Effect of Coulomb Friction and Stiction on Force Control. *Proceedings of the IEEE International Conference on Robotics and Automation*, pp. 883–889.

Whitney, D.E. 1976. Force Feedback Control of Manipulator Fine Motions. *Proceedings of the Joint Automatic Control Conference*, pp. 687–693.

Whitney, D.E. 1985. Historical Perspective and State of the Art in Robot Force Control. *Proceedings of the IEEE International Conference on Robotics and Automation*, pp. 262–268.

Wlassich, J.J. 1986. *Nonlinear Force Feedback Impedance Control*. Master's thesis, M.I.T. Department of Mechanical Engineering.

Yabuta, T., Chona, A.J., and Beni, G. 1988. On the Asymptotic Stability of the Hybrid Position/Force Control Scheme for Robot Manipulators. *Proceedings of the IEEE International Conference on Robotics and Automation*, pp. 338–343.

Youcef-Toumi, K. and Li, D. 1987. Force Control of Direct Drive Manipulators for Surface Following. *Proceedings of the IEEE International Conference on Robotics and Automation*, pp. 2055–2060.

Robot Planning and Control Via Potential Functions

Daniel E. Koditschek [1]
Center for Systems Science
Department of Electrical Engineering
Yale University
New Haven, CT 06520

1 Introduction

There mingle in the contemporary field of robotics a great many disparate currents of thought from a large variety of disciplines. Nevertheless, a largely unspoken understanding seems to prevail in the field to the effect that certain topics are conceptually distinct. In general, methods of task planning are held to be unrelated to methods of control. The former belong to the realm of geometry and logic whereas the latter inhabit the the earthier domain of engineering analysis; geometry is usually associated with off-line computation whereas everyone knows that control must be accomplished in real-time; the one is a "high level" activity whereas the other is at a "low level". This article concerns one circle of ideas that, in contrast, intrinsically binds action and intention together in the description of the robot's task. From the perspective of task planning, this point of view seeks to represent abstract goals via a geometric formalism which is guaranteed to furnish a correct control law as well. From the perspective of control theory, the methodology substitutes reference dynamical systems for reference trajectories. From the point of view of computation, less is required off-line, while more is demanded of the real-time controller. From the historical perspective, these techniques represent the effort of engineers to avail themselves of natural physical phenomena in the sythesis of unnatural machines.

---

[1] This work was supported in part by the National Science Foundation under grant no. DMC-8505160

This article reviews the historical context and some of the more important contemporary practitioners of the potential field based methodology for robot planning and control. In Section 2, the example of PD feedback for a one degree of freedom mechanical system is used to illlustrate the basic idea in a trivial context. Namely, we interpret the proportional gain as representative of a one dimensional "planning system" based upon the geometry of cost functions. The derivative gain allows us to "embed" the limiting behavior of the planning system in the two dimensional physical plant. In Section 3 we sketch the history of the control theoretic aspect of this methodology — the progress of total energy conceived as a Lyapunov function — focusing upon the important work of Arimoto and colleagues. Finally, in Section 4 we suggest the planning capabilities of this point of view by focusing on the pioneering work of Khatib in the domain of obstacle avoidance, and touching upon Hogan's re-interpretation in the domain of tasks requiring contact with the environment.

2 Example: An End-Point Task for a Simple Robot

Consider the "one-degree-of-freedom" robot — a single (revolute or prismatic) joint whose position, q, and velocity, $\dot{q}$, are measured exactly and instantaneously by a perfect sensor, actuated by a motor which delivers exact torque, τ, according to the user's instantaneous command — and its dynamical equation, given by Newton's second law as

$$M\ddot{q} = \tau, \qquad (1)$$

where M is the mass (or moment of inertia in the revolute case) of the robot link. Suppose the robot has been given the task of moving to a point, q^*, and remaining there. One might imagine splitting the task up into a "high level" geometric problem — find a curve in the jointspace, $c(t)$, which ends up at q^* — and a "low level" control problem — find a contol law, $\tau(t)$, which "forces" the robot to "track" the commanded behavior, $q(t) \rightarrow c(t)$. In much of the robotics literature, these two problems are solved independently. In the methodology under consideration, they are solved at the same time.

2.1 Geometry: Hill Climbing

The geometric aspect of this task may be represented by the following optimization problem. Conceive of a "cost" function on the jointspace, φ, which assigns a scalar value to every position, vanishing uniquely at the "target", $\varphi(q^*) = 0$, and growing larger farther away. For example, the quadratic function, which we shall come to call a "Hook's Law" cost,

$$\varphi \triangleq \frac{1}{2} K_P \left[q - q^* \right]^2,$$

would do nicely for all positive values of the scalar K_P. If φ is continuously differentiable then it has a well defined negative gradient system,

$$\dot{q} = -grad\ \varphi. \tag{2}$$

Solutions of this differential equation follow the "fall line" of the "hill" defined by the cost function — i.e. at every position, the velocity of any solution curve is specified by the directional derivative of φ. If q^* is a local minimum of φ it follows that all solutions of system (2) which originate in some neighborhood of that point, define curves which lead to that point. If, in addition, q^* is the *only* extremum of φ, then *every* solution curve leads there. For example, in the case of the quadratic cost function, φ_{HL}, the gradient system works out to be the scalar linear time invariant differential equation,

$$\dot{q} = -K_P \left[q - q^* \right].$$

Since q^* is a minimum, and the only extremum of φ to boot, it follows that this gradient system generates a solution to the geometric "find path" problem from *any* starting position.

2.2 Control: Energy Dissipation

Faced with the particular problem of navigating from some initial position, q_0, to the target, q^*, one might now define a reference trajectory, $c(t)$, by solving the gradient differential equation, (2), for the initial condition, q_0, and then attempt a tracking control. Instead, we will re-interpret the cost function, φ, as a *potential function*, and introduce a control law which achieves the desired result with no recourse to explicit solutions of the original gradient system, (2).

If we are to interpret φ as a potential function, we must form the total energy by taking its sum with the kinetic energy,

$$\eta \triangleq \frac{1}{2} \dot{q}^T M \dot{q} + \varphi,$$

and then apply the Lagrangian formalism,

$$\left[\frac{d}{dt} \partial/\partial \dot{q} - \partial/\partial q \right] \left(\frac{1}{2} \dot{q}^T M \dot{q} - \varphi \right) = f_{ext}$$

(where f_{ext} denotes all non-conservative forces) to obtain the Newtonian law of motion,

$$M\ddot{q} + grad\,\varphi = f_{ext}.$$

If f_{ext} represents the effect of a dissipative force, say a Rayleigh damper, $f_{ext} = -K_D\dot{q}$, where K_D is positive, then the total energy must decrease:

$$\dot{\eta} = -K_D\dot{q}^2.$$

In consequence, it seems intuitively plausible, and will be made rigorously clear below, that $(q, \dot{q})$ converges to $(q^*, 0)$ from some neighborhood of that point in phase space — the space of positions and velocities — as long as q^* is a minimum of φ.

The final equation of motion resulting from this formulation is

$$M\ddot{q} + K_D\dot{q} + grad\,\varphi = 0. \tag{3}$$

It is clear that (1) may be made to look like (3) by assignment of the control law,

$$\tau \triangleq -K_D\dot{q} - grad\,\varphi.$$

Thus, if q^* is a local minimum of φ, then, under the influence of this control law, our robot is guaranteed to approach $(q^*, 0)$ asymptotically from any initial state, $(q_0, \dot{q}_0)$, which is sufficiently close to $(q^*, 0)$. In the particular case of a Hook's Law spring potential, φ_{HL}, this control strategy corresponds exactly to the time honored "proportional-derivative" feedback control strategy of linear systems theory,

$$\tau \triangleq -K_D\dot{q} - K_P\,[q - q^*],$$

with the familiar closed loop dynamics,

$$M\ddot{q} + K_D\dot{q} + K_P\,[q - q^*] = 0.$$

3 Control: Dissipative Mechanical Systems

The modern history of this idea might be said to begin with the discovery by the American engineering community of the work of Lyapunov [26]. The specific application to the class of mechanical systems seems to have been rediscovered on several different occasions by a variety of researchers in a variety of subdisciplines: Section 3.1 offers an illustrative (and certainly not exhaustive) sketch of this development. [1] Credit for first introducing these results into the robotics literature is due Arimoto and colleagues: Section 3.2 presents an overview of their central contributions.

3.1 Lyapunov Theory and Total Energy

Two hundred years ago, Lagrange showed that a conservative physical system has stable behavior with respect to any minimum of its potential energy. Roughly one century later, Lord Kelvin argued that the addition of a dissipative field would induce asymptotic stability of any minimum state. Since our understanding of the relation between gradient planning systems and robot dynamics stems entirely from these ideas, it is worth spending a little time to sketch a brief history of their subsequent development and discovery by engineers.

3.1.1 Lyapunov's Direct Method

In his doctoral dissertation of 1892, Lyapunov unified the asymptotic analysis of linear systems based upon eigenvalues with the energy-based analysis of mechanical (generally speaking, nonlinear) systems discovered by Lagrange and Lord Kelvin. The so-called "direct method" provides a remarkably simple test for stability of an equilibrium state, x^*, of a general dynamical system,

$$\dot{x} = f(x), \qquad f(x^*) = 0. \qquad (1)$$

Namely, one studies the derivative of a scalar valued function, $V(x)$, along trajectories of (1) in the neighborhood of x^*. Notice that

$$\dot{V}[x(t)] = grad\ V \cdot \dot{x} = grad\ V \cdot f(x),$$

hence no explicit knowledge of the trajectories of (1) (which are, in general, unobtainable for nonlinear systems) is required in order to apply the theory. Lyapunov showed that if V is *positive definite* at x^* — that is, takes positive values on a neighborhood, vanishing only at that point — and if $-\dot{V}$ is positive definite as well, then x^* is asymptotically stable.

While this powerful technique was developed and refined by a variety of Soviet mathematicians over the following five decades [5,27], it remained virtually ignored in the west. In the nineteen fifties, Lyapunov's method gained the attention of the American mathematician Solomon Lefschetz [25]. His colleague, Lasalle, discovered a fundamental relationship between Lyapunov functions and limit sets of dynamical systems which proves to be of primary importance for mechanical systems.

---

[1] Of course, the central idea of energy dissipation is to be found in physics texts as well [1][Prop. 3.7.17].

A glance back at the time derivative, $\dot{\eta}$, in Section 2.2 shows that the total energy is not a *strict* Lyapunov function: this means that its derivative vanishes (in this case, at every zero velocity state) even though the trajectory has not arrived at the desired equilibrium state. In general, if V is positive definite and $-\dot{V}$ is merely non-negative, then one may deduce stability (solutions originating near the equilibrium, x^* remain near that point) but not necessarily asymptotic stability (nearby solutions not only remain near, but eventually end up at x^*). Thus, the total energy appears to have a significant flaw for purposes of deducing asymptotic attraction to a nearby minimum of the potential energy: indeed, Lord Kelvin's original argument [41][§345] ignores this flaw. In his simple but illuminating book with Lefschetz, Lasalle [24] presented his "Invariance Principle", which demonstrated in the context described above that all solutions eventually end up in an *invariant* subset of the set $\dot{V} \equiv 0$. An invariant set of (4) is characterized by the property that all solutions originating there, remain there for all time. Thus, with a little more effort, Lyapunov theory may be used to make rigorous the intuitive arguments of Lord Kelvin.

3.1.2 Application to Problems in Mechanical Engineering

In conjunction with the advocacy of these ideas by Lefschetz came the the awareness of their relevance to engineering problems. The weakness of Lyapunov's method is that it provides information only if a "candidate function", V, is available: the construction of such functions remains something of a black art in general. In particular, however, when the dynamical system (4) has a specific structure, then it is often possible to say a great deal. This is the situation for linear systems — Lyapunov had already proven the existence of quadratic functions, V, in that case — as well as for mechanical systems where, as we have seen, the total energy provides the obvious choice. Kalman and Bertram published an influential article in 1960 describing the relevance of Lyapunov functions for control applications [13], leading to an avalanche of interest of which the present application are represents just one stream. Credit for first making use of total energy as a Lyapunov function for a specific engineering application involving mutually constrained bodies would appear to be due Pringle.

In his 1966 paper [31], Pringle first presents a succinct historical review and then defines his notion of a "connected system." This is a single rigid body together with a set of point masses interconnected via holonomic constraints. There are three theorems: the first expresses Lyapunov's direct method; the second presents the application to mechanical systems taken from the Soviet literature cited above, together with the appropriate appeal to Lasalle's invariance principle; the final theorem states that the block matrix

$$\begin{bmatrix} R & \frac{1}{2}\epsilon C \\ \frac{1}{2}\epsilon C^{\mathrm{T}} & \alpha P \end{bmatrix}$$

may be made positive definite, regardless of the properties of C, by a sufficiently small choice of ϵ relative to α as long as R, P are positive definite square matrices. This last theorem is used in conjunction with the Lyapunov arguments to determine the asymptotic behavior of his connected systems — considered as linked rigid bodies by passing to the limiting case of increasingly large potential forces — in the presence of passive dampers.

3.2 Arimoto's Use of Total Energy for Robot Systems

The first clear exposition of the use of total energy in robotics was contributed by Arimoto. Over the better part of a decade, he and his colleagues have offered numerous extensions and refinements of these ideas. An increasing number of researchers (this author among them) have had the experience of reporting a "new" result in the application of Lyapunov theory to robotics and subsequently discovering the same or similar ideas in papers written several years earlier by Arimoto. This section attempts to present an overview of his work in the area by concentrating on two important papers.

3.2.1 The Central Idea

In constrast to the situation two decades ago treated by Pringle who analyzed passive arrangements of dampers and springs, the advance of technology in robotics lends the designer complete freedom (at least in theory) to choose an arbitrary force law at each degree of freedom. Arimoto's contribution was to argue that computationally very simple force laws based upon the principles of energy dissipation discussed so far could accomplish rather sophisticated end-point tasks. The first exposition of this point of view seems to be his 1981 paper with Takegaki [39].

The authors observe that existing potential fields may be exactly cancelled and replaced by artificial fields in a manipulator whose joints are all actuated by user commanded torques. They next introduce a Rayleigh damping term and argue, as in our simple example of Section 2.2, that the total energy — the robot's physical kinetic energy added to the artificial potential function — must decrease. By invoking LaSalle's invariance principle, they obtain their desired asymptotic stability result. Since the Lagrangian formulation of dynamics obtains from variational principles, the authors are able to assert the optimality of the Rayleigh damping scheme with respect to a quadratic performance index based upon it. They next choose a quadratic artificial potential function: since the stability result requires only that the potential function be positive definite at the desired rest point, they point out that simple decoupled PD feedback will suffice.

The paper now shifts to applications involving workspace based planning. A quadratic artificial potential function in work space leads to a feedback law whose proportional term involves the workspace errors multiplied by the transposed jacobian of the forward kinematics. A certain amount of unnecessary confusion is added by sticking with the Hamiltonian rather than Lagrangian formulation of the dynamics: the former requires analysis involving the inverse jacobian whereas the latter would not. Holonomic constraints in the workspace are handled by a local technique based upon the implicit function theorem which yields tangent information to the jointspace based torque inputs. This necessitates a different feedback structure for each region of the task giving rise to the use of time varying potentials for which no stability proof is offered.

Subsequent to the publication of Arimoto's first paper, independent derivations of the same result were reported by Van der Schaft [35] and this author [22]. The latter development proceeded from the Lagrangian rather than Hamiltonian formulation of robot arm dynamics, thus the author was led to discover a structural aspect of the Coriolis terms which had not been reported in Arimoto's work. This "skew-symmetric" structure was later clarified and exploited by Slotine and Li [36] in the construction of a globally stable adaptive controller for robot arms.

3.2.2 Extensions and Problems

The important 1984 paper by Arimoto and Miyazaki [38], receives an independent review in the same issue of this journal. For the purposes of the present article, that paper provides a convenient point of departure for discussion of extensions and open problems in this domain. In the effort to gain a stability result for the traditional PID control algorithm from linear systems theory, the authors are forced to confront and "fix" the weakness of Lord Kelvin's original result: namely, the lack of a negative definite derivative which necessitates appeal to LaSalle's Invariance Principle.

They modify the total energy function

$$V_E(x) = x^T \begin{bmatrix} K_P & 0 \\ 0 & M \end{bmatrix} x$$

by adding a cross term which is bilinear in position and velocity,

$$V_{Ari}(x) = x^T \begin{bmatrix} K_P & \alpha M \\ \alpha M & M \end{bmatrix} x$$

resulting a locally strict Lyapunov Function,

$$\dot{V}_{Ari} = x^T \begin{bmatrix} K_P & \alpha M \\ \alpha M & M \end{bmatrix} \begin{bmatrix} 0 & I \\ -M^{-1}K_P & -M^{-1}K_D \end{bmatrix} x$$

$$= -x^T \begin{bmatrix} \alpha K_P & \frac{1}{2}\alpha K_D \\ \frac{1}{2}\alpha K_D & K_D - \alpha M \end{bmatrix} x.$$

For the purposes of this illustrative linear example, both V and $\dot{V}$ may be made sign definite via small enough choice of the parameter, α: a global stability argument results. In the general situation, the presence of the Coriolis forces results in a cubic term in $\dot{V}$ (it is linear in the position error and quadratic in the velocity error) which, though necessarily sign indefinite, is dominated by $K_D - \alpha M$ for small enough position error magnitudes. Thus V_{Ari} yields only a local stability argument: the designer must insure that the initial conditions originate in a neighborhood of the desired state small enough to guarantee that decreasing V implies the error remains less than a magnitude proportional to $\|K_D - \alpha M\|$. The alternative — enlarging the domain of attraction — may be accomplished only by recourse to larger derivative gains, K_D.

Tracking. The notion of tracking appears to be out of context in a review article concerning robot task encoding via reference dynamics. Actually, the use of a "moving Hook's Law Potential" — that is, a PD-based tracking method — serves as the first instance of generalized moving potential fields. For example, it would be very nice to use the obstacle avoidance schemes discussed below to navigate amidst moving environments. Two years ago, this author [21] and Bayard and Wen [42] announced independent work leading to the proof of exponential stability of PD compensated nonlinear mechanical systems. This construction afforded a proof of global boundedness of

PD-based tracking schemes, and reasonable estimates of the rate of convergence to the bound. Both results were based upon a modified version of total energy which yielded a locally strict Lyapunov function. At the following year's IEEE Conference on Robotics and Automation, Arimoto quietly pointed out that he had obtained the same result several years earlier [3]. The three independent papers offer strikingly similar variants of the total energy — all with the same local limitations discussed above. In fact, they are all similar to the original modification Arimoto had introduced in 1984.

Very recently, this author has developed a further modification of the total energy which yields a strict Lyapunov function whose extent is global[17]. Moreover, unlike the earlier modifications, which assume a Hook's Law Spring Potential, the new variant is defined for a large class of potential laws. This opens the desirable possibility of combining time variation with the complex artificial potentials to be discussed in the last Section of this article.

Internal Model Principle. Arimoto's extension of Lord Kelvin's ideas in the 1984 paper expand the analogy between linear second order systems and nonlinear mechanical systems from PD to PID techniques. In classical linear systems theory, the introduction of integral feedback action is associated with tracking a reference (or rejecting a disturbance) which is known structurally, but only up to an unknown set of parameters. For example, the addition of an integrator is dictated by the presence of a step input; two cascaded integrators are required in the presence of a ramp input; and so on. These ideas from the classical theory were formalized by Wonham and colleagues as "The Internal Model Principle" [8].

Since the most popular reference trajectories in the robotics literature are generated by polynomial spline techniques — that is, they may be modeled as outputs of unforced linear time invariant dynamical reference systems whose dimension is specified by the degree of the polynomial — the internal model principle is immediately applicable to Cartesian robot arms. Its extension to nonlinear mechanical systems would be of great significance for practical robot control.

Transient Analysis. Even though the desired asympotic behavior is assured, it has been pointed out on many occasions, originally in Arimoto's 1981 paper [39], that the transients resulting from potential field control of nonlinear mechanical systems may be poorly behaved. More generally, given a kinetic energy, it seems very important to develop a theory that matches the dissipative term to the particular form of the potential, φ, in order to ensure that the closed loop mechanical system give rise to a "sufficiently accurate" copy of the gradient "planning system" trajectories. In the simple example of Section 2 we already have an intuitive idea of what this means. The "best" match between (2) and (3) arguably obtains from the assignment of gains for the latter which produces a critically damped system whose natural frequency is the time constant of the former. It seems important to generalize this notion in the nonlinear context.

4 Planning: Gradient Dynamical Systems

The idea of using "potential functions" for the specification of robot tasks was pioneered by Khatib [14] in the context of obstacle avoidance, and further advanced by fundamental work of Hogan [12] in the context of force control. The methodology was developed independently by Arimoto in Japan [28], and by Soviet investigators as well [30]. Of course, the possibility of solving complex problems by resort to analog (or iterated discrete approximations of analog) methods of computation has a much older history which we touch upon briefly in Section 4.1 before turning to Khatib's contributions in Section 4.2.

4.1 Hill Climbing via Analog Computation

Possibly the most common use of gradient vector fields in engineering occurs in the context of root finding: given a function, $g(x)$, find those points which attain a particular value. For example, in robotics, g might likely represent the kinematic transformations. Recent work of Smale [37] and Hirsch and Smale [10] has established variants of the iterated Newton-Raphson technique which succeed for almost all initial points, in the search space. They point out that the gradient algorithm — climbing down the "error hill" — is a much older variant with poorer properties. Yet recourse to hill climbing has two other advantages that make it so attractive in the present context. First, peaks as well as valleys may be encoded: this idea leads to the notion of repelling sets of obstacles. Second, as we have seen in the previous section, the dissipative mechanical systems make natural analog computers for integrating gradient vector fields.

Of course, the appeal to analog computation has an old and established history. Even after the domination of electronic computation by digital technology in the sixties, serious engineering effort has been expended upon analog technology for special purpose problems of data acquisition and manipulation, e.g. for pulse code modulation devices [4], or image sensory processing [7,6]. More general classes of combinatorially complex problems have been approached from this point of view in recent years. Work of Hopfield and Tank [40] has demonstrated the possibility of finding good sub-optimal solutions to the travelling salesman problem (in the class np-complete) by formulating them in terms of a scalar potential function and implementing their solution using "neural nets" consisting of many processing elements endowed with elementary analog capability. Kirkpatrick and colleagues [16] have generated a great deal of interest in a class of objective functions denoted "simulated annealing" (because of their origin in models of physical cooling processes) used, again, to find good sub-optimal solutions to the problem of VLSI circuit device placement (also np-complete). It should be mentioned that the gradient method has a long history in systems science as well as the general world of applied mathematics. For instance, the parameter adjustment algorithms of adaptive control and estimation schemes may be seen as a gradient solution to a set of linear algebraic equations [29].

Unlike most nonlinear differential equations, gradient vector fields have particularly simple dynamical behavior. Systems possessed of isolated equilibrium states come with the guarantee that any bounded solution must asymptotically approach one of them travelling orthogonal to the level lines of the original "objective" function [11]. Thus, a careful construction methodology for potential functions brings the promise of good transient shaping in the planning system. Of course, our true interest is in an on-line implementation of the gradient algorithm implicit in the motion of the robot arm itself. Work of this author has established the global similarity in limiting behavior between the planning system and the ultimate closed loop mechanical system [22,17]. Yet, as mentioned at the end of the previous section, it remains to determine the relationship between the transients.

4.2 Khatib's Method of Obstacle Avoidance

There are certain ideas which are so natural and compelling that it becomes hard to distinguish a particular individual responsible for their introduction to a community researchers. Happily, in the case of artificial potential fields — surely the exemplar of such a compelling idea — it seems clear that credit for its introduction to the robotics community is due Khatib.

4.2.1 FIRAS

In his original 1978 paper with Le Maitre [14], Khatib suggested the desirability of bringing more "adaptibility" to the control level, while maintaining a relatively lean (hence, rapidly executable), yet dynamically meaningful maneuvering algorithm, for robots in cluttered environments. He proposed adopting the general paradigm of Renaud, who had suggested the relationship between cost function gradients and mechanical systems in a doctoral dissertation two years earlier [32], to the specific problem of obstacle avoidance.

Assuming that each obstacle is described as the zero level surface of a known scalar valued analytic function, $f(x, y, z) = 0$, Khatib formed a local inverse square potential law: it goes to infinity as the inverse square of f near the obstacle, and gets cut off at zero at some positive level surface, $f(x, y, z) = f_0$, presumably "far enough" away from the obstacle, as determined by the designer. A particle moving according to Newton's Laws in such a potential field would clearly never hit this obstacle. Khatib further observed that the sum of the gradients is the gradient of the sum: thus adding up the potential laws for many obstacles would result in a single function under whose influence the particle could not hit any obstacle.

Since the objects to be maneuvered are not point masses, but rigid links of a kinematic chain, Khatib proposed identifying a number of distinguished points on the manipulator body, and subjecting each of them to the potential field of the obstacles. This induces a potential function on the joint space: the techniques of stable hill climbing via Lagrangian dynamics described above are immediately applicable. To complete the methodology, Khatib suggested adding obstacle potentials representing the joint limits to the induced workspace function. Thus was born FIRAS — "force inducing an artificial repulsion from the surface" — in 1978.

4.2.2 Problems and Extensions

In the decade since its introduction, the idea of using artificial potential functions for robot task description and control has been adopted or reintroduced independently by a growing number of researchers [28,2,30]. There are several conceptual problems with the potential function methodology, of which the more important include the following.

Terrain Shaping Khatib [15] observed that the use of analytic level surfaces to prescribe obstacle boundaries and neighboring regions becomes difficult when the surface is complex because the relative locus of the concentric level surfaces may be poorly behaved. His alternative — reliance upon distance functions computed online — represents an interesting point of departure but would surely give rise to equally difficult problems in an effort to obtain theoretical understanding of the properties of such trajectories given the documented trickiness of set distances — see the work of Gilbert and Johnson [9], for example.

It seems equally plausible to imagine a more careful examination of the construction of implicit representations of solid body surfaces via scalar valued functions. Of course, the study of representations of solid bodies is a vast discipline of its own. While those researchers seem to favor explicit parametrizations rather than implicit representations [33], it seems certain that increased contact between the two communities would result in enhanced understanding for both. In particular, a careful listing of desiderata for such functions — both with respect to their use in robot navigation as well as in computer models of solid bodies — would seem to be essential before they are rejected out of hand.

Rigid Bodies While it is most natural to construct obstacle avoidance potentials on the workspace, existing methods are actually more suited to configuration space where the exact state of the robot at any instant is represented by a single point. As discussed above, work space boundaries (obstacle surfaces) are plausibly available from a CAD/CAM model: their "inverse images" under the kinematic transformation representing any particular robot are not. Thus practitioners of the methodology must rely upon heuristics. A creditable example is proposed by Khatib [15] who places his "points subjected to potential" forces on a specific locus — a line through the body — of each link in the kinematic chain. While this seems like a reasonable procedure, and the designer is guaranteed that the particular points in question will not collide with obstacles, there is no such guarantee for the entire link itself. Thus, there is a crying need for provably correct procedures which build obstacle avoidance potentials on the configuration space of rigid bodies rather than simply point robots.

Spurious Minima Even given a good description of the configuration space, a central difficulty with this technique has been the possibility of extra minima. As has been asserted, gradient vector fields have particularly simple dynamical behavior. Systems possessed of isolated equilibrium states come with the guarantee that any bounded solution must asymptotically approach one of them [11]. However, there is no generally applicable method for determining the particular equilibrium state achieved from a particular initial condition. All points in some open neighborhood of a local minimum will approach that point, whereas the set of points which approach a saddle has measure zero, and the set of points which approach a local maximum is empty. Thus, if the problem formulation gives rise to an objective function with more than one local minimum a sub-optimal solution is guaranteed for all initial conditions in some set of non-zero measure whose location is generally impossible to characterize analytically.

This has prompted the author to spend a growing amount of attention focused on the problem of how to construct "good" potential functions on manifolds with boundary [20]. We have been led to define the notion of a *navigation function* [34] — a refinement of the potential function which includes the condition of a single minimum (at the desired destination), a uniform finite height on all the boundaries, along with several other technical requirements — and have constructed them on a variety of spaces. Thirty year old topological results of S. Smale assure the existence of smooth navigation functions on any smooth space with boundaries [18]: it is now up to the engineering community to actually construct them.

4.3 Hogan's Impedance Functions

It seems useful to include in this article a brief look at some work which demonstrates that the potential function approach to unified task description and control is not limited to task domains involving a purely geometric environment. Fundamental work by Hogan [12] advances persuasive arguments for encoding general manipulation tasks in the form of "impedances". Impedances and admittances are formal relationships between the force exerted on the world at some cartesian position and the motion variables - displacement, velocity, acceleration, etc. - at that position with respect to some reference point (or "virtual position" in Hogan's terminology [12]). He argues that for purposes of modeling manipulation tasks, the kinematic and dynamical properties of a robots's contacted environment must be understood as admittances - systems for which the relationship operates as a function describing a specified displacement for any input force. Arguing, further, that physical systems may only be coupled via port relationships which match admittances to impedances, and that robots can violate physics no more than any other objects with mass, he arrives at the conclusion that the most general model of manipulation is the specification of an impedance - a system which returns force as a function of motion. By construing motions relative to a virtual position as defining tangent vectors at that position, Hogan notes that an impedance may be defined in terms of a scalar valued function on the cross product of two copies of the tangent space at each virtual position whose gradient co-vector determines the relationship between motion and force. Thus, an impedance may be reinterpreted as the gradient co-vector field of an "objective function", whose fall lines specify the desired dynamical response of the robot end-effector in response to infinitesimal motions imposed by the world. In this context, unlike the other gradient vector field task definitions, it is intended a priori that the dynamics be second order - i.e. define changes of velocity (force) rather than changes of position.

of view, while highlighting the significant contributions of three researchers: Arimoto, Hogan, and Khatib. A more elaborate treatment of these ideas may be found in the author's recent tutorial articles [19,23].

References

[1] Ralph Abraham and Jerrold E. Marsden. *Foundations of Mechanics*. Benjamin/Cummings, Reading, MA, 1978.

[2] J. R. Andrews and N. Hogan. Impedance control as a framework for implementing obstacle avoidance in a manipulator. In David E. Hardt and Wayne J. Book, editors, *Control of Manufacturing and Robotic Systems*, pages 243–251, A.S.M.E., Boston, MA, 1983.

[3] S. Arimoto and F. Miyazaki. Asymptotic stability of feedback controls for robot manipulators. In *Proceedings 1st IFAC Symposium on robot Control*, Barcelona, Spain, 1985.

[4] K. W. Cattermole. *Principles of Pulse Code Modulation*. American Elsevier, NY, 1969.

[5] N. G. Chetaev. *The Stability of Motion*. Pergammon, New York, 1961.

[6] E. R. Fossum. *Charge-Coupled Analog Computer Elements and Their Application to Smart Image Sensors*. PhD thesis, Yale University, New Haven, Ct., 1984.

[7] Eric R. Fossum and Richard C. Barker. A linear and compact charge-coupled charge packet differencer/replicator. *IEEE Transactions on Electron Devices*, ED-31(12):1784–1789, Dec 1984.

[8] B.A. Francis and W. M. Wonham. The internal model principle for multivariable regulators. *Appl. Math. & Optimization*, 2(2):170–194, 1975.

[9] E. Gilbert and D. W. Johnson. Distance functions and their application to robot path planning in the presence of obstacles. *IEEE Journal of Robotics and Automation*, RA-1(1):21–30, Mar 1985.

[10] M. W. Hirsch and S. Smale. On Algorithms for Solving $f(x) = 0$. *Comm. Pure and Appl. Math*, XXXII:281–312, 1979.

[11] Morris W. Hirsch and Stephen Smale. *Differential Equations, Dynamical Systems, and Linear Algebra*. Academic Press, Inc., Orlando, Fla., 1974.

[12] Neville Hogan. Impedance control: an approach to manipulation. *ASME Journal of Dynamics Systems,Measurement, and Control*, 107:1–7, Mar 1985.

[13] R.E. Kalman and J.E. Bertram. "Control systems analysis and design via the 'second method' of Lyapunov". *Journal of Basic Engineering*, :371–392, June 1960.

[14] O. Khatib and J.-F. Le Maitre. Dynamic control of manipulators operating in a complex environment. In *Proceedings Third International CISM-IFToMM Symposium*, pages 267–282, Udine, Italy, Sep 1978.

[15] Oussama Khatib. Real time obstacle avoidance for manipulators and mobile robots. *The International Journal of Robotics Research*, 5(1):90–99, Spring 1986.

[16] S. Kirkpatrick, C. D. Gelatt, and M. P. Vecchi. *Optimization by Simulated Annealing*. Research Report RC 9355 (#41093), IBM Thomas J. Watson Research Center, Yorktown Heights, NY, Apr 1982.

[17] D. E. Koditschek. *A Strict Global Lyapunov Function for Mechanical Systems*. Technical Report 8707, Center for Systems Science, Yale University, November 1987.

[18] D. E. Koditschek and E. Rimon. *Navigation Functions for Robot Obstacle Avoidance*. Technical Report 8803, Center for Systems Science, Yale University, January 1988.

[19] Daniel E. Koditschek. Automatic planning and control of robot natural motion via feedback. In Kumpati S. Narendra, editor, *Adaptive and Learning Systems: Theory and Applications*, pages 389–402, Plenum, 1986.

[20] Daniel E. Koditschek. Exact robot navigation by means of potential functions: some topological considerations. In *IEEE International Conference on Robotics and Automation*, pages 1–6, Raleigh, NC, Mar 1987.

[21] Daniel E. Koditschek. High gain feedback and telerobotic tracking. In *Workshop on Space Telerobotics*, page , Jet Propulsion Laboratory, California Institute of Technology, Pasadena, CA, Jan 1987.

[22] Daniel E. Koditschek. Natural motion for robot arms. In *IEEE Proceedings 23rd Conference on Decision and Control*, pages 733–735, Las Vegas, Dec 1984.

[23] Daniel E. Koditschek. Robot control systems. In Stuart Shapiro, editor, *Encyclopedia of Artificial Intelligence*, pages 902–923, John Wiley and Sons, Inc., 1987.

[24] J. P. Lasalle and S. Lefschetz. *Stability by Lyapunov's Direct Method with Applications*. Academic, New York, 1961.

[25] Solomon Lefschetz. *Differential Equations: Geometric Theory*. Dover, NY, 1977.

[26] A. M. Lyapunov. *Problème Gén éral de la Stabilité du Mouvement*. Princeton University, Princeton, NJ, 1949.

[27] I. G. Malkin. *The Theory of Stability of Motion*. US Atomic Energy Commission AEC-TR-3352, Wash. DC, (translation).

[28] Fumio Miyazaki and S. Arimoto. Sensory feedback based on the artificial potential for robots. In *Proceedings 9th IFAC*, Budapest, Hungary, 1984.

[29] K. S. Narendra and L. S. Valavani. Stable adaptive observers and controllers. *Proceedings of the IEEE*, 64(8), 1976.

[30] V. V. Pavlov and A. N. Voronin. The method of potential functions for coding constraints of the external space in an intelligent mobile robot. *Soviet Automatic Control*, (6), 1984.

[31] Ralph Pringle, Jr. On the stability of a body with connected moving parts. *AIAA*, 4(8):1395–1404, Aug 1966.

[32] M. Renaud. *Contribution à l' étude de la modelisation des systèmes mécaniques articulés*. Thèse de Docteur-Ingénieur, Univ. Toulouse, December, 1976.

[33] A.A.G. Requicha and H. B. Voelcker. Solid modeling: current status and reserach directions. *IEEE Computer Graphics and Applications*, 3(10):25–37, Oct 1983.

[34] E. Rimon and D. E. Koditschek. Exact robot navigation using cost functions: the case of spherical boundaries in $\mathbf{E}^n$. In *IEEE International Conference on Robotics and Automation*, page , Philadelphia, PA, Apr 1988.

[35] A. J. Van Der Schaft. *Stabilization of Hamiltonian Systems*. Memo 470, Technische Hogeschool Twente, Twente, Netherlands, Jan 1985.

[36] Jean-Jacques E. Slotine and Weiping Li. On the adaptive control of robot manipulators. In *Proc. ASME Winter Annual Meeting*, page , Anaheim, CA., Dec 1986.

[37] S. Smale. The fundamental theorem of algebra and complexity theory. *Annals of Mathematics*, 4(1):1–36, Bull. AMS.

[38] M. Takegaki and S. Arimoto. Stability and robustness of pid feedback control for robot manipulators of sensory capability. In *Robotics Research, First International Symposium*, MIT Press, 1984.

[39] Morikazu Takegaki and Suguru Arimoto. A new feedback method for dynamic control of manipulators. *ASME Journal of Dynamics Systems, Measurement, and Control*, 102:119–125, 1981.

[40] David W. Tank and John J. Hopfield. Simple "Neural Optimization Networks": an a/d converter, signal decision circuit, and a linear programming circuit. *IEEE Transactions on Circuits and Systems*, CAS-33(5):533–541, May 1986.

[41] Sir W. Thompson and P. G. Tait. *Treatise on Natural Philosophy*. University of Cambridge Press, 1886, Cambridge.

[42] John T. Wen and David S. Bayard. *Robust Control for Robotic Manipulators Part I: Non-Adaptive Case*. Technical Report 347-87-203, Jet Propulsion Laboratory, Pasadena, CA, 1987.

Adaptive Trajectory Control of Manipulators

Jean-Jacques E. Slotine
Nonlinear Systems Laboratory
Massachusetts Institute of Technology
Cambridge, MA 02139, U.S.A.

and

John J. Craig
SILMA Inc.
1601 Saratoga-Sunnyvale Rd.
Cupertino, CA 95014, U.S.A.

1. INTRODUCTION

The development of effective adaptive controllers represents an important step towards versatile applications of high-speed and high-precision robots. Even in a well structured industrial setting like the proverbial, though elusive, "factory of the future", robots still have to face uncertainty on the parameters describing the dynamic properties of the grasped load, such as moments of inertia or exact position of the center of mass in the end-effector. Since these parameters are difficult to compute or measure for geometrically complex objects, they limit the potential for robots to accurately manipulate objects of size and weights similar to their own, as the human arm routinely does. The control performance of conventional industrial robot controllers, and even that of more recent approaches (such as the computed torque method) may become unsatisfactory due to parameter uncertainties and variations. This sensitivity is especially severe in high speed operations or when controlling direct-drive robots, for which no gear reduction is available to mask effective inertia variations. Two classes of approaches have been actively studied to maintain the performance of the manipulators in the presence of parameter uncertainties: robust control (*e.g.*, [Slotine, 1985; Ha and Gilbert, 1987; Spong and Vidyasagar, 1987]) and adaptive control. An advantage of the adaptive approach is that the accuracy of a manipulator carrying unknown loads improves with time, because the adaptation mechanism keeps extracting parameter information from tracking errors, so that adaptive controllers potentially hold the promise of consistent performance in the face of very large load variations.

Adaptive control of linear time-invariant single-input single-output systems has been extensively studied, and a number of globally convergent controllers have been derived. Extensions of the results to nonlinear or multivariable systems have rarely been achieved. Yet, in the case of robot manipulators, which represent an important and unique class of nonlinear, time-varying, multi-input multi-output dynamic systems, similar global convergence properties can indeed be obtained.

It seems possible to discern two phases in the history of adaptive robot control research, an "approximation" phase (roughly 1979-1985) and a "linear parametrization" phase (after 1985). In the approximation phase, due to the complicated nonlinear, time-varying and coupled robot dynamics, researchers had to rely on restrictive assumptions or approximations of one kind or another for adaptive control design and analysis, e.g., linearization of robot dynamics, decoupling assumption for joint motions, or slow variation of the inertia matrix. The pioneering papers of this first phase included [Dubowsky and DesForges, 1979], [Horowitz and Tomizuka, 1980], [Koivo, *et al.*, 1981], and [Balestrino *et al.*, 1981]. [Hsia, 1986] presents a reasonably complete review of these early methods, most of which are troubled either by lack of stability proofs, due to the simplifying assumptions on the robot model, or by the excessive computational burden resulting from the large number of parameters to be updated. The explicit introduction in adaptive robotic control research of the linear parametrization of robot dynamics, inspired by earlier results developed in the context of parameter estimation [Nicolo and Katende, 1982; An, *et al.*, 1985; Khosla and Kanade, 1985], represents a turning point.

Based on this possibility of selecting a proper set of equivalent parameters such that the manipulator dynamics depend linearly on these parameters, research on adaptive robot control can now take full consideration of the nonlinear, time-varying and coupled nature of robot dynamics. The proposed adaptive controllers can be classified into three categories: direct, indirect, and composite adaptive controllers. The direct adaptive controllers (e.g., [Craig, *et al.*, 1986], [Slotine and Li, 1986], [Bayard and Wen, 1987], [Sadegh and Horowitz, 1987], [Koditschek,

1987]) use tracking errors of the joint motion to drive parameter adaptation. The indirect adaptive controllers (e.g., [Middleton and Goodwin, 1986], [Hsu, *et al*, 1987], [Li and Slotine, 1988]), on the other hand, use prediction errors on the filtered joint torques to generate parameter estimates to be used in the control law. The composite adaptive controllers ([Slotine and Li, 1987d, 1988a]) use both tracking errors in the joint motion and prediction errors on the filtered torques to drive parameter adaptation.

We shall now briefly discuss each of these approaches. Recall that, in the absence of friction or other disturbances, the dynamics of a rigid manipulator (with the load considered as part of the last link) can be written as

$$\mathbf{H}(\mathbf{q})\ddot{\mathbf{q}} + \mathbf{C}(\mathbf{q},\dot{\mathbf{q}})\dot{\mathbf{q}} + \mathbf{g}(\mathbf{q}) = \tau \qquad (1)$$

where $\mathbf{q}$ is the $n \times 1$ vector of joint displacements, τ is the $n \times 1$ vector of applied joint torques (or forces), $\mathbf{H}(\mathbf{q})$ is the $n \times n$ symmetric positive definite manipulator inertia matrix, $\mathbf{C}(\mathbf{q},\dot{\mathbf{q}})\dot{\mathbf{q}}$ is the $n \times 1$ vector of centripetal and Coriolis torques, and $\mathbf{g}(\mathbf{q})$ is the $n \times 1$ vector of gravitational torques. The adaptive controller design problem is as follows: given the desired trajectory $\mathbf{q}_d(t)$, with the joint positions and velocities being measured, and with some or all the manipulator parameters being unknown, derive a control law for the actuator torques, and an adaptation (or estimation) law for the unknown parameters, such that the manipulator joint position $\mathbf{q}(t)$ closely track the desired trajectory $\mathbf{q}_d(t)$. If the adaptive controller is designed such that for any unknown load/link parameters and any errors in initial joint position and velocity, the tracking error $\mathbf{q}(t) - \mathbf{q}_d(t)$ converge to zero, the adaptive controller is said to have global tracking convergence.

2. DIRECT ADAPTIVE CONTROLLERS

The *direct* adaptive controllers, first investigated by [Craig, *et al.*, 1986], use tracking errors of the joint motion to drive the parameter adaptation. In this class of adaptive controllers, the predominant concern of the adaptation laws is to reduce the tracking errors.

[Craig, et al, 1986, 1987] propose an adaptive controller based on computed-torque control and show its global convergence, including robustness to a class of unmodelled effects which can be *a priori* bounded. The authors' desire to strictly maintain the structure of the computed-torque control scheme brings about the need to use acceleration measurements, and to invert the estimated inertia matrix as part of the algorithm. A sufficient excitation result is also presented which allows proposed trajectories to be tested for their ability to excite all equivalent parameters of the system.

In [Slotine and Li, 1986, 1987c, 1987e], the global tracking convergence of a new adaptive feedforward-plus-PD controller is established and its good performance is demonstrated experimentally. The algorithm only requires joint positions and velocities to be measured, and avoids the inversion of the estimated inertia matrix. The approach also avoids the difficulties linked to the SPR (strictly positive real) requirement in traditional adaptive control by taking advantage of the inherent positive definiteness of the manipulator's inertia matrix.

Subsequent related work appears in [Bayard and Wen, 1987], [Sadegh and Horowitz, 1987], and [Koditschek, 1987].

Two properties of the manipulator dynamics (1) can be used for direct adaptive control design. The first, exploited in all the adaptive controllers discussed in this paper, is the linear parametrization property, *i.e.*, the fact that each of the individual terms of the left hand side of (1), and therefore the whole robot dynamics, is linear in terms of a suitably selected set of robot and load parameters [Khosla and Kanade, 1985; An, et al, 1985]. Second, as pointed out by [Takegaki and Arimoto, 1981; Koditschek, 1984], the matrices $\mathbf{H}$ and $\mathbf{C}$ in (1) are not independent; specifically, using a proper definition of the matrix $\mathbf{C}$ (only the *vector* $\mathbf{C}\dot{\mathbf{q}}$ is uniquely defined) the matrix $(\dot{\mathbf{H}} - 2\mathbf{C})$ is *skew-symmetric*, as can be simply shown [Slotine and Li, 1987c]. This second property, which reflects conservation of energy, is exploited in [Slotine and Li, 1986, 1987c].

Note that the number of equivalent parameters may be much smaller than that of physical parameters, since the equivalent parameters consist of nonlinear combinations of the physical parameters. Therefore, some physical parameters may be unidentifiable. This does not represent a difficulty from a control point of view, since only the equivalent parameters affect the dynamics and control.

Since the parameters of the load change each time a new payload is picked up while the parameters of manipulator links are the same, in practice only the parameters of the load (10 parameters in general motion, namely, load mass, 3 parameters for mass center location, 6 moments of inertia) need to be estimated. In most applications, it is reasonable to estimate the parameters of the links beforehand to reduce on-line computation. The controllers can also be made to be robust to time-varying disturbances, such as friction. The approach can also be extended easily to hybrid force/position control and external adaptive control of passive mechanisms [Slotine and Li, 1987a]. Issues of persistency of excitation are detailed in [Slotine and Li, 1987b].

3. INDIRECT ADAPTIVE CONTROLLERS

The *indirect* adaptive manipulator controllers, pioneered by [Middleton and Goodwin, 1986], use prediction errors on the filtered joint torques to generate parameter estimates to be used in the control law. The predicted torque error consists of the difference between the actual applied torque vector, and the value of the torque vector necessary to support the current motion as computed from the current parameter estimates and the values of $\mathbf{q}$, $\dot{\mathbf{q}}$, and $\ddot{\mathbf{q}}$. Using the *filtered* predicted torque error permits such algorithms to avoid the need for acceleration measurements.

For the indirect adaptive controllers, the predominant concern of the adaptation is to extract information about the true parameters from the prediction errors, with no *direct* concern to adapt the parameters so that the tracking errors converge to zero. [Middleton and Goodwin, 1986, 1988] show the global tracking convergence of their adaptive controller, which is composed of a modified computed-torque controller and a modified least-square estimator. The computation of their adaptive controller again requires inversion of the estimated inertia matrix. The indirect adaptive controller in [Li and Slotine, 1988] has global asymptotic tracking convergence in general, and exponential convergence in the presence of persistent excitation, while avoiding the requirement of inertia matrix inversion by using a different modification of the computed-torque controller.

Most of the indirect algorithms (as well as [Craig, et al., 1986, 1987]) have to assume (or to develop procedures to guarantee) that the estimated inertia matrix remains positive definite in the course of adaptation. If only the load is to be estimated, a simple projection approach can be used to maintain this positive definiteness while preserving convergence properties, as the convexity result of [Li and Slotine, 1988] shows.

Indirect controllers allow the vast parameter estimation literature to be used to select time-variations of the adaptation gains, while in principle direct controllers are limited to constant adaptation gains.

4. COMPOSITE ADAPTIVE CONTROLLERS

For complex nonlinear control problems such as adaptive manipulator control, global tracking convergence does not necessarily imply *exponential* tracking convergence. In other words, the algorithms do not by themselves guarantee *time-constants* within which tracking convergence is achieved and maintained. Now in adaptive controllers, intuitively, convergence time-constants must depend on how demanding (or "exciting") the desired trajectory is: the algorithm may easily follow an initially simple trajectory, and apparently converge, but later exhibit tracking errors if the trajectory becomes more exciting. Conversely, if the trajectory is initially exciting, then the algorithm will have no difficulty following a simpler trajectory later on: the initial excitation was sufficiently demanding for tracking convergence to necessarily imply exact parameter convergence, and therefore subsequently "exact" performance.

Such exponential convergence, with known convergence rates depending on the excitation of the desired trajectory, can be achieved by "composite" adaptive controllers. Composite adaptive controllers, studied in [Li and Slotine, 1987; Slotine and Li, 1987d, 1988a], use *both* tracking errors in the joint motion *and* prediction errors in the predicted filtered torque to drive the parameter adaptation. They are based on the observation that since parameter uncertainty is reflected in both the tracking error (or output error) and the prediction error (or input error), it is desirable to extract parameter information from both sources, thereby achieving a mixture of direct and indirect approaches. This not only permits full use of available information sources, but also offers an automatic way of modulating the adaptation gain according to the excitation of the desired trajectories, yielding exponential convergence of the tracking and estimation errors in the presence of persistent excitation, with known convergence rates depending on the degree of excitation of the desired trajectory. These controllers also allow standard parameter estimation techniques to be exploited.

5. CONCLUDING REMARKS

Adaptive manipulator control has been evolving rapidly in the past few years, and is the subject of active research. Current research directions include issues of computational efficiency (e.g. [Walker, 1988; Niemeyer and Slotine, 1988]), design of exciting desired trajectories (e.g., [Armstrong, 1987]), robustness to unmodelled dynamics (such as structural resonant modes, joint flexibility, or neglected actuator dynamics), and extensions to compliant motion control.

REFERENCES

An, C.H., Atkeson, C.G. and Hollerbach, J.M., 1985. Estimation of inertial parameters of rigid body links of manipulators, *I.E.E.E. Conf. Decision and Control*, Fort Lauderdale.
Armstrong, B., 1987. On Finding 'Exciting' Trajectories for Identification Experiments involving Systems with Non-Linear Dynamics, *IEEE Conference on Robotics and Automation*, Raleigh, NC.
Asada, H., and Slotine, J.J.E., 1986. Robot Analysis and Control, *John Wiley and Sons*.
Balestrino, A., DeMaria, G., and Sciavicco, L., 1981. Adaptive Control of Robotic Manipulators, *AFCET Congres Automatique*, Nantes, France.
Bayard,D.S., and Wen,J.T., 1987. Simple Adaptive Control Laws for Robotic Manipulators, *Proceedings of the Fifth Yale Workshop on the Applications of Adaptive Systems Theory*.

Craig, J.J., Hsu, P., and Sastry, S., 1986. Adaptive Control of Mechanical Manipulators, *I.E.E.E. Int. Conf. Robotics and Automation*, San Francisco.

Craig, J.J., Hsu, P., and Sastry, S., 1987. Adaptive Control of Mechanical Manipulators, *Int. J. Robotics Research*, 6(2).

Dubowsky, S., and DesForges, D., 1979. The Application of Model-Referenced Adaptive Control to Robotic Manipulators, *ASME Journal of Dynamic Systems Measurement and Control*.

Ha I.J, and Gilbert E.G., 1987. Robust Tracking in Nonlinear Systems, *I.E.E.E. Trans. Autom. Control*, AC-32, 9.

Horowitz, R., and Tomizuka, M., 1980. An Adaptive Control Scheme for Mechanical Manipulators -- Compensation of Nonlinearity and Decoupling Control, *ASME Paper No. 80-WA/DSC-6*.

Hsia, T.C., 1986. Adaptive Control of Robot Manipulators - A Review, *I.E.E.E. Int. Conf. Robotics and Automation*, San Francisco. and Control, Prentice Hall.

Hsu, P., Sastry, S. , Bodson, M. and Paden, B. 1987. Adaptive Identification and Control of Manipulators With Joint Acceleration Measurements, *I.E.E.E. Int. Conf. Robotics and Automation*, Raleigh, NC.

Khosla, P.,and Kanade, T., 1985. Parameter Identification of Robot Dynamics, *I.E.E.E. Conf. Decision and Control*, Fort Lauderdale.

Koditschek, D.E., 1987. Adaptive Techniques for Mechanical Systems, *Proceedings of the Fifth Yale Workshop on the Applications of Adaptive Systems Theory*.

Koivo, A., 1981. Control of Robotic Manipulator with Adaptive Controller, *IEEE Conference on Decision and Control*, San Diego, Ca.

Li, W., and Slotine, J.J.E., 1987. Parameter Estimation Strategies for Robotic Aplications. *A.S.M.E. Winter Annual Meeting*, Boston, MA.

Li, W., and Slotine, J.J.E., 1988a. Indirect Adaptive Robot Control, *5th I.E.E.E. Int. Conf. Robotics and Automation*, Philadelphia.

Li, W., and Slotine, J.J.E., 1988b. An Indirect Adaptive Robot Controller, to be published in *Systems and Control Letter*.

Middleton, R.H. and Goodwin, G.C. , 1986. Adaptive Computed Torque Control for Rigid Link Manipulators, 25th *I.E.E.E. Conf. on Dec. and Contr.*, Athens, Greece.

Middleton, R.H., and Goodwin, G.C., 1988. Adaptive Computed Torque Control for Rigid Link Manipulators, *Systems and Control Letters*.

Nicolo, F., and Katende, 1982. *IASTED Conf. Robotics and Automation*, Lugano.

Niemeyer, G., and Slotine, J.J.E, 1988. Performance in Adaptive Manipulator Control, *I.E.E.E. Conf. Decision and Control*, Austin.

Sadegh, N., and Horowitz, R., 1987. Stability Analysis of an Adaptive Controller for Robotic Manipulators. *I.E.E.E. Int. Conf. Robotics and Automation*, Raleigh, NC.

Slotine, J.J.E.,, 1985. The Robust Control of Robot Manipulators, *Int. J. Robotics Research*, 4(2).

Slotine, J.J.E., and Li, W., 1986. On The Adaptive Control of Robot Manipulators, *A.S.M.E. Winter Annual Meeting*, Anaheim, CA.

Slotine, J.J.E., and Li, W., 1987a. Adaptive Strategies in Constrained Manipulation, *I.E.E.E. Int. Conf. Robotics and Automation*, Raleigh, NC.

Slotine, J.J.E. and Li, W., 1987b. Theoretical Issues In Adaptive Manipulator Control. In *the Proceedings of the Fifth Yale Workshop on Applications of Adaptive Systems Theory*.

Slotine, J.J.E., and Li, W., 1987c. On the Adaptive Control of Robot Manipulators, *Int. J. Robotics Res.*, vol. No.3

Slotine, J.J.E., and Li, W., 1987d. Adaptive Manipulator Control: A New Perspective, *I.E.E.E. Conf. Decision and Control*, Los Angeles.

Slotine, J.J.E., and Li, W., 1987e. Adaptive Manipulator Control: A Case Study, *I.E.E.E. Int. Conf. Robotics and Automation*, Raleigh, NC.

Slotine, J.J.E., and Li, W., 1988a. Composite Adaptive Manipulator Control, to be published in *Automatica*.

Slotine, J.J.E., and Li, W., 1988b. Adaptive Manipulator Control: A Case Study, *I.E.E.E. Trans. Autom. Control*, 33-11.

Spong, M.W., and Vidyasagar, M., 1987. Robust Linear Compensator Design For Nonlinear Robotic Control. *I.E.E.E. J. Robotics and Automation*, RA-3, 4.

Takegaki, H., and Arimoto, S., 1981. Adaptive Control of Manipulators, *Int. J. Contr.*, vol. 34, No. 2

Walker, M., 1988. An Efficient Algorithm for the Adaptive Control of a Manipulator, *I.E.E.E. Conference on Robotics and Automation*, Philadelphia, PA.

Compliance and Force Control for Computer-Controlled Manipulators[1]

Matthew T. Mason

Reviewed by

Haruhiko Asada
Department of Mechanical Engineering
Massachusetts Institute of Technology
Cambridge, MA 02139

Compliant motion control is a critical issue in advanced manipulation where robots interact with the environment mechanically. Assembly, grinding, and many other important tasks are performed through mechanical interactions between robots and their environment. A number of researchers have then addressed issues concerning force control and compliant motion control in the past. Useful control methods including bilateral servo (Inoue 1971), damping control (Whitney 1977), hybrid control (Mason 1981), impedance control (Hogan 1985), and operational space approach (Khatib 1987) have been developed. Competent technologies such as force sensors, the RCC hand, and direct-drive motors have also made significant progress in the past decade.

These technologies have enabled us to build force-controlled robots. However, force control and compliant motion control have been used very little in industry so far. Applications of force control and compliant motion control to real manipulative tasks are still unsatisfactory. The reason for this is that we are lacking knowledge and experience about the application of force and compliant motion control. We do not know in detail how to use them in order to perform a given task. To achieve a task goal, we need to find an appropriate strategy and reduce the task goal to stepwise motion commands that can be performed by the robot. This is an intricate process which needs insight and reasoning about the task as well as some knowledge about the

---

[1] *Transactions of IEEE Sys. Man. Cybern.*, Vol. 11-SMC, 1981.

robot. There exist significant obstacles between technologies available today and what we want to do with them.

In 1981, Mason discovered a useful guideline which provides a theoretical basis for generating a task strategy for force and compliance control (Mason 1981). According to his theory, one can designate an appropriate control mode, either position or force, to each of the coordinate axes. Given geometric constraints, the control mode is chosen in such a way that the robot movement does not conflict with the geometric constraints. His theory also provides a guideline for designating reference inputs to each of the position and force control systems.

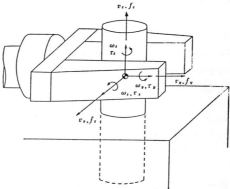

Fig. 1 Peg-and-Hole Problem

Let us consider the task of pulling a peg out of a hole as shown in Fig. 1. Obviously, the admissible motion of the peg is limited to translation in the Z direction and rotation about the Z axis. Namely, the linear and angular components of velocity of the peg must satisfy

$$v_x = 0, \quad v_y = 0, \quad \omega_x = 0, \quad \omega_y = 0 \qquad (1)$$

On the other hand, the admissible components of force and torque to be applied to the peg must satisfy

$$f_z = 0, \quad \tau_z = 0 \qquad (2)$$

in order to maintain the quasi-static balance of force and torque. Note that friction between the peg and the hole is ignored. The above equations provide conditions that the robot has to obey when the peg is constrained by the hole. These are referred to as Natural Constraints according to Mason.

The structure of the robot control system must be consistent with the Natural Constraints. If one forms a position control loop and designates a position command to one of the constrained axes described by eq.1, a conflict occurs between the natural constraint and the position control loop. The movement of the robot is prevented by the natural constraint, and thereby the position command can never be accomplished without breakage of the peg or the hole. Similarly, one cannot designate a force command to the axes described by eq.2. From these observations, the control modes that may not conflict with the natural constraints are determined as

$$\text{position mode}: v_z, \omega_z$$

$$\text{force mode}: f_x, f_y, \tau_x, \tau_y$$

One can designate arbitrary input commands to the above variables without conflicting with the natural constraints. The input commands should be chosen so that the given task goal may be accomplished. For the peg-and-hole problem in Fig. 1,

$$v_z = V > 0, \qquad \omega_z = 0 \tag{3}$$

$$f_x = 0, \qquad f_y = 0, \qquad \tau_x = 0, \qquad \tau_y = 0 \tag{4}$$

where V is an appropriate positive value. The above equations are referred to as Artificial Constraints in accordance with Mason.

Let us discuss the significance of Mason's result and formulate the relationship between natural and artificial constraints in a generic form. We begin by considering a six-dimensional vector space V^6 associated with the end effector motion, e.g., translation and rotation in the three dimensions. Let V_a be the subset of V^6 that consists of all admissible velocity vectors and V_c be the subset of all admissible force vectors. An important proposition derived from Mason's theory is the orthogonality relationship between the admissible motion space V_a and the admissible force space V_c. Namely, the space V_a is always the orthogonal complement of space V_c. This is obviously held for the example that we discussed above. The following is a formal proof for a general case.

Let $\delta \mathbf{q} \in V^6$ be an arbitrary virtual displacement that meets the geometrical constraints. Note that $\delta \mathbf{q}$ is involved in space V_a; $\delta \mathbf{q} \in V_a$. Let $\mathbf{Q} \in V^6$ be an arbitrary force and moment acting on the rigid body carried around by the robot. The rigid body is in quasi-static equilibrium if, and only if, the following virtual work vanishes for an arbitrary virtual displacement $\delta \mathbf{q}$.

$$\delta Work = \mathbf{Q}^T \delta \mathbf{q} \qquad \forall \delta \mathbf{q} \in V_a \qquad (5)$$

Since the space V_c consists of the force $\mathbf{Q}$ that satisfies the quasi-static balance condition, that is, eq. 5, V_c is the orthogonal complement of V_a.

$$V_c \perp V_a \qquad (6)$$

	Kinematics	Statics
Natural Constraints	Vc $\perp$ Velocities=0	Va Forces=0
	$\perp$	$\perp$
Artificial Constraints	Va Velocities =Specified	$\perp$ Vc Forces =Specified

Fig. 2 Natural and Artificial Constraints

Fig. 2 shows the relationship between the natural and artificial constraints as well as the one between kinematic and static conditions.

Note that once either V_a or V_c is identified, one can find all the subspaces in which natural and artificial constructions are designated. The result shown in Fig. 2 is the condition that the task planner must go through in order to generate an appropriate control strategy. Usually, we perform this step of investigation by reasoning. The significance of Mason's theory is to provide a theoretical guideline that eliminates a part of the reasoning in the task planning.

Mason's theory has provided a theoretical framework for planning compliant motion control. Nevertheless, we have many other parameters involved in the control system that are not uniquely determined by Mason's theory. For example, command inputs involved in artificial constraints were determined in an intuitive or heuristic manner. Control modes are not necessarily either pure position or pure force control, but can be other types. Stiffness control and damping control as well as impedance control can be employed without contradicting Mason's theory. Detailed planning of many parameters and choices of control schemes are issues for future research.

References

Hogan, N., 1985. Impedance Control: An Approach to Manipulation. *ASME J. of Dynamic Systems, Measurement and Control*, Vol. 107-1, pp. 1–22.

Inoue, H., 1971. A Computer-Controlled Bilateral Manipulator. *Bulletin of JSME*, Vol. 14, pp.69–76.

Khatib, O., 1987. A Unified Approach for Motion and Force Control of Robot Manipulators: The Operational Space Formulation. *Trans. of IEEE Robotics and Automation*, Vol. RA3-1, pp. 43–53.

Mason, M.T., 1981. Compliance and Force Control for Computer-Controlled Manipulators. *Trans. of IEEE Sys. Man. Cybern.*, Vol. 11-SMC, pp. 418–432.

Whitney, D.E., 1977. Force Feedback Control of Manipulator Fine Motions. *ASME J. of Dynamic Systems, Measurement and Control*, Vol. 99, pp. 91–97.

Stability and Robustness of PID Feedback Control for Robot Manipulators of Sensory Capability[1]

Sugaru Arimoto and Fumio Miyazaki

Reviewed by

John J. Craig
Silma Inc.
1601 Saratoga-Sunnyvale Road
Cupertino, CA 95014

This paper is concerned with proving the stability of a very simple class of robot control schemes similar to those found in industrial robot controllers today. Although the control scheme is simple, due to the complicated nature of robot manipulator dynamics the overall system is not simple, and so the work presented is by no means trivial. Whereas many published analyses of robot control algorithms resort to linearization and treatment by linear methods, Arimoto and Miyazaki's stability proof deals directly with the nonlinear manipulator equations of motion. Their proof is only for the case of stability about a stationary desired position, and so does not deal with stability about a trajectory. Also, the work deals solely with the stability problem, rather than providing results related to performance of the system in a transient situation. Despite this somewhat limited scope, the paper is of interest because the topic treated is of such a fundamental nature, and needed to be answered first before research on the harder problems could be undertaken on solid footing.

The paper can be viewed as an extension of the earlier work by Takegaki and Arimoto (1981), with the principal extension being the inclusion of an integral term in the control law. As in the earlier work, this analysis assumes rigid body dynamics and that the gravity term in the dynamic equation which can be cancelled exactly (or is not present, as in a space application). The first result of the paper is an elegant proof using Lyapunov theory of the global stability of a manipulator controlled with the surprisingly simple controller given by

$$\tau_i = -a_i \dot{\theta}_i - b_i(\theta_i - \theta_{di}) + g_i(\Theta) \qquad (1)$$

---
[1] M. Brady and R.P. Paul (editors), "Robotics Research, First International Symposium", MIT Press, 1984.

where τ_i is the torque command to the i-th joint, θ_i is the i-th joint variable, θ_{di} is its desired value, $g_i(\Theta)$ is the torque at the i-th joint due to gravity, and a_i and b_i are gains. It is important to note that the stability analysis is for the case of a fixed desired position, rather than a time varying one, i.e a trajectory. The proof is the same as in (Takegaki and Arimoto, 1981) and similar proofs for passive mechanical systems seem to have been known more than a century ago (Lagrange, 1788; Thompson and Tait 1886).

The authors then proceed with a somewhat involved proof of the stability of the control law given by

$$\tau_i = -a_i\dot{\theta}_i - b_i(\theta_i - \theta_{di}) - c_i \int_0^t (\theta_i - \theta_{di})dt + g_i(\Theta) \qquad (2)$$

where a_i, b_i, and c_i are appropriately chosen gains. In classical control engineering, this control law is known as the PID (proportional, integral, derivative) controller. Note that some PID controllers use the derivative of position error, rather than the derivative of sensed position, and the last term for gravity compensation is not typical. Because the closed loop system is nonlinear and contains an integral term, this global stability result is quite interesting. Unfortunately, global stability of the PID scheme is not shown, but rather, the result is local in nature.

While the paper is generally concerned with a somewhat idealized model of manipulator dynamics, one section of the paper goes beyond this. The authors make use of the theory of singular perturbations to show stability of the joint space PID controller in the the presence of some additional dynamics (motor inductance or shaft flexibility, for example).

In another section of the paper the control of a robot in task-oriented coordinates (or *operational* space) is considered. Once again, the authors show that a quite simple PID controller also stabilizes such a system. The controller proven stable is given by

$$\Gamma = -A\dot{\Theta} - J(\Theta)^T [B(X - X_d) + C \int_0^t (X - X_d)dt] + G(\Theta) \qquad (3)$$

where Γ is the $n \times 1$ vector of joint torques (n is the number of joints), X is the manipulator position in the operational space, $J(\Theta)$ is the manipulator Jacobian, $G(\Theta)$ is the $n \times 1$ vector of torques due to gravity, and A, B, and C are gain matrices. The stability proof is restricted to the subset of operational space where $J(\Theta)J(\Theta)^T$ is nonsingular, and again, due to the inclusion of the integral term, only local stability is shown.

Arimoto and Miyazaki's work shows that the fundamental goal of closed loop stability of a position controlled manipulator can be acheived with a very simple control algorithm. Almost every industrial robot operating today uses a control law similar to (2), and this paper provides a theoretical proof of its stability. Hence, this paper is important because in some sense it shows why today's robots work. Further, the authors show that even if the control is performed in an operational space, rather than in the manipulator joint space, it is still true that a simple controller suffices to guarantee stability (except at singular points).

While Arimoto and Miyazaki have shown that a simple controller can acheive stability, obtaining good performance is another issue. Particularly in a complex nonlinear system, a stability result such as the one shown provides little or no information regarding performance of the system. In fact, the performance of such a control system may be so poor as to render the system useless. In the case of current industrial robots, performance problems are often seen, and improved performance would certainly bring benefits.

High performance robot controllers will require more advanced control techniques. Much of the literature published on manipulator control in the last two decades is concerned with exploring various methods to obtain higher performance robot control systems. These methods generally involve dynamic compensation and nonlinear control techniques as contrasted with the simple fixed gain techniques analyzed by Arimoto and Miyazaki. Among the methods which have been proposed are those which make use of a more or less complete dynamic model of the manipulator, sliding mode control, adaptive control, and learning control. Stability analysis and performance evaluation of robot control systems is an active area of current research. The work of Arimoto and Miyazaki is an important contribution to this on-going effort.

References

Takegaki, M., and Arimoto, S. 1981 (June). A New Feedback Method for Dynamic Control of Manipulators. *ASME Journal of Dynamic Systems, Measurement, and Control*, Vol. 102.

Lagrange, J. 1788. Mechanique Analytique. *Gauthier-Villars*, Paris.

Thompson, W. and Tait, P.G. 1886. Treatise on Natural Philosophy. *University of Cambridge Press*, Cambridge.

Satellite-Mounted Robot Manipulators – New Kinematics and Reaction Moment Compensation[1]

R. W. Longman, R. E. Lindberg, and M. F. Zedd

Reviewed by

Gerd Hirzinger
DFVLR
8037 Wessling
West Germany

The use of robots in space (e.g. for the assembly of space structures or the repair of satellites) will be among the most challenging applications of robotics. Therefore, space technology may become one of the major drivers for the development of more intelligent robots. However there are a couple of problems that do not occur in ground-based systems and which, surprisingly enough, have not been thoroughly tackled in robotics research literature. One of the most important gaps of this type is closed by this paper. It deals with the problem that a robot arm mounted on a satellite and executing commanded motions produces a motion of its carrier and thus the end-effector would miss a free-flying target. The paper presents a new kind of robot kinematics that adjusts the joint angle commands to account for the base motion, and at the same time derives the satellite attitude disturbances that may be fed in a feedforward manner into the vehicles's attitude control system.

The paper makes one important but realistic assumption: the base vehicle has an attitude control system that is not able to compensate for translational disturbances but can compensate for rotational ones (e.g. by using three reaction wheels). This assumption simplifies the problem in an essential way: it reduces the otherwise coupled nonlinear differential equations involving the complete history of robot motion to a *nonstandard* kinematic problem. Proper initialization has to take into account the initial joint positions and the robot load as

---
[1] The International Journal of Robotics Research, Vol. 6, No. 3, 1987.

well as the definition of an inertial frame, with respect to which the overall system's center of mass remains constant. The solution of this new kinematic problem for a spherical-polar robot with five rotational joints and one prismatic joint is demonstrated in the paper. The authors indicate that there will be a future, more complete version of this work which will include elbow robots like the space shuttle manipulator.

An equally interesting part of the paper is concerned with calculating the reaction forces and moments acting on the base vehicle. First the base moments and forces for an inertially fixed robot are derived. The next step takes into account that the base is translationally free, but rotationally fixed. Thus, modified terms for the base moments and forces are derived. In the last step these base forces and moments are used to compute the torques that have to be compensated for (e.g. by reaction wheels) in order to guarantee the assumption of rotational stability. Thus a clever loop of presumptions and conclusions is closed. Clearly the base forces contribute to the moments as these force vectors in general will not pass through the vehicle's center of gravity. However if the compensation device produces a true couple or torque, as is the case with reaction wheels, it may be mounted anywhere on the vehicle.

Impressive simulation examples showing motions, forces, and moments are given for the case of a spherical-polar manipulator. Numerical examples show that it might be necessary to provide small moment control gyros instead of reaction wheels in order to produce the necessary reaction torques.

The technique developed in this paper provides an effective tool for the development of free-flying telerobot systems and constiutes an important contribution to research in the general area of space robotics.

Optimal Trajectory Planning for Industrial Robots[1]

R. Johanni and F. Pfeiffer

Reviewed by

Nenad M. Kircanski
M. Pupin Institute
POB 15, 11000 Beograd
Yugoslavia

In this paper a new method for obtaining an optimal velocity profile for a prescribed path of the end-effector is presented. The end effector path is specified in world coordinates. Constraints on the joint torques, joint velocities and on velocity of the end effector are taken into account.

The authors compared their work with the method proposed by Dubowsky and Schiller (1985) for obtaining optimal robot trajectories, Hollerbach's method (1986) for dynamic scaling of manipulator trajectories, Shin and McKay's method (1985) for minimum-time control of robotic manipulators, and several other methods.

The end effector path was specified by a scalar parameter s measuring the arclength of the path. In the case of point-to-point motion the authors use cubic splines to describe the corresponding continuous path.

Given the position and orientation vectors of the gripper along the end-effector path, the authors recognized that after applying inverse kinematics, the joint coordinates could also be expressed in s coordinates. Since the entire dynamic system model for robots with n

---

[1] *Robotersysteme*, Springer-Verlag, No. 3, 1987 (in German). A similar version of the paper was published in english as: Pfeiffer, F. and Johanni, R., "A Concept for Manipulator Trajectory Planning," in *IEEE Conference on Robotics and Automation*, San Francisco, 1986.

degrees of freedom depends on joint coordinates, velocities, and accelerations, the authors expressed the dynamic model in s coordinates

$$A_i(s)\frac{d}{ds}(\dot{s}^2) + B_i(s)\dot{s}^2 + C_i(s) = T_i;\qquad(1)$$

where A_i, B_i, and C_i are the coefficients which depend on s, T_i is the driving torque/force applied to the joint i.

The model form (1) appeared to be very suitable for optimization of velocity profile along a given end-effector path. Now, the authors introduced a performance index of integral type and the constraints on joint torques (1), joint velocities, and the end effector velocity. The following performance indices are discussed: $f_1 = 1/\text{abs}(\dot{s})$ which corresponds to time-minimum control, and $f_2 = f_1(k_1 + k_2\Sigma c_i T_i^2)$ which describes a combined time-energy consumption criterion.

The authors propose an algorithm to solve the time optimization problem without using classical mathematical optimization tools. They noticed that the constraints upon joint torques represent an polygon area in the $x = \frac{d}{ds}(\dot{s}^2)$, $y = \text{abs}(\dot{s})$ plane. Obviously, different polygons correspond to different values of s. Now, it was easy to introduce the end effector velocity constraint $\text{abs}(\dot{s})_{\max}$ as a new line in the $x - y$ plane, and to get the maximum $\ddot{s}$ given the parameter s. Thus, upon finding the extreme values of velocity and acceleration in $x - y$ plane for each s along a specified path, the time-optimization problem was solved.

The general optimization problem was not considered in detail in this paper, although the authors do mention dynamic programming as one of the possible methods. Although Vukobratovic and Kircanski (1982) and Shin and McKay (1985) applied this method in optimal trajectory planning, the paper of Johanni and Pfeiffer seems to be more practical since the dynamic programming can be very time consuming and numerically inefficient.

While the authors have shown that a simple optimization technique can be used for solving the time-minimum problem given a specified end-effector path, obtaining the optimal trajectories in general is another issue. In particular, trajectory planning for robots with redundant kinematic chains working in a constrained environment will require more advanced optimization techniques. Finally, real-time optimization and self-learning based on specified criteria are goals of future robotics research. The optimization of robot trajectories and other path-planning tasks represent an active area of research. The work of Johanni and Pfeiffer is an important contribution to this ongoing effort.

References

Dubowsky, S. and Schiller Z. 1985. Optimal Dynamic Trajectories for Robotic Manipulators. Proceedings of the 5th CISM-IFTOMM Symposium on the Theory and Practice of Robots and Manipulators, London: Kogan Page.

Hollerbach, J. M. 1986. Dynamic Scaling of Manipulator Trajectories. ASME Journal of Dynamic Systems, Meas., Contr. 106, pp. 102-106.

Shin, K. G. and McKay, N. D. 1985. Minimum-time Control of Robotic Manipulators with Geometric Path Constraints. IEEE trans. Automat. Contr. 30, pp. 531- 541.

Vukobratovic, M. and Kircanski M. 1982. A Method for Optimal Synthesis of Manipulation Robot Trajectories. ASME Journal of Dynmaic Systems, Contr. 104, pp. 188-194.

Shin, K. G. and McKay, N. D. 1986. A Dynamic Programming Approach to Trajectory Planning of Robotic Manipulators, IEEE Trans. Automat. Contr. 31, pp. 491-500.

Extended Kinematic Path Control of Robot Arms[1]
Evgeny Krustev and Ljubomir Lilov

Reviewed by

Simeon Patarinski
Department of Instrumentation and Robotics
Institute of Mechanics and Biomechanics
Bulgarian Academy of Sciences
Block 4, Acad G. Bonchev St.
1113 Sofia, Bulgaria

This paper proposes a general and unified approach for planning continuous path motions of robot manipulators which explicitly considers the restrictions imposed on the joint variables and velocities.

The authors define "continuous path motion" as the relation between time instants and the points (representing the position and the orientation of the end-effector) of a prescribed path (Krustev and Lilov 1986). Path planning starts from a prescribed path which is treated as a purely geometrical object with an arbitrary parametrization in "operational space" (Khatib 1987). Planning then proceeds by defining the desired time-based motion of the end-effector, and, using kinematic transformations, results in the joint variables as functions of time. This is a rather general statement of the problem, including inverse kinematics evaluations under inequality type restrictions and giving a unified notion of the manipulator motion in both joint and manipulation variables, "operational space coordinates."

Two basic assumptions are made:

A1. The mapping $\mathbf{F}(.)$ from the configuration space

$$\mathbf{Q}^O = \mathbf{q}: \ \mathbf{a}^O < \mathbf{q} < \mathbf{b}^O \in \mathbf{R}^n;$$

---
[1] Robotica, Vol. 5, pp. 45-53, 1987.

where n is the number of degrees of mobility of the robot arm, $\mathbf{q} \in \mathbf{R}^n$ is the joint variable vector, $\mathbf{a}^O, \mathbf{b}^O \in \mathbf{R}^n$, and the relation "less than" acts component-wise, to the manipulation space

$$\mathbf{X}^O = \mathbf{x} : \mathbf{x} = \mathbf{F}(\mathbf{q}), \ \mathbf{q} \in \mathbf{Q}^O \in \mathbf{R}^m;$$

where $m \leq \min(6, n)$ is the dimension of the task, $\mathbf{x} \in \mathbf{R}^m$ is the manipulation variable vector (Hanafusa, Yoshikawa, and Nakamura 1981) and

$$\mathbf{x} = \mathbf{F}(\mathbf{q}); \tag{1}$$

is the geometric model of the robot arm has full rank m almost everywhere along the prescribed path. This assumption (called by Krustev and Lilov the "displacement feasibility assumption") insures the existence of a continuous joint trajectory $\boldsymbol{\Psi} \in \mathbf{R}^n$, given a continuous path $\boldsymbol{\Gamma} \in \mathbf{X}^O$. That is, the singular configurations are completely ignored by a proper generation of the desired path.

A2. The vectors $\mathbf{a}^O$ and $\mathbf{b}^O$ are constant - in a more general case their components could be subject to a functional relationship.

The first assumption is very common in robotics (there are even methods proposed (Klein 1985) which utilize kinematic redundancy for singularity avoidance) and does not represent a major restriction. The second assumption is central to the approach developed in the paper - fortunately, while not completely general, it is often satisfied by typical industrial and research robot manipulators. Despite these restrictions, the topics dicussed are of fundamental interest and the results obtained are quite general from both theoretical and practical points of view.

The paper is an extension of the earlier work by Krustev and Lilov on kinematic path control of robot manipulators (Krustev and Lilov 1986), which took the physical limits on the joint positions and velocities into consideration. The problems are stated and treated in an extremely formal way that allows development of constructive and efficient computational procedures for continuous path motion planning and kinematic control. These include some specific problems such as minimum time motion and motion with a prescribed velocity profile for the end-effector.

Denote by $\gamma(\lambda) : \Lambda \Longrightarrow \mathbf{X}^O$ the prescribed path (involving, in the general case of $m = 6$, both the position and the orientation of the end-effector), where some parametrization is introduced by $\lambda \in \Lambda = [\lambda^O, \lambda^f]$. Then, every re-parameterization function

$$\lambda = \lambda(t); \qquad (2)$$

where $t \in \mathbf{T} = [t^O, t^f]$ is the time, defines an unique motion $\gamma = \gamma(t)$ of the end-effector along the prescribed path, and conversely, given some $\gamma(t)$ and $\mathbf{T}$, there always exists a unique function (2). This shows, that the motion of the end-effector along the prescribed path is invariant with respect to its particular parametrization. Thus, the planning of continuous path motions can be performed in two independent stages:

S1. Resolve the geometric model (1) of the robot arm to define the respective joint variables $\mathbf{q} = \mathbf{q}(\lambda)$.

S2. Consider the restrictions

$$\mathbf{q} \in \mathbf{Q}^O; \qquad (3)$$

imposed on the joint variables and introduce proper scaling so that

$$\dot{\mathbf{q}} \in \mathbf{Q}^1; \qquad (4)$$

where $\mathbf{Q}^1 = \dot{\mathbf{q}} : \mathbf{a}^1 < \dot{\mathbf{q}} < \mathbf{b}^1 \in \mathbf{R}^n$ are constant vectors.

Let $\mathbf{q}^* = [\mathbf{q}, \lambda]^T \in \mathbf{Q}^O \times \Lambda$ and $\mathbf{F}^*(\mathbf{q}^*) = \mathbf{F}(\mathbf{q}) - \gamma(\lambda)$. Each function $\mathbf{q}^*(t)$, for which $\mathbf{F}^*(\mathbf{q}^*(t)) = \mathbf{O_n}$, where $\mathbf{O_n} \in \mathbf{R}^n$ is the null vector, is called "an admissible motion" (Krustev and Lilov 1986). It follows from assumption A1, that all admissable motions lie over a smooth manifold. In the first stage (S1), their equations are obtained (after linearization of $\mathbf{F}^*(.)$ along the prescribed path $\mathbf{\Psi}$) in the form

$$\dot{\mathbf{q}}^* = K\mathbf{u}^*; \qquad (5)$$

where $\mathbf{u}^* = [\mathbf{u}^T, \dot{\lambda}]^T$ is a piecewise continuous vector-valued function ("kinematic control"), and

$$K = \begin{bmatrix} I_n - J^+J & A^{-1}J^T(JA^{-1}J^T)^{-1}\partial\gamma/\partial\lambda \\ O_n & 1 \end{bmatrix};$$

where $I_n \in \mathbf{R}^{n,n}$ is the unit matrix, J is the Jacobian matrix of $\mathbf{F}(.)$, and J^+ is its pseudoinverse. By $A = A(\mathbf{q})$ the Riemann metric on $\mathbf{\Psi}$ is defined (or A can be simply thought of as a positive definite weighting matrix (Whitney 1972)). The ordinary differential equation (5) uniquely and completely describes the joint motions that provide the desired continuous path motion of the end-effector for both kinematically redundant and non-redundant robot manipulators. Any ordered pair $(\mathbf{q}^*(t), \mathbf{u}^*)$ of admissible motion $\mathbf{q}^*(t)$ and kinematic control $\mathbf{u}^*$, satisfying equation (5) is referred to as a solution of the continuous path motion planning problem.

Thus stage S1 is completed - the desired motion of the end-effector along the prescribed path $\mathbf{\Psi}$ is transformed into equivalent joint motions.

With regard to the kinematic path control approach developed by Krustev and Lilov, it should be noted that the great majority of methods proposed during the last two decades for inverse kinematics evaluation and redundancy resolution use some kind of pseudo-inversion of the Jacobian matrix (a brief review is made in (Klein and Huang 1983)), and are generally some form of resolved motion rate control proposed by Whitney (1969). However, they all consider the continuous path motion of the end-effector and the path itself (with its particular parametrization) as being identical. First of all, this does not allow the prescribed path $\mathbf{\Gamma}$ and the respective joint trajectory $\mathbf{\Psi}$ to be considered entirely and, secondly, does not provide constructive means for satisfying the restrictions of equations (3) and (4).

If, in the second stage (S2) of motion planning, the admissible motion $\mathbf{q}^*(t)$ intersects the extended configuration space $\mathbf{Q}^O \times \mathbf{\Lambda}$, some ideas from variable structure systems theory are applied. A switching surface $\mathbf{\Sigma}$ is synthesized by a smooth deformation of an elliptical cylinder inscribed within $\mathbf{Q}^O \times \mathbf{\Lambda}$. This cylinder can approximate the extended configuration space with any prescribed accuracy. Then a sliding mode $\mathbf{u}^\# = \mathbf{u}^* - \sigma$ is introduced over the switching surface, where $\sigma(\mathbf{q}^*)$ is a vector-valued function defined by the normal vector to $\mathbf{\Sigma}$. Thus, equation (5), where $\mathbf{u}^\#$ is substituted for the kinematic control $\mathbf{u}^*$

$$\dot{\mathbf{q}}^\# = K\mathbf{u}^\#; \qquad (6)$$

defines an admissible motion $\mathbf{q}^\#(t)$ that satisfies the restriction (3).

To satisfy the restriction (4) upon the joint velocities, Krustev and Lilov introduce a scaling factor $\nu = \max a > 0 : a\mathbf{q} \in \mathbf{Q}^1$, where $\mathbf{q}$ is determined by (6). The admissible motion that satisfies both restrictions (3) and (4) is then given by the solution of the equation

$$\dot{\tilde{\mathbf{q}}} = \min(1,\nu)K\mathbf{u}^{\#}. \tag{7}$$

It should be noted here, that both admissible motions $\mathbf{q}^{\#}(t)$ and $\tilde{\mathbf{q}}(t)$ do not necessarily provide the desired continuous path motion of the end-effector, even in case of kinematically redundant manipulators – the accuracy estimation is still an open question.

To illustrate the results, Krustev and Lilov consider an example of an articulated manipulator with five degrees of mobility, under a prescribed profile of the end-effector's velocity (of the common trapezoidal type). The admissible motions, satisfying the restriction (3) and (4), and the sliding mode are shown by numerical simulation.

The replacement of fixed automation with advanced robot systems, implies ever growing requirements upon their performance. To meet these requirements a more general and better understanding of the essential kinematic and dynamic features of a robot manipulator, as well as effective computational methods for motion planning and real-time control are needed. The extended kinematic path control approach of Krustev and Lilov is an important contribution to this field.

References

Hanafusa, H., Yoshikawa, T., and Nakamura, Y. 1981. Analysis and Control of Articulated Arms with Redundancy. Proc. 8th IFAC World Congress, Vol. XIV, Pergamon Press, pp. 78-83.

Khatib, O. 1987. A Unified Approach for Motion and Force Control of Robot Manipulators: The Operational Space Formulation. IEEE Journal of Robotics and Automation, Vol. RA-3 No. 1, pp 43-53.

Klein, C.A., and Huang, C.H. 1983. Review of Pseudoinverse Control for Use with Kinematically Redundant Manipulators. IEEE Trans. Systems, Man, Cybern., Vol. SMC-13, No. 2, pp.245-250.

Klein, C.A. 1985. Use of Redundancy in the Design of Robotic Systems. Proc. 2nd Int. Symp. Robotics Research MIT Press, pp. 207-214.

Krustev, E., and Lilov, L. 1986. Kinematic Path Control of Robot Arms. Robotica, Vol. 4, pp. 107-116.

Whitney, D.E. 1969. Resolved Motion Rate Control of Manipulators and Human Prosthesis. IEEE Trans. Man-Machine Systems, Vol.MMS-10, No.2 pp. 47-53.

Whitney, D.E. 1972 (Dec.). The Mathematics of Coordinated Control of Prosthetic Arms and Manipulators. ASME Journal of Dynamic Systems, Meas., and Control, Vol. 94, Series G, pp. 303-309.

Experiments in Force Control of Robotic Manipulators
James A. Maples and Joseph Becker

Reviewed by

Kenneth Salisbury
Artificial Intelligence Laboratory
Massachusetts Institute of Technology
Cambridge, MA 02139

The authors of this paper address the problem of force control of robotic manipulators from two points of view. First, they present a classification scheme for force control algorithms and second, they discuss their own implementation of several such algorithms. Their work addresses a particular manipulator configuration, one which employs electric motors with position sensors located at the motor and force sensors located at the wrist. While other configurations may ultimately yield better performance (notably those with torque sensors at the motor), the one they explore is a configuration frequently discussed in the literature and one which has been the subject of many control implementations. The work is particularly relevant in that it suggests a practical method for retrofitting existing position controlled manipulators so that they may perform force control. One of the significant inferences they make is that by using a high gain inner position or velocity control loop one may achieve greatly increased disturbance rejection.

Their classification scheme considers two criteria for for distinguishing force control systems: the coordinate system in which errors are computed and the type of inner control loop. Systems which compute the position and force error in Cartesian (or operational) space are identified as Cartesian based systems and those which compute the errors in joint space are identified as joint based systems. While in both cases the same sensors ultimately are used to derive the errors, the organization of computation, and to some degree its efficiency, is affected by the choice of error frame. They suggest that more efficient computations may be made in joint space systems, while more geometrically precise control of compliance behavior may be achieved with Cartesian systems. In the hardware configuration they employ, the force errors are available from a wrist force sensor in Cartesian coordinates while the position errors are derived from joint sensors. Although commands are typically in terms of Cartesian specifications, the actuators are inherently in joint coordinates and thus complex servo-rate transformations between the two coordinate systems are unavoidable.

The second criterion they identify for classification is the type as servo loop used as the innermost loop. This may be either a force, velocity or position loop. They conclude that more robust control systems may be achieved with high gain inner control loops servoing the position or velocity of the actuator. This is consistent with the notion that sensors physically proximate to the actuators are less susceptible to errors in unmodeled dynamics and therefore permit higher loop gain to be achieved. Conversely, a force sensor located at the distal end of the arm gives information about joint torques which is corrupted by complex arm dynamics, mechanism non-linearities and changing geometry and therefore requires lower gain for stability. Present-day actuators are poor force sources and the transmissions they employ often introduce additional errors in force output. Thus, from the point of view of precision, it is appropriate to have the force sensors at the point of ultimate force application (i.e. at the wrist). Future improvements in torque controllable actuators and direct-drive or low-friction transmissions will likely improve this situation to the point where high gain force control loops may be usable as the inner loops and perhaps be supplemented with additional endpoint sensing for better precision.

Rather than address the problem modeling the arm's configuration dependent dynamics and gravity disturbances, the authors suggest that the use of high gain inner position control loops and force sensors is a more expedient and, for the moment, practical approach to arm control. This has been the practical approach taken by many industrial robot controllers for position control. While the authors show that this simple approach can be extended to achieve force control, they do not conclusively argue against the advantages of careful dynamic modeling. With increasingly powerful computational facilities it becomes easier to take advantage of our knowledge of varying arm dynamics to improve the performance and robustness of arm controllers. This is particularly important in force controllers which act in moving frames of reference and at dynamically significant speeds.

In the second portion of the paper the authors discuss at length their experience with a Cartesian based controller with an inner position control loop. They show real data demonstrating the success of their implementation on the AdeptOne robot. In addition, they discuss some of the important implementation details involved in combining force and position control with a stiffness or hybrid controller. They bring to light the advantage of using a dual rate controller with appropriately higher servo rates at the inner position loop. One minor shortcoming is that they do not explicitly discuss the underlying dynamic model of the system upon which their compensation was based.

Overall the paper clarifies the differences among a large class of force control schemes and presents a lucid and insightful account of the implementation of several such schemes. They present real performance data, welcome in a field replete with speculative results. Although convincing in their argument for inner loop position control with present hardware, they do not address the potential gains to be achieved with truly force controllable hardware. For example, if their system is subjected to impacts, the force errors will be large because the disturbance frequency content will exceed the bandwidth of the force controller. By comparison a back-drivable system with low-mass torque source actuators will yield to such impacts as though it were a pure mass. However, until such mechanisms are available, the approach suggested by Maples and Becker is likely to be of significant utility. While this paper does not put to rest the controversy between Cartesian and joint based controllers. (particularly where dynamic effects are significant) it does give important structure and insight into the issues.

Adaptive Control of Mechanical Manipulators[1]
John J. Craig, Ping Hsu, and S. Shankar Sastry

Reviewed by

Tsuneo Yoshikawa
Automation Esearch Laboratory
Faculty of Engineering
Kyoto University
Uji, Kyoto 611, Japan

This paper proposes a new parameter-adaptive control scheme for manipulators that can be regarded as a combination of a nonlinear, model-based scheme which is known as the *computed torque method* and an adaptation scheme for unknown parameters of the dynamic model. A unique feature of the proposed scheme is that while maintaining this physically reasonable structure, it has been proven stable in the full, nonlinear setting using the positive real lemma and Lyapunov stability theory.

The computed torque method ideally yields a controller that suppresses disturbances and tracks desired trajectories uniformly in all configurations of the manipulator. However, this method must usually overcome two difficulties. One is that quick computation of the dynamic model is necessary. The other is that an accurate dynamic model should be available. The first difficulty has been decreased due to some recent work on efficient computational algorithms and due to advancement of microelectronics. The paper is intended to address the second difficulty from the viewpoint of adaptive control.

Although many studies have been done on adaptive control of manipulators, most of them are based on standard model reference adaptive control theory or self- tuning regulator theory for linear, time-invariant systems. Thus these results are only valid under the assumption that the nonlinear effect of the manipulator is negligible.

The proposed adaptive control scheme is based on the joint error dynamics derived from use of the computed torque method with the estimated dynamic parameters and the desired joint acceleration. A

---
[1]International Journal of Robotics Research, Vol. 6, No. 2, pp. 16-17, Summer 1987.

rough sketch of the scheme is as follows. The manipulator dynamics is assumed to be given by

$$T = M(\Theta)\ddot{\Theta} + Q(\Theta,\dot{\Theta}) = W(\Theta,\dot{\Theta},\ddot{\Theta})P; \qquad (1)$$

where T is the $n \times 1$ joint torque (or force) vector, Θ is the $n \times 1$ joint position, M is the mass matrix, Q represents centrifugal, Coriolis, gravitational, and frictional forces. P is the unkown parameters, and W is the coefficient matrix. The proposed control law is given by

$$T = W(\Theta,\dot{\Theta},\ddot{\Theta}^*)\widehat{P} = \widehat{M}(\Theta)\ddot{\Theta}^* + \widehat{Q}(\Theta,\dot{\Theta}); \qquad (2)$$
$$\ddot{\Theta}^* = \ddot{\Theta}_d + K_v\dot{E} + K_pE; \qquad (3)$$

with

$$E = \Theta_d - \Theta;$$

where $\widehat{P}$ is the estimate of P, Θ_d is the desired joint trajectory, and K_v and K_p are diagonal gain matrices (see Figure 1). Then the tracking error E satisfies the system's error equation

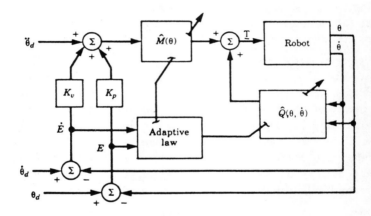

Figure 1. The controller with an adaptive element.

$$\ddot{E} + K_v\dot{E} + K_pE = \widehat{M}^{-1}W(\Theta,\dot{\Theta},\ddot{\Theta}^*)(P - \widehat{P}). \qquad (4)$$

Further, let the filtered servo error E_1 be given by

$$E_1 = \dot{E} + \Psi E. \qquad (5)$$

where Ψ is a diagonal matrix such that the transfer function from to E_1 is strictly positive real. Then it can be shown by using the positive real lemma and Lyapunov stability theory that the parameter adaptation law

$$\hat{P} = \Gamma W^T \widehat{M}^{-1} E_1. \tag{6}$$

makes E and $\dot{E}$ converge to zero under the condition that Γ is an diagonal matrix with positive diagonal elements and $\widehat{M}$ remains positive definite and invertible. In order to insure the positive definiteness of $\widehat{M}$, the authors propose to give upper and lower bounds on each element of $\hat{P}$.

After establishing the adaptation law, the authors give a persistent excitation condition on the matrix W_d which is the W function evaluated along the desired trajectory

$$\alpha I_r \leq \int_{t_0}^{t_0+\rho} W_d^T W_d dt; \quad \text{for some } \alpha > 0; \tag{7}$$

where t_0 is the initial time, I_r is the $r \times r$ identity matrix, and ρ is the duration of trajectory control. If this condition is met, the estimate $\hat{P}$ will converge to the true value P. Using condition (7) a preplanned trajectory can be tested to verify if it provides persistent exitation in an off-line manner.

They also show that the presence of bounded disturbances does not result in loss of stability or unbounded estimates. In this case, although servo errors do not converge asymptotically to zero, they converge to a bounded region near zero. However, in the case of bounded disturbances, the persistent excitation condition can only be stated in terms of the matrix W evaluated along the actual trajectory.

Experimental results for the two major links of the Adept One are given. Although these results indicate that the adaptive controller is still not outperforming Adept's fixed controller, this experiment is quite a good contribution in the sense that it encourages research of adaptive control for manipulators in the expectation that the adaptive controller will outperform the conventional controllers in the near future.

Further study is warranted on a few points. The reviewer found that it may sometimes be difficult to make the width of the bounds on $\hat{P}$ large. Also it would be much better if the acceleration signal $\ddot{\Theta}$ were not necessary in the adaptation law. These points are now being studied by several researchers (Hsu et al, 1987; Slotine and Li, 1987; Li and Slotine 1988) and will eventually be settled. In spite of these points to be improved, the reviewer believes that the paper shows a definite step toward practical implementation of adaptive control schemes to manipulators and other mechanical systems.

References

Hsu, P., Bodson, M., Sastry, S., and Paden, B. 1987. Adaptive Identification and Control of Manipulators Without Using Joint Accelerations. IEEE International Conference on Robotics and Automation, Raleigh.

Slotine, J.J. and Li, W. 1987. On the Adaptive Control of Robot Manipulators. International Journal of Robotics Research, 6(3).

Li, W. and Slotine, J.J. 1988. Indirect Adaptive Robot Control. IEEE International Conference on Robotics and Automation, Philadelphia.